Cerebrovascular Diseases: From Pathogenesis to Treatment

Cerebrovascular Diseases: From Pathogenesis to Treatment

Guest Editors

Adrià Arboix
Teresa Gasull
Francisco Purroy

MDPI

Basel • Beijing • Wuhan • Barcelona • Belgrade • Novi Sad • Cluj • Manchester

Guest Editors
Adrià Arboix
Department of Neurology
Universitat de Barcelona
Barcelona
Spain

Teresa Gasull
Department of Neurosciences
Germans Trias i Pujol
Research Institute (IGTP)
Badalona
Spain

Francisco Purroy
Neurology Department
Universitat de Lleida
Lleida
Spain

Editorial Office
MDPI AG
Grosspeteranlage 5
4052 Basel, Switzerland

This is a reprint of the Special Issue, published open access by the journal *Current Issues in Molecular Biology* (ISSN 1467-3045), freely accessible at: https://www.mdpi.com/journal/cimb/special_issues/S00EE5WM08.

For citation purposes, cite each article independently as indicated on the article page online and as indicated below:

Lastname, A.A.; Lastname, B.B. Article Title. *Journal Name* **Year**, *Volume Number*, Page Range.

ISBN 978-3-7258-6426-3 (Hbk)
ISBN 978-3-7258-6427-0 (PDF)
https://doi.org/10.3390/books978-3-7258-6427-0

Contents

About the Editors

Adrià Arboix

Adrià Arboix is an honorary professor at the University of Barcelona (Catalonia, Spain). He is an Advisory Member of the Research Committee of the Sagrat Cor University Hospital of Barcelona and former President of the Catalan Society of Neurology. He research has resulted in several books and manuals and he is the author of more than 250 peer-reviewed scientific papers and serves as a reviewer for dozens of SCI/SSCI journals. His current research focuses on cerebrovascular diseases, lacunar strokes, acute strokes and vascular cognitive impairment.

Teresa Gasull

Teresa Gasull holds a master's degree and a PhD in Neuroscience. Following a Ramón y Cajal appointment at the Spanish National Research Council (CSIC), she founded the Cellular and Molecular Neurobiology Group at the Germans Trias i Pujol Research Institute (IGTP) in 2006 and has led it ever since. Her research focuses on stroke, with a strong translational approach to neurovascular disorders, cerebral iron dysregulation, and multimodal imaging of ischemic and hemorrhagic injury. Using preclinical rodent and large-animal models, her team develops innovations in therapies, stratification biomarkers, predictive tools, and machine-learning-based neurobehavioral assessment. She is a member of the national stroke network RICORS-ICTUS and of the IGTP AGAUR Neurosciences group, has received Bioemprendedor XXI and CaixaImpulse innovation awards, has led 17 competitive research projects and industry collaborations, holds two stroke-related patents, and serves on the external committee of the IGTP/HGTP Biobank.

Francisco Purroy

Francisco Purroy is a neurologist with over 20 years of experience in the field of cerebrovascular disease. He is Full Professor at the University of Lleida and Head of the Neurology Department, as well as Director of a multidisciplinary research group focused on clinical neurosciences and stroke. He has authored more than 200 scientific publications and has participated in major national and international clinical trials. He is the principal leader of the REMOTE-CAT academic trial investigating remote ischemic conditioning for neuroprotection, and served on the steering committee of the CONVINCE trial. He is also a member of the Spanish Stroke Research Network (RICORS-ICTUS) funded by the Instituto de Salud Carlos III. In addition, he has contributed to the development of European Stroke Organisation clinical guidelines, including those on transient ischemic attack. His career integrates clinical practice, academic leadership, and translational research, with a strong commitment to improving stroke care and prevention.

current issues in molecular biology

MDPI

Editorial

Editorial for Special Issue "Cerebrovascular Diseases—From Pathogenesis to Treatment: Opportunities and Challenges"

Teresa Gasull [1], Francisco Purroy [2] and Adrià Arboix [3,*]

[1] Cellular and Molecular Neurobiology (CMN) Research Group at Germans Trias i Pujol Research Institute, 08916 Badalona, Catalonia, Spain; tgasull@igtp.cat
[2] Neurology Department, Stroke Unit, Hospital Universitari Arnau de Vilanova de Lleida, Universitat de Lleida, Biomedical Research Institute of Lleida, 25198 Lleida, Catalonia, Spain; fpurroy.lleida.ics@gencat.cat
[3] Cerebrovascular Division, Department of Neurology, Hospital Universitari del Sagrat Cor, Universitat de Barcelona, 08916 Barcelona, Catalonia, Spain
* Correspondence: aarboix@quironsalud.es; Tel.: +34-93-4948940

Stroke remains a leading cause of death and disability and has a complex pathophysiology [1,2]. Increasing evidence suggests that the brain is highly sensitive to even short-duration ischemia and that multiple mechanisms are involved in the tissue damage that results from cerebral ischemia. Ischemic stroke initiates a cascade of events, including ATP depletion, ionic dysregulation, increased release of glutamate, excess production of free radicals, edema, and inflammation; all of these events eventually contribute to cell death. In contrast, in intracerebral hemorrhage, the compression and destruction of brain tissue by a hematoma are the primary causes of brain injury, but inflammation, the coagulation response, and released hemoglobin toxicity play pivotal roles as well. Cell death after stroke has mainly been attributed in the past to necrosis or apoptosis, but recent reports show the involvement of other newly described forms of cell death [3–8].

This Special Issue, entitled "Cerebrovascular Diseases—From Pathogenesis to Treatment: Opportunities and Challenges", aims to highlight some of the most novel pathophysiological and molecular aspects of acute stroke, emphasizing neurological biomarkers; cerebrovascular risk factors; the molecular and clinical implications of sleep-related breathing disorders for treating and recovering from acute stroke; therapeutic approaches targeting neuroinflammation, oxidative stress, and endothelial dysfunction in cerebral small vessel disease; as well as the glycoprotein IIIa PlA1/A2 polymorphism's role in cerebrovascular disease risk, areas in which recent advances have been made. However, despite significant advancements, several controversies persist in the field of acute stroke mechanisms, pathophysiology, and treatment.

Following an excellent response from researchers around the world, the contributions listed below come from nine countries (China, Malaysia, Russia, the USA, Romania, the UK, Türkiye, Spain, and South Africa), reflecting the truly global and collaborative nature of contemporary cerebrovascular research.

Twelve manuscripts were submitted for consideration for the Special Issue, and all were subject to the rigorous *Current Issues in Molecular Biology* review process. In total, ten papers were ultimately accepted for publication and inclusion in this Special Issue (see list of contributions).

In this Special Issue, five original articles (contributions 1, 2, 4, 8, and 10) and five updated review papers (contributions 3, 5–7, and 9) were included (Table 1). In this Issue, we have collated a diverse set of research articles that highlight the importance of understanding the molecular signaling pathways involved in stroke.

Table 1. Analysis of the published contributions included in the present Special Issue "Cerebrovascular Diseases—From Pathogenesis to Treatment: Opportunities and Challenges".

N#*	Authors	I/H	Stroke Relevance	Topic	Characteristics of the Article	Conclusions
1	Lin C-I et al.	H	Hemorrhagic stroke pathophysiology (intraventricular and subarachnoid hemorrhage)	Role of the transmembrane-4L-six family (TM4SF1) in embryonic blood vessel development.	Experimental study	This study reveals that the timeliness of reaching an optimal TM4SF1 expression level to support TMED formation for endothelial cell and mesenchymal stem cell signaling and intercellular interactions is of critical importance during embryonic blood vessel development.
2	Coadă et al.	I/H	Cerebrovascular risk factors	Role of the glycoprotein IIIa PIA1/A2 polymorphism in the risk of cerebrovascular diseases.	Systematic search and meta-analysis	The carrier status for the PIA2 polymorphism was significantly associated with a marginally elevated risk of ischemic stroke. No significant association could be found with hemorrhagic stroke.
3	Che Mohd Nassir et al.	I/H	Chronic cerebrovascular dysfunctions that follow COVID-19	Dipeptidyl peptidase IV's role in post-COVID-19 cerebrovascular complications.	Narrative review	Dipeptidyl peptidase IV is a main mediator in metabolic–inflammatory cascades leading to the aggravation of post-COVID-19 fatigue and cerebrovascular complications. Its participation in glucose metabolism, immune regulation, and endothelial dysfunction, therefore, places it at the heart of the chronic cerebrovascular outcomes that accompany COVID-19 survivors, such as neurovascular damage, stroke, and cognitive decline.
4	Obukhova et al.	I	Cerebrovascular risk factors (blood lipid spectrum)	Analysis of the major parameters of the blood lipid spectrum in atherosclerosis and brain tumors, as well as assessing any correlations of these indicators with the Ki-67 proliferative activity index of the cells involved.	Comparative clinical and biological study	Significant differences were only found in the parameter values between the atherosclerosis sufferers and the control group for their ratios of neutral lipids to cholesterol. Of the short-chain fatty acids, butyric acid is of the greatest interest due to the significant differences in its levels in the blood of patients with meningiomas and gliomas from the control group. However, these parameters cannot serve as a preoperative non-invasive predictor of the Ki-67 mitotic index level, as no significant differences between them were found for low- and high-grade anaplasia of brain tumors.
5	Uscamaita et al.	I/H	Sleep-related breathing disorders as a cerebrovascular risk factor	Molecular and clinical implications of sleep-related breathing disorders on treating and recovering from acute stroke.	Scoping review	The relationship between sleep-related breathing disorders and cardiovascular conditions is significant, encompassing stroke (ischemic and hemorrhagic), hypertension, myocardial infarction, post-myocardial infarction heart failure, and atherosclerosis. The prompt recognition of sleep apnea in acute stroke patients could be important for possible therapy and forecasting potential complications and long-term outcomes.

Table 1. *Cont.*

N#*	Authors	I/H	Stroke Relevance	Topic	Characteristics of the Article	Conclusions
6	Yılmaz and Bayraktutan	I	Therapeutic approach in cerebral small vessel disease	Therapeutic approaches targeting neuroinflammation, oxidative stress, and endothelial dysfunction in cerebral small vessel disease.	Updated review	A list of drugs were analyzed to prevent neuroinflammation and oxidative damage and effectively maintain endothelial function and blood–brain barrier integrity, including anti-anginal drugs, acetylcholine esterase inhibitors, β-hydroxy β-methylglutaryl-CoA reductase inhibitors, lithium drugs, phosphodiesterase inhibitors, oral antihyperglycemic drugs, and tetracycline antibiotics. However, it is necessary to improve the understanding of the molecular pathologies underlying cerebral small vessel disease and explore potential pharmacological interventions through well-designed preclinical studies targeting neuroinflammation, oxidative stress, and endothelial dysfunction.
7	Teodoru M. et al.	I	Role of HSP47 in thrombotic disorders	Relevant in vitro and in vivo studies examining platelet aggregation, collagen–HSP47 interactions, and thrombus formation.	Systematic literature review	Reveals the emerging significance of heat shock protein 47 (HSP47) in thrombotic disorders, particularly deep vein thrombosis (DVT) and atherosclerosis. HSP47 facilitates platelet–collagen interactions critical for thrombus formation and vascular stability in vulnerable plaques.
8	Yuan et al.	I	Therapeutic approach of using isoliquiritigenin in ischemic stroke	An integrated network pharmacology, molecular docking, molecular dynamics simulation, and experimental validation study was performed to investigate the potential mechanism of isoliquiritigenin in treating ischemic stroke.	Experimental study	The report emphasizes that isoliquiritigenin may exert an anti-ischemic stroke effect by regulating APP, ESR1, MAO-A, and PTGS2, determined according to the combined results of network pharmacology, molecular docking, molecular dynamics simulation, and experimental validation.
9	Naicker and Khathi	I	Cerebrovascular risk factors (prediabetes and stroke)	Links between prediabetes and stroke.	Systematic literature review	Individuals with prediabetes may show alterations in inflammatory (IL-6), coagulation (D-dimer), and neurovascular (S100B, GFAP, NSE) markers, though some of these findings are inconsistent, reflecting early pathophysiological changes that may increase stroke risk.
10	Naicker and Khathi	I	Cerebrovascular risk factors (prediabetes and stroke)	Relevance of prediabetes as a cerebrovascular risk factor.	Comparative clinical and biological study	Inflammatory, coagulation, and neurovascular biomarkers, particularly S100B, may indicate early stroke risk in prediabetes.

I: ischemic stroke; H: hemorrhagic stroke; N#*: number of contributions.

Examining the latest evidence on this topic, a total of six articles analyzed cerebral infarctions (contributions 4 and 6–10), while ischemic and hemorrhagic strokes were analyzed in three studies (contributions 2, 3, and 5) and hemorrhagic stroke alone was analyzed in another study (contribution 1).

Emphasis was placed on the role of transmembrane 4 L six family member 1 (TM4SF1) in embryonic blood vessel development and hemorrhagic stroke pathophysiology (contribution 1) and glycoprotein IIIa PlA1/A2 polymorphism in the risk of acute cerebrovascular diseases (contribution 2). Also examined were the role of the dipeptidyl peptidase IV in post-COVID-19 cerebrovascular complications (contribution 3) and the analysis of the major blood lipid spectrum parameters in atherosclerosis and brain tumors as well as the assessment of correlations between these indicators and the Ki-67 proliferative activity index of the cells involved (contribution 4). The role of HSP47 in thrombotic disorders was evaluated (contribution 7). The molecular and clinical implications of sleep-related breathing disorders on treating and recovering from acute stroke (contribution 5) and the therapeutic approaches targeting neuroinflammation, oxidative stress, and endothelial dysfunction in cerebral small vessel disease (contribution 6) were also reviewed. Finally, a therapeutic approach using isoliquiritigenin (a type of chalcone that is widely found in medicinal plants of the Fabaceae family) in ischemic stroke was evaluated (contribution 8), and the relevance of prediabetes as a potential cerebrovascular risk factor was also evaluated (contributions 9 and 10).

The goal of this Monographic Issue is to provide a critical overview of the underlying factors involved in stroke-related brain injury, as well as the recent advances and controversies in this topic. Lin C.-I. et al., in their experimental study (contribution 1], demonstrated that the timeliness of reaching an optimal level of transmembrane 4 L six family member 1 (TM4SF1) expression to support TMED formation (TM4SF1-enriched microdomain) for endothelial cell and mesenchymal stem cell signaling and intercellular interactions is of critical importance during embryonic blood vessel development, preventing intraventricular and subarachnoid hemorrhage.

In a systematic search and meta-analysis, Coadă et al. (contribution 2) critically emphasized that the carrier status for the PIA2 polymorphism was significantly associated with a marginally elevated risk of ischemic stroke, though no significant association was found with hemorrhagic stroke.

Che Mohd Nassir et al., in their narrative review (contribution 3), showed that dipeptidyl peptidase IV is a main mediator in metabolic–inflammatory cascades, leading to the aggravation of post-COVID-19 fatigue and cerebrovascular complications. Its participation in glucose metabolism, immune regulation, and endothelial dysfunction, therefore, places it at the heart of the chronic cerebrovascular outcomes that accompany COVID-19 in survivors, such as neurovascular damage, stroke, and cognitive decline.

In their comparative study, Obukhova et al. (contribution 4) showed that there were only significant differences in the parameter values between atherosclerosis sufferers and the control group in terms of their ratios of neutral lipids to cholesterol. Of the short-chain fatty acids, butyric acid was of greatest interest due to the significant difference in its levels compared to the control group in the blood of patients with meningiomas and of those with gliomas. However, these parameters cannot serve as preoperative, non-invasive predictors of the Ki-67 mitotic index level, as no significant differences were found between them for low- and high-grade anaplasias of brain tumors.

Moreover, the scoping review by Uscamaita et al. (contribution 5) showed that the relationship between sleep-related breathing disorders and cardiovascular conditions is significant, encompassing stroke (ischemic and hemorrhagic), hypertension, myocardial infarction, post-myocardial infarction heart failure, and atherosclerosis. Promptly recognizing sleep apnea in acute stroke patients could be important for guiding therapy and forecasting potential complications and long-term outcomes.

Yılmaz and Bayraktutan (contribution 6), using an updated review, analyzed therapeutic approaches targeting neuroinflammation, oxidative stress, and endothelial dysfunction in cerebral small vessel disease and concluded that there is a need to deepen the understanding of the molecular pathologies underlying cerebral small vessel disease and explore potential pharmacological interventions through well-designed preclinical studies targeting neuroinflammation, oxidative stress, and endothelial dysfunction.

In their systematic literature review, Teodoru et al. (contribution 7) remarked that HSP47, a key collagen-specific molecular chaperone, plays a vital role in collagen maturation, folding, and stabilization in humans. Its activity is crucial for maintaining the stability and assembly of collagen types essential to connective tissue health. While primarily functioning within the endoplasmic reticulum, HSP47's influence extends beyond cellular protein synthesis, impacting adiposity regulation and cardiovascular health.

Lastly, Yuan et al. (contribution 8) emphasized that isoliquiritigenin may exert an anti-ischemic stroke effect through regulating APP (amyloid precursor protein), ESR1 (estrogen receptor alpha), MAO-A (monoamine oxidase type-A), and PTGS2 (prostaglandin-endoperoxidase synthase).

The links between prediabetes and acute stroke were analyzed by Naicker and Khathi (contributions 9 and 10).

Overall, the research presented in this Special Issue underscores the complex nature of stroke and the importance and necessity of a multidisciplinary research approach to advance the understanding of stroke pathophysiology [9–18].

In summary, in this Special Issue, the authors highlight a number of research directions and topics with the aim of studying the molecular and cellular mechanisms of acute stroke. Improving our understanding of the underlying molecular mechanisms of cerebrovascular diseases is imperative for further integrating innovative drug, cell, and gene therapy approaches into carefully controlled clinical trials, and will promote significant advances in a field with many unmet needs.

We hope that the diverse perspectives and methodological innovations showcased in this Special Issue will be both stimulating and valuable to our readers, inspiring collaborative efforts to advance the field of molecular biology in cerebrovascular research.

Open lines of research include the pathophysiological and molecular aspects of acute stroke in octogenarians; differences between men and women; and comparing cardioembolic versus atherothrombotic strokes or lacunar versus non-lacunar infarctions—all of which are important and challenging subsets of acute cerebrovascular diseases that should be evaluated in further investigations.

We would like to extend our sincere thanks to all the authors, peer reviewers, and editors who contributed to this Special Issue.

Author Contributions: Writing, reviewing, and editing, A.A.; reviewing and editing, T.G. and F.P. This article is dedicated to the memory of Josep Lluís Martí-Vilalta. All authors have read and agreed to the published version of the manuscript.

Funding: The authors received no funding for this Editorial.

Acknowledgments: Thank you to all contributors, peer reviewers, and editors of *CIMB*, an MDPI journal, for their excellent collaboration and support.

Conflicts of Interest: The authors declare no conflicts of interest.

List of Contributions:

1. Lin, C.-I.; Merley, A.; Wada, H.; Zheng, J.; Jaminet, S.-C.S. Transmembrane-4 L-Six Family Member-1 Is Essential for Embryonic Blood Vessel Development. *Curr. Issues Mol. Biol.* **2024**, *46*, 13105–13118. https://doi.org/10.3390/cimb46110781.
2. Coadă, C.A.; Lupu, M.; Florea, I.; Di Constanzo, S.; Coluccelli, S.; Şimon, I. Association of Glycoprotein IIIa PlA1/A2 Polymorphism with Risk of Stroke: Updated Meta-Analysis. *Curr. Issues Mol. Biol.* **2024**, *46*, 5364–5378. https://doi.org/10.3390/cimb46060321.
3. Che Mohd Nassir, C.M.N.; Che Ramli, M.D.; Jaffer, U.; Abdul Hamid, H.; Mehat, M.Z.; Mohamad Ghazali, M.; Kottakal Cheriya, E.N. Neurological Sequelae of Post-COVID-19 Fatigue: A Narrative Review of Dipeptidyl Peptidase IV-Mediated Cerebrovascular Complications. *Curr. Issues Mol. Biol.* **2024**, *46*, 13565–13582. https://doi.org/10.3390/cimb46120811.
4. Obukhova, L.; Shchelchkova, N.; Medyanik, I.; Yashin, K.; Grishin, A.; Bezvuglyak, O.; Abdullaev, I. Brain Tumor-Induced Changes in Routine Parameters of the Lipid Spectrum of Blood Plasma and Its Short-Chain Fatty Acids. *Curr. Issues Mol. Biol.* **2025**, *47*, 228. https://doi.org/10.3390/cimb47040228.
5. Uscamaita, K.; Parra Ordaz, O.; Yazbeck Morell, I.; Pla, M.G.; Sánchez-López, M.-J.; Arboix, A. From Molecular to Clinical Implications of Sleep-Related Breathing Disorders on the Treatment and Recovery of Acute Stroke: A Scoping Review. *Curr. Issues Mol. Biol.* **2025**, *47*, 138. https://doi.org/10.3390/cimb47030138.
6. Yılmaz, H.; Bayraktutan, U. Cerebral Small Vessel Disease: Therapeutic Approaches Targeting Neuroinflammation, Oxidative Stress, and Endothelial Dysfunction. *Curr. Issues Mol. Biol.* **2025**, *47*, 232. https://doi.org/10.3390/cimb47040232.
7. Teodoru, M.; Stoia, O.-M.; Vladoiu, M.-G.; Tonch-Cerbu, A.-K. The Role of HSP47 in Thrombotic Disorders: Molecular Mechanism and Therapeutic Potential. *Curr. Issues Mol. Biol.* **2025**, *47*, 283. https://doi.org/10.3390/cimb47040283.
8. Yuan, H.; Hou, Y.; Jiao, Y.; Lu, X.; Liu, L. An Integrated Network Pharmacology, Molecular Docking, Molecular Dynamics Simulation, and Experimental Validation Study to Investigate the Potential Mechanism of Isoliquiritigenin in the Treatment of Ischemic Stroke. *Curr. Issues Mol. Biol.* **2025**, *47*, 627. https://doi.org/10.3390/cimb47080627.
9. Naicker, Y.; Khathi, A. Pathophysiological Links Between Stroke and Prediabetes: A Systematic Review. *Curr. Issues Mol. Biol.* **2025**, *47*, 854. https://doi.org/10.3390/cimb47100854.
10. Naicker, Y.; Khathi, A. A Retrospective Assessment of Changes in Stroke Risk-Related Biomarkers in Individuals with Prediabetes from Durban, South Africa: Preliminary Findings. *Curr. Issues Mol. Biol.* **2025**, *47*, 884. https://doi.org/10.3390/cimb47110884.

References

1. Kleindorfer, D.O.; Towfighi, A.; Chaturvedi, S.; Cockroft, K.M.; Gutierrez, J.; Lombardi-Hill, D.; Kamel, H.; Kernan, W.N.; Kittner, S.J.; Leira, E.C.; et al. Guideline for the prevention of stroke in patients with stroke and transient ischemic attack; A guideline from the American Heart Association/American Stroke Association. *Stroke* **2021**, *52*, e364–e467. Correction in *Stroke* **2021**, *52*, e483–e484. https://doi.org/10.1161/STR.0000000000000383. [CrossRef] [PubMed]
2. Arboix, A.; Massons, J.; Oliveres, M.; García-Eroles, L.; Titus, F. Análisis de 1.000 pacientes consecutivos con enfermedad cerebrovascular aguda. Registro de patología cerebrovascular de La Alianza-Hospital Central de Barcelona. [An analysis of 1000 consecutive patients with acute cerebrovascular disease. The registry of cerebrovascular disease of La Alianza-Hospital Central of Barcelona]. *Med. Clin.* **1993**, *101*, 281–285. (In Spanish) [PubMed]
3. Alsbrook, D.L.; Di Napoli, M.; Bhatia, K.; Biller, J.; Andalib, S.; Hinduja, A.; Rodrigues, R.; Rodriguez, M.; Sabbagh, S.Y.; Selim, M.; et al. Neuroinflammation in Acute Ischemic and Hemorrhagic Stroke. *Curr. Neurol. Neurosci. Rep.* **2023**, *23*, 407–431. [CrossRef] [PubMed]
4. Sang, H.; Cao, Z.; Du, J.; Nguyen, T.N.; Saver, J.L.; Mao, A.; Nogueira, R.G.; Tao, Z.; Zhou, S.; DEVT and RESCUE BT Trial; et al. Investigators. Intravenous Tirofiban Versus Alteplase Before Endovascular Treatment in Acute Ischemic Stroke: A Pooled Analysis of the DEVT and RESCUE BT Trials. *Stroke* **2024**, *55*, 856–865. [CrossRef] [PubMed]
5. Pawluk, H.; Kołodziejska, R.; Grześk, G.; Kozakiewicz, M.; Woźniak, A.; Pawluk, M.; Kosinska, A.; Grześk, M.; Wojtasik, J.; Kozera, G. Selected Mediators of Inflammation in Patients with Acute Ischemic Stroke. *Int. J. Mol. Sci.* **2022**, *23*, 10614. [CrossRef] [PubMed]

6. Cullell, N.; Gallego-Fábrega, C.; Cárcel-Márquez, J.; Muiño, E.; Llucià-Carol, L.; Lledós, M.; Martín-Campos, J.M.; Molina, J.; Casas, L.; Almeria, M.; et al. ICA1L is associated with small vessel disease: A proteome-wide association study in small vessel stroke and intracerebral haemorrhage. *Int. J. Mol. Sci.* **2022**, *23*, 3161. [CrossRef] [PubMed]
7. Azarbarzin, A.; Zinchuk, A.; Peker, Y.; Sands, S.A. Reply to: Heart Rate Response in Obstructive Sleep Apnea: A Clue to Reveal Cardiovascular Benefit from Continuous Positive Airway Pressure? *Am. J. Respir. Crit. Care Med.* **2022**, *206*, 1181–1182. [CrossRef] [PubMed]
8. Capmany, R.P.; Arboix, A.; Casañas-Muñoz, R.; Anguera-Ferrando, N. Specific cardiac disorders in 402 consecutive patients with ischaemic cardioembolic stroke. *Int. J. Cardiol.* **2004**, *95*, 129–134. [CrossRef] [PubMed]
9. D'aes, T.; Marlier, Q.; Verteneuil, S.; Quatresooz, P.; Vandenbosch, R.; Malgrange, B. Re-Evaluating the Relevance of the Oxygen-Glucose Deprivation Model in Ischemic Stroke: The Example of Cdk Inhibition. *Int. J. Mol. Sci.* **2023**, *24*, 7009. [CrossRef] [PubMed] [PubMed Central]
10. Uscamaita, K.; Ordaz, O.P.; Morell, I.Y.; Pla, M.G.; Sánchez-López, M.-J.; Arboix, A. From Molecular to Clinical Implications of Sleep-Related Breathing Disorders on the Treatment and Recovery of Acute Stroke: A Scoping Review. *Curr. Issues Mol. Biol.* **2025**, *47*, 138. [CrossRef] [PubMed]
11. Agarwal, N.; Lewis, L.D.; Hirschler, L.; Rivera, L.R.; Naganawa, S.; Levendovszky, S.R.; Ringstad, G.; Klarica, M.; Wardlaw, J.; Iadecola, C.; et al. Current Understanding of the Anatomy, Physiology, and Magnetic Resonance Imaging of Neurofluids: Update From the 2022 "ISMRM Imaging Neurofluids Study group" Workshop in Rome. *J. Magn. Reson. Imaging* **2024**, *59*, 431–449. [CrossRef] [PubMed] [PubMed Central]
12. Yılmaz, H.; Bayraktutan, U. Cerebral Small Vessel Disease: Therapeutic Approaches Targeting Neuroinflammation, Oxidative Stress, and Endothelial Dysfunction. *Curr. Issues Mol. Biol.* **2025**, *47*, 232. [CrossRef] [PubMed]
13. Rehman, S.; Nadeem, A.; Akram, U.; Sarwar, A.; Quraishi, A.; Siddiqui, H.; Malik, M.A.J.; Nabi, M.; Haq, I.U.; Cho, A.; et al. Molecular Mechanisms of Ischemic Stroke: A Review Integrating Clinical Imaging and Therapeutic Perspectives. *Biomedicines* **2024**, *12*, 812. [CrossRef] [PubMed]
14. Segura, B.; Jurado, M.Á.; Freixenet, N.; Bargalló, N.; Junqué, C.; Arboix, A. White matter fractional anisotropy is related to processing speed in metabolic syndrome patients: A case-control study. *BMC Neurol.* **2010**, *10*, 64. [CrossRef] [PubMed] [PubMed Central]
15. Sharma, R.; Lee, K. Advances in treatments for acute ischemic stroke. *BMJ* **2025**, *389*, e076161. [CrossRef] [PubMed]
16. Kumar, S.; Chou, S.H.-Y.; Smith, C.J.; Nallaparaju, A.; Laurido-Soto, O.J.; Leonard, A.D.; Singla, A.K.; Leonhardt-Caprio, A.; Stein, D.J.; American Heart Association Stroke Council; et al. Addressing Systemic Complications of Acute Stroke: A Scientific Statement from the American Heart Association. *Stroke* **2025**, *56*, e15–e29. [CrossRef] [PubMed]
17. McDermott, C.M.; Seminer, A.; Comer, S.; Mulihano, A.; Reddin, C.; Judge, C.; Costello, M.; Canavan, M.; Smyth, A.; Hurley, P.; et al. Discrimination of ischemic versus hemorrhagic stroke type by presenting symptoms or signs: A systematic review and meta-analysis. *J. Stroke Cerebrovasc. Dis.* **2025**, *34*, 108413. [CrossRef] [PubMed]
18. Babić, A.; Bonifačić, D.; Komen, V.; Kovačić, S.; Mamić, M.; Vuletić, V. Blood Biomarkers in Ischemic Stroke Diagnostics and Treatment-Future Perspectives. *Medicina* **2025**, *61*, 514. [CrossRef] [PubMed] [PubMed Central]

current issues in molecular biology

Article

Transmembrane-4 L-Six Family Member-1 Is Essential for Embryonic Blood Vessel Development

Chi-Iou Lin [1,2,†], Anne Merley [1,3,†], Hiromi Wada [1,4], Jianwei Zheng [1,5] and Shou-Ching S. Jaminet [1,6,*]

1 Center for Vascular Biology Research, Department of Pathology, Beth Israel Deaconess Medical Center, Harvard Medical School, Boston, MA 02215, USA; clin@riverview.org (C.-I.L.); anne_merley@brown.edu (A.M.); hwada@ric.u-tokyo.ac.jp (H.W.); zhengjianwei@bjtth.org (J.Z.)
2 Anesthesiology Department, Riverview Hospital, Noblesville, IN 46060, USA
3 Center for Animal Resources and Education, Brown University, Providence, RI 02912, USA
4 Isotope Science Center, The University of Tokyo, Tokyo 113-0032, Japan
5 Department of General Surgery, TianTan Hospital, Capital Medical University, Beijing 100070, China
6 Angiex Inc., Cambridge, MA 02140, USA
* Correspondence: sjaminet@angiex.com
† These authors contributed equally to this work.

Abstract: Transmembrane-4 L-six family member-1 (TM4SF1) is a small cell surface glycoprotein that is highly and selectively expressed on endothelial cell and mesenchymal stem cell surfaces. TM4SF1 regulates cellular functions by forming protein complexes called TMED (TM4SF1-enriched microdomains) that either recruit growth-factor activated proteins and internalize them via microtubules to distribute the recruited molecules intracellularly or support the formation of nanopodia for intercellular interactions extracellularly. Through a genetically manipulated mouse model for global *Tm4sf1* gene knockout, we demonstrate here that TM4SF1 is essential for blood vessel development. *Tm4sf1*-null embryos fail to develop blood vessels and experience lethality at E9.5. *Tm4SF1*-heterozygous embryos are smaller in body size during early embryonic development, and almost half die in utero due to intracranial hemorrhage in the intraventricular and subarachnoid space, which becomes apparent by E17.5. Surviving *Tm4SF1*-heterozygotes do not display overt phenotypic differences relative to wild type littermates postnatally. Together, these studies demonstrate that TM4SF1, through its molecular facilitator and nanopodia formation roles in TMED, intimately regulates blood vessel formation during embryonic development.

Keywords: TM4SF1; endothelial cell; mesenchymal stem cell; knockout mice; blood vessel development; intraventricular and subarachnoid hemorrhage

1. Introduction

Transmembrane-4 L-six family member-1 (TM4SF1) is a small plasma membrane glycoprotein of tetraspanin topology but distinct homology from the thirty-three genuine tetraspanins [1,2]. TM4SF1 was originally discovered as a tumor-associated antigen [3,4] that is also expressed at low levels by a normal vascular endothelium [5–7]. Subsequently, we demonstrated that TM4SF1 is expressed at high levels in cultured endothelial cells originating from blood or lymphatic vessels [6,8], endothelial progenitor cells, and in bone marrow-derived mesenchymal stem cells [9], as well as in vivo in an endothelium undergoing pathological angiogenesis in both experimental models of angiogenesis in mice [6] and in human cancers [6,7]. Our studies further demonstrated that the depletion of TM4SF1 in endothelial cells through knockdown (i) prevents the cellular polarization necessary for endothelial cell migration and proliferation in vitro [6,8]; (ii) blocks intercellular interactions and junction formation in cultured endothelial cells [8]; and (iii) inhibits vessel maturation in pathological angiogenesis in vivo [6].

TM4SF1 regulates biological and cellular activities through its ability to form 100–300 nm diameter protein complexes called TMEDs (TM4SF1-enriched microdomains) on the surfaces of cells that express high levels of TM4SF1 in vitro [6–9] and the tumor endothelium of human cancers in vivo [7]. Each TMED contains three to fourteen TM4SF1 molecules and functions as a membrane dock that recruits membrane proteins and brings along their intracellular membrane-proximal signaling components [7,8]. Through a novel and as of yet unelucidated mechanism, TMEDs internalize from the cell surface via microtubules with an ultimate destination in the nuclear compartment [7] and distribute recruited proteins from the plasma membrane to intracellular locations. TMEDs also facilitate the adhesion of endothelial cells to the matrix by recruiting integrin-α5β1 [6], a process that enables the formation of nanopodia: thin, long, membrane-lined channels within which actin microfilaments extend and/or retract during cell movement, and which facilitate the homotypic and heterotypic intercellular interactions for stable vessel formation [6,8–10].

In this study, we investigated TM4SF1's functional role in normal blood vessel development using a global knockout strategy in mice. We report that *Tm4sf1*-null embryos are avascular and experience lethality at E9.5, while *Tm4sf1*-heterozygous embryos are smaller in body size during early embryonic development, and about half die in utero due to an intracranial hemorrhage, which becomes apparent in the intraventricular and subarachnoid space by E17.5. Together these studies reveal that TM4SF1, through its ability to form TMED, intimately regulates blood vessel formation during embryonic development.

2. Materials and Methods

2.1. Anti-TM4SF1 Antibodies

Anti-TM4SF1 antibodies were the same as used in our earlier studies: (1) the anti-mouse TM4SF1 antibody 2A7A (a chimeric antibody with the variable region of a rabbit anti-mouse TM4SF1 monoclonal antibody and the constant region of a human IgG1) [11] and (2) the anti-human TM4SF1 antibody 8G4 [IgG1, mouse monoclonal [9,10]; MilliporeSigma (Burlington, MA, USA), catalog number MABC1723].

2.2. Cells and Cell Culture

All seven human primary endothelial cells (ECs), HUVEC (human umbilical vein EC), HLMEC (lung microvascular EC), HDMEC, (human dermal microvascular EC), HCAEC (human coronary artery EC), HPAEC (human pulmonary artery EC), ECFCs (endothelial colony forming cells), HMVEC-dLy (human dermal microvascular lymphatic EC), as well as HBdSMC (smooth muscle cells, bladder), HDF (fibroblast, dermal), HemaLP (epidermal melanocytes), and bone marrow-derived mesenchymal stem cells were purchased along with their respective media from Lonza (Walkersville, MD, USA), and used within six passages in all experiments conducted in this study. Three cell lines, 3T3 (fibroblast, mouse embryo), HEK393 (epithelial cells, kidney), and MS1 (mouse SV1 immortalized islet EC), were acquired from ATCC (American Type Culture Collection, Gaithersburg, MD, USA), cultured in their respective media suggested by ATCC, and used within five passages after being awakened from liquid nitrogen. Total RNAs for mouse leukocytes were extracted from fresh mouse blood while it was purchased from Thermo Fisher (Waltham, MA, USA) for human leukocytes.

2.3. Flow Cytometry

Experimental procedures were described in detail previously [10]. Briefly, 1×10^6 freshly harvested cells (HPAEC, HDF, and HEK293) were washed in cold PBS, suspended in 1 mL of cold blocking buffer (PBS/2% FBS) that contained 100 ng of 1st antibody 8G4 or mouse IgG1 (Sigma, Saint Louis, MO, USA), and incubated at 4 °C for 30 min with occasional agitation. Cells were centrifuged (500× *g* for 5 min), followed by 100 ng/mL 2nd antibody (Alexa-488 labeled donkey anti-human or mouse IgG; Thermo Fisher) for another 30 min at 4 °C with washing 3× with cold PBS in between the incubations. Cell suspensions were analyzed with FACScan (Becton Dickinson, San Jose, CA, USA). A total of 10^4 events were collected for each analysis.

2.4. Generation of TM4SF1 Knockout Mice

Four key steps were involved to generate the *Tm4sf1* knockout mice: (1) we produced the linearized *Tm4sf1* genome targeting construct pM253/PUH-PDH/Neo (8461 bp), (2) introduced the construct into embryonic stem cells (ESCs) as described below, (3) identified mouse W4/129S6 ESCs that had *Tm4sf1* genome Ex3 to Ex5, targeted and replaced with loxP-Neo-loxP through homologous recombination, and (4) implanted the ESC clones into Fv129 mice (Charles River Laboratory, Shrewsbury, MA, USA). A schematic summary of the generation of the knockout mice is detailed in Supplementary Figure S2. All PCR primers used in *Tm4sf1* knockout mice are listed in Supplementary material Table S1. The Institutional Animal Care and Use Committee at Beth Israel Deaconess Medical Center, Boston approved all animal experiments (protocol #100-2011 and 111-2014).

RP24-402K15 chloramphenicol resistant Bac clone that contains mouse *Tm4sf1* locus (acquired from NIH) was used as the source of the mouse *Tm4sf1* genomic sequence. Four constructs were made to generate a *Tm4sf1* gene targeting construct (Supplementary Figure S2): **(i) pM253/UH-DH (5.7 kb) plasmid**. PCR amplified UH (657 bp; 5′ Ex3 flanking region) and DH (634 bp; reside in intron 6) DNA fragments from RP24-402K15, cut with Xba1 and ligated to generate UH-DH, followed by Kpn1 and HindIII, which were cut for insertion into the pM253 vector (provided by Dr. W.C. Aird [12,13]) to produce pM253/UH-DH (ampicillin resistant). **(ii) pM253/UH-HR-DH (13.5 kb) gene repair vector**. pM253/UH-DH plasmid was linearized and UH-DH arms separated through an Xba1 cut and electroporated into RP24-402K15/EL-250 bacteria for homologous recombination (HR) to bring in *Tm4sf1* Ex3 to Ex6 (9013 bp) genome segment to the plasmid. The plasmid was confirmed via PCR through UP and DP primer sets. **(iii) PUH/loxP-Neo-loxP/PDH (2.85 kb) cassette**. PCR amplified the PUH (463 bp; 5′ Ex3 flanking region adjacent to Ex3) and PDH (587 bp; 3′ Ex5 intron region) DNA fragments from RP24-402K15, respectively; they were cut with Bsa1 and BsmB1 and inserted on either side of the loxP-Neo-loxP cassette. **(iv) pM253/PUH/loxP-Neo-loxP/PDH (12.9 kb) targeting plasmid**. RP24-402K15/EL-250 bacteria that contain pM253/UH-HR-DH gene repair vector were transfected with PUH/loxP-Neo-loxP/PDH cassette to generate pM253/PUH/loxP-Neo-loxP/PDH plasmid through HR (neomycin resistant). The removal of 2345 bp long Ex3 to Ex5 was then confirmed via PCR using TM-UPH-Neo forward and reverse primers. The 8461 bp targeting construct was then released via Kpn1 and HindIII, cut and transfected W4/129S6 ESCs (Taconic Biosciences, Albany, NY, USA). G418 antibiotic (Thermo Fisher) selection at 350 ug/mL was initiated 24 h following targeting and neomycin-resistant ESC clones were screened for successful homologous recombination with PCR using the genotyping primer sets.

Positive *Tm4sf1* −/+ ES clones were amplified and submitted to the BIDMC transgenic core facility for injection into blastocysts. Briefly, targeted ESC clones were microinjected into C57BL/6-derived blastocyst stage embryos which were then transplanted into the uteri of recipient C57BL/6 females (Charles River Laboratory) to generate chimeric mice. Resulting chimeric males were bred with C57BL/6 females to obtain mice heterozygous for the inactivated loci. The heterozygous F1 male were back-crossed with C57BL/6 females to generate the F2 heterozygous male and repeated for 8 generations to stabilize the chromosome before conducting the *Tm4sf1*-heterozygotes male and female inbred mating to obtain homozygous embryos.

2.5. PCR Genotyping

Genotyping was conducted using chromosomal DNA extracted from either yolk sac or a 2 mm tail clip from P14 to P21 postnatal mice. PCR analyses were performed with primers specific for the wild type and targeted alleles. Tissues were submerged in Gitschier Buffer in the presence of β-mercaptoethanol and proteinase-K to release chromosomal DNA. PCR conditions as described by the manufacturer of AmpliTaq Polymerase (Perkin-Elmer Cetus,

Norwalk, CT, USA) were used to generate fragments from wild type (596 bp) and targeted (397 bp) alleles on 1% agarose gels.

2.6. Multi-Gene Transcriptional Profiling (MGTP) Approach to Quantitative Real-Time PCR (qPCR)

MGTP is a qPCR technique that efficiently quantifies mRNA copies per cell [14] and is performed by MGTP core. Total RNA was prepared using the RNeasy kit with DNase-I treatment (Qiagen, Chatsworth, CA, USA) and cDNA using random primers and SuperScript III (Thermo Fisher). Total RNAs for both a human aorta and vein were purchased from BioChain (Newark, CA, USA). For each data point, mean ± SD were calculated from three different samples from three separate experiments. PCR reactions for each cDNA sample were performed in duplicate. Transcript abundances were normalized/10^6 18S-rRNA copies to an approximate number of transcripts/cell [14]. All MGTP primer sequences used in the study are listed in Supplementary Table S2.

2.7. Mouse Organ Harvest and Immunohistochemical (IHC) Staining

Experimental procedures were described in detail in previous studies [15]. Briefly, mouse tissues were immediately placed in 4% paraformaldehyde (PFA) after harvesting from 8 week old C57BL/6 mice, and cryostat sections were immunostained with anti-mouse TM4SF1 antibody 2A7A followed by donkey anti-human HRP-conjugated secondary antibodies (Thermo Fisher). Normal human liver section was prepared from surgically removed liver transplant by Dr. Lawrence Brown at BIDMC's Pathology Dept and placed in 4% paraformaldehyde followed by OCT mounting and cryostat sectioning for staining with mouse anti-human TM4SF1 antibody 8G4, with subsequent anti-mouse HRP-conjugated 2nd antibodies (Thermo Fisher). All IHC images were representative selections from at least three separate sections and were taken by Dr. Harold Dvorak. For wholemount mouse embryo and retina staining, Alexa488-conjugated 2A7A was used and images were captured via Keyence BZ-9000E microscope (Itasca, IL, USA).

2.8. Embryo Images, Videos and Immunostaining

All bright field images or videos of embryos harvested from different developmental stages were taken from live cam attached Wild Photo Mikroskop M400 stereo photomicroscope (Martin Microscope, Easley, SC, USA). Wholemount staining was carried out as described [12]. Briefly, E9.5 embryos were fixed for 4 h in 4% PFA in PBS at 4 °C, rinsed in PBST (0.1% Tween-20/PBS), and incubated with Alexa488-conjugated anti-mouse TM4SF1 antibody 2A7A in blocking buffer (PBST/2% FBS) for two nights at 4 °C. Fluorescence images were then acquired via Keyence microscope after thoroughly washing the embryos in PBST overnight at 4 °C. For *Tm4s1*-histology staining, 4% PFA fixed embryos were cryoprotected overnight in 20% sucrose/PBS, embedded in OCT (Tissue-Tek), sectioned (6 μm), and stained with the 2A7A antibody, followed by HRP-conjugated anti-human 2nd antibody and Eosin counterstaining. For H&E staining, embryos were preserved in 50% ethanol before being processed for FFPE for H&E staining at the BIDMC's tissue processing core.

2.9. Statistical Analysis

Each experiment was repeated at least three times and the statistical difference between groups is presented as mean ± standard deviation. Analysis was performed using GraphPad Prism 10.0 software. The significance of differences between groups was assessed by Student's *t*-test; p values < 0.05 were considered statistically significant.

3. Results

3.1. TM4SF1 Expression Is Largely Limited to Endothelial Cells In Vitro and In Vivo

We surveyed TM4SF1 expression in cultured endothelial cells and six non-endothelial cell and non-tumor cell types that originated from both human and mouse tissues. qPCR revealed that human TM4SF1 expression in non-endothelial cells, with the exception of mesenchymal stem cells (MSCs, bone marrow derived), was either very low as in smooth muscle cells (HBdSMCs, bladder) and fibroblasts (HDF, dermal), or was not detectably expressed as in kidney epithelial cells (HEK293), epidermal melanocytes (HemaLP), and white blood cells (WBCs) (Supplementary Figure S1). TM4SF1 expression in cells of mouse origin such as MS1 (SV40-immortalized mouse islet endothelial cells), 3T3 (embryonic fibroblasts), and mWBCs was consonant with TM4SF1 expression in the corresponding human cell lineages (Supplementary Figure S1A).

IHC staining in tissue sections prepared from six different mouse tissues (liver, lung, heart, kidney, brain, and retina) provided further evidence that TM4SF1 expression was limited to the vascular endothelium, and that expression was highest in the endothelium of arteries, followed by veins and capillaries (Figure 1A). Strong TM4SF1 expression is also seen in kidney glomeruli and in the choroid plexus epithelium of the lateral ventricle as well as the subependymal cell layer in the mouse brain (Figure 1A). Similar TM4SF1 localization was demonstrated in both mouse and human tissues; for example, the human liver exhibits arterial > venous > capillary endothelial staining (Figure 1B). In agreement, qPCR demonstrated that TM4SF1 gene expression in RNA from the human aorta was 4.8-fold greater than that from the pulmonary vein ($p = 0.0026$; Figure 1C).

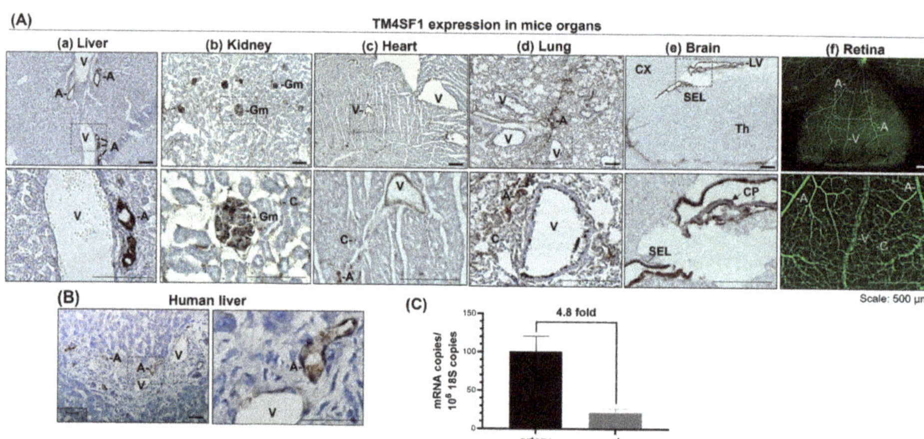

Figure 1. TM4SF1 expression in the endothelium *in vivo*. TM4SF1 IHC staining with HRP-conjugated 2nd antibody and eosin was conducted in tissue sections prepared from (**A**) 8 week old C57BL/6 mouse organs [liver (**a**), kidney (**b**), heart (**c**), lung (**d**), and brain (**e**)] and (**B**) human liver. Wholemount staining was performed in postnatal day 18 C57BL/6 mouse retina (**f**) with Alexa488-conjugated anti-mouse TM4SF1 antibody 2A7A. Representative IHC staining in both (**A**,**B**) tissue sections showed that TM4SF1 expression was largely limited to blood vessels with a tendency for the highest TM4SF1 expression to be in arteries (A), intermediate expression in veins (V), and the lowest expression in capillaries (C). Positive TM4SF1 expression was also seen in kidney glomeruli (Gm) and in the choroid plexus (CP; black arrow), the epithelium of lateral ventricles (LV), along with the subependymal cell layer (SEL) in the brain; no expression was observed in the cortex (CX) and thalamus (Th). (**C**) qPCR using total RNA (normalized against 10^6 18S copies) acquired from three different human aorta and pulmonary vein samples showed that TM4SF1 expression was 4.8-fold higher in the artery than in the vein ($p = 0.0026$).

3.2. *Tm4sf1*-Knockout Mice Are Embryonic Lethal at Embryonic Day 9.5 (E9.5)

The mouse *Tm4sf1* gene is encoded on the minus strand of chromosome-3 from nucleotides 57,105,910 to 57,089,531 and contains seven exons (Ex). The protein coding region begins in Ex3 and ends in Ex7 and encodes 202 amino acids (Supplementary Figure S2A). We used standard gene targeting strategies to generate *Tm4sf1*-knockout mice; *Tm4sf1*'s first three coding exons (Ex3 to Ex5) start with an initiation codon ATG, and our targeting strategy was designed to completely eliminate TM4SF1 protein translation by deleting Ex3 to EX5 (Supplementary Figure S2B).

Genotypes of progeny of intercrossed *Tm4sf1*-heterozygous mice from E9.5 to weaning age (WA) (postnatal day 14 to day 21) are presented in Table 1. At WA, 154 pups were accounted to 30 litters, 74 pups were heterozygous (het; +/−), 80 were wild type (wt; +/+), and zero were knockout (ko; −/−) (Table 1). The ratio of +/− to +/+ was 0.93 at WA, substantially different from the expected Mendelian ratio of 2. The 5.1 pups/litter was also below the normal average of 8 pups/litter (the normal distribution for wt:het:ko for an eight pup litter would be 2:4:2). These results indicate that all *Tm4sf1*-knockout mice died in utero; of the *Tm4sf1*-heterozygous, (1) working from the expected litter size, only 62% were born live (74 pups out of an expected 120 pups for the total of 30 litters); or alternatively, (2) working from Mendelian ratios, about 48% (74 pups out of 154 pups for the total of 30 litters with wt:het ratio of 0.93 instead of 2.0).

Table 1. Genotypes of progeny from *Tm4sf1*-heterozygous intercrosses.

	# of Embryos (% over Total #)									
Age	**+/+**	**+/− [a]**	**+/− [b]**	**+/− [c]**	**−/−**	**Total #**	**Litters**	**#/Litter**	**Viable −/−**	**+/− [a]: +/+**
E9.5	34 (29.6)	63 (54.8)	0 (0.0)	0 (0.0)	16 (13.9)	113	12	9.4	0	1.85
E10.5	22 (30.1)	41 (56.2)	2 (0.0)	0 (0.0)	10 (13.7)	75	9	8.3	0	1.86
E12.5	33 (34.0)	58 (59.8)	0 (0.0)	0 (0.0)	6 (6.2) §	97	10	9.7	0	1.76
E14.5	25 (39.1)	29 (45.3)	4 (6.3)	2 (3.1)	4 (6.3) §	64	8	8.0	0	1.16
E16.5	27 (40.3)	26 (38.8)	7 (10.4)	1 (1.5)	6 (9.0) §	67	8	8.4	0	0.96
E18.5	34 (39.1)	31 (35.6)	19 (21.8)	3 (3.4)	0 (0.0)	87	11	7.9	0	0.91
WA	80 (51.9)	74 (48.1)	0 (0.0)	0 (0.0)	0 (0.0)	154	30	5.1	0	0.93

E, embryonic day; **WA**, weaning age for genotyping; **a**, embryos without visible hemorrhagic blood vessel or live at the WA; **b**, embryos with visible hemorrhagic brain; **c**, necrotic dead embryo; **§**, embryos were deteriorating or being absorbed.

Although only 48–62% of the *Tm4sf1*-heterozygous reached WA, a near Mendelian ratio of 1:2:1 (wt:het:ko) was observed at E9.5 to E12.5 (Table 1). However, none of the homozygous *Tm4sf1*-null embryos were viable with a vital heartbeat; see representative movies taken from E9.5 *Tm4sf1*-wild type (Supplementary Movie S1) and *Tm4sf1*-knockout (Supplementary Movie S2) littermate embryos. Representative E9.5 embryos further demonstrated that unlike *Tm4sf1*-wild type embryos (Figure 2A(a)), *Tm4sf1*-knockout embryos lacked visible blood vessels in the yolk sac and embryo body in bright field images and lacked TM4SF1 positively stained blood vessels in immunofluorescence wholemount (Figure 2A(b)). E10.5 littermates provided further evidence of the inability of *Tm4sf1*-knockout embryos to generate vasculature (Supplementary Figure S3). A representative mid-sagittal section of a wild type E10.5 embryo revealed that all major vessels stained positively for TM4SF1, as did the cephalic mesenchyme, the condensing mesenchyme in the head (Supplementary Figure S4).

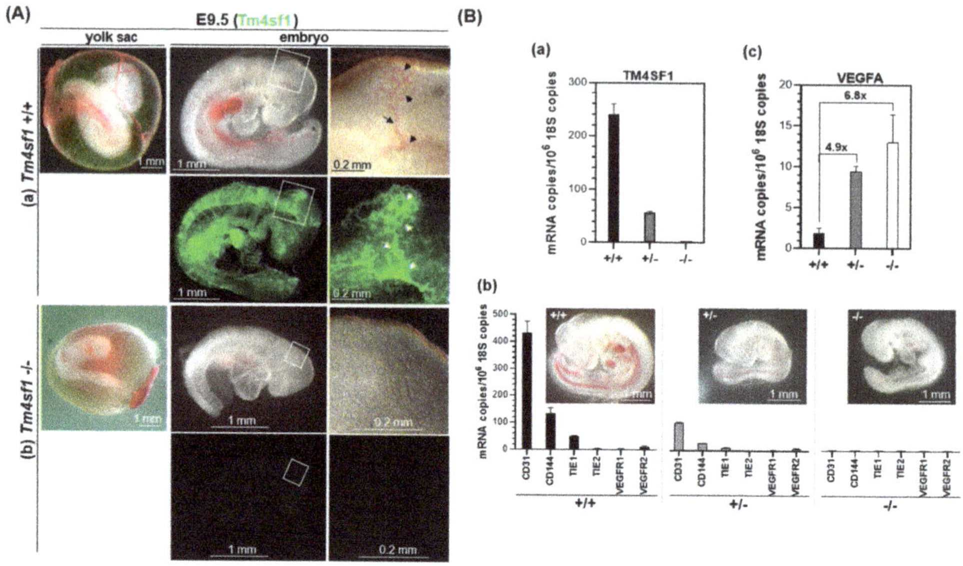

Figure 2. *Tm4sf1*-knockout mice are embryonically lethal at E9.5. (**A**) Representative E9.5 bright field embryo images of littermates show developing blood vessels in both the yolk sac and embryo in *Tm4sf1* +/+ (**a**) but not in *Tm4sf1* −/− (**b**). Wholemount embryo staining using Alexa488-conjugated 2A7A demonstrated TM4SF1 protein localization in the *Tm4sf1* +/+ embryo vasculature (white arrows indicate the same locations as the black arrows in bright field images), but no 2A7A staining occurred in the *Tm4sf1* −/− embryo. (**B**) qPCR was conducted on total RNA prepared from E9.5 littermate embryos (three each of +/+, +/−, and −/− from three different litters). (**a**) Representative RNA quantification demonstrated that *Tm4sf1* −/− embryos, unlike *Tm4sf1* +/+ and +/− embryos, did not express the *Tm4sf1* gene and that (**b**) the known vascular markers CD31, CD144, TIE1, TIE2, VEGFR1, and VEGFR2 were also too low to be detected; and, (**c**) in comparison to *Tm4sf1* +/+ embryos, VEGFA expression was 6.8-fold ($p = 0.00003$) higher in *Tm4sf1* −/− embryos and 4.9-fold ($p = 0.0014$) higher in *Tm4sf1* +/− embryos. (**b**) The representative inset littermate embryo images show an absence of blood vessels in the −/− embryos, and a small body size in the +/− embryo in comparison to the +/+.

In accordance with yolk sac genotyping, the qPCR performed on the total RNA extracted from E9.5 embryos demonstrated the absence of detectable *Tm4sf1* gene expression in the *Tm4sf1*-knockout embryos (Figure 2B(a)). *Tm4sf1*-knockout embryos also were deficient in blood vessel markers including CD31, CD144, TIE1, TIE2, VEGFR1, and VEGFR2 (Figure 2B(b)). Expression of TM4SF1 and the six vascular markers in *Tm4sf1*-heterozygous embryos was less than half of that in their *Tm4sf1*-wild type littermates (Figure 2B(b)). These differences in vascular gene expression were accompanied by differences in embryo phenotype: representative E9.5 embryo images from the same litter showed that the *Tm4sf1*-heterozygous embryo exhibited a smaller body size than their wild type littermates, and the *Tm4sf1*-knockout embryo lacked blood vessels (Figure 2B(b), embryo image inset). VEGFA expression was, respectively, 6.8-fold and 4.9-fold higher in the *Tm4sf1*-knockout and *Tm4sf1*-heterozygous embryos than in wild type embryos (Figure 2B(c)).

3.3. Tm4sf1-Heterozygous Embryos Were Smaller in Size, and Approximately Half Evolved Significant Brain Hemorrhage

Abnormalities in *Tm4sf1*-heterozygous embryos were clearly noted starting at about E14.5, when many developed brain hemorrhage (Figure 3; Table 1). Representative E15.5 littermate embryo images revealed that *Tm4sf1*-heterozygous displayed a smaller body size than their wild type littermate, with one of the two *Tm4sf1*-heterozygous exhibiting brain hemorrhage (Figure 3A, blue arrow), while other regions of the body, including the yolk sac, appeared to be normal.

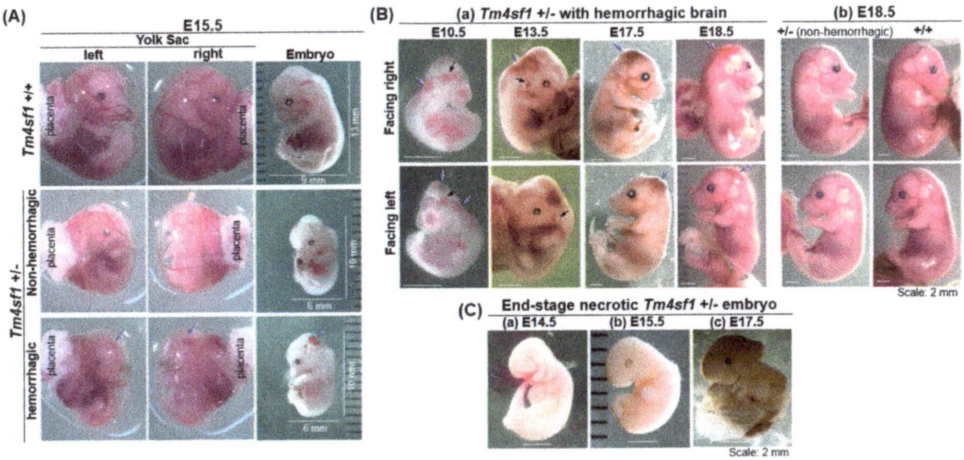

Figure 3. *Tm4sf1*-heterozygous embryos can develop lethal brain hemorrhage. Representative images from the sequential developmental stages of mouse embryos portray brain hemorrhages that occurred in some *Tm4sf1* +/− embryos. (**A**) Two yolk sac images were taken with the placenta positioned at left or right. E14.5 *Tm4sf1* +/− embryos had a smaller body size than their wild type littermate, and one of the two *Tm4sf1* +/− embryos displayed hemorrhage in the head (blue arrows). No obvious hemorrhage appeared in other parts of the embryo body including the yolk sac. (**B**) The hemorrhage can be seen as early as E10.5 and became more apparent over the subsequent days of development (blue arrows), with some embryos also showing a vascular defect near the jugular vein (black arrows). Brain hemorrhages did not affect the embryo growth rate; at E18.5, *Tm4sf1* +/− embryos with and without brain hemorrhage were comparable in body size and had nearly caught up in size to *Tm4sf1* +/+ embryos. (**C**) Deceased *Tm4sf1* +/− embryos were noted during embryo harvests at (**a**) E14.5, (**b**) E15.5, and (**c**) E17.5.

Representative images of developing embryos provide further evidence that visible vascular defects in *Tm4sf1*-heterozygous embryos were largely confined to the head (Figure 3B, blue arrow) with some also seen in the vicinity of the jugular vein (Figure 3B, black arrows). Occasionally, some remnants of dead *Tm4sf1*-heterozygous embryos were identified during embryo harvest (Figure 3C). This is consistent with the progressively declining ratio of non-hemorrhagic *Tm4sf1*-heterozygous to wild type embryos from the expected 2:1 Mendelian ratio and observed as 1.16 at E14.5, 0.91 at E18.5, and 0.93 at WA. These data imply that lethal vascular defects can happen before E18.5 in *Tm4sf1*-heterozygous embryos.

A higher resolution bright field image of a representative E17.5 *Tm4sf1*-heterozygous embryo that exhibited brain hemorrhage demonstrates a lack of integrity in the forebrain–forebrain (fb-fb) and forebrain–midbrain (fb-mb) junctions and an accumulation of blood around the third and fourth ventricles (Figure 4A(c)). The transverse H&E section of the hemorrhagic *Tm4sf1*-heterozygous embryo showed that all four ventricles (lateral left and right, third, and fourth) in the head along with the subarachnoid space were filled with blood (Figure 4B(c)), but no hemorrhage was observed in the cortex (Figure 4B(c)). The *Tm4sf1*-heterozygous littermate without brain hemorrhage resembled wild type embryos and showed normal integrity in the fb-fb and fb-mb regions (Figure 4A(a,b)), without blood accumulation in the ventricles or subarachnoid space (Figure 4B(a,b)).

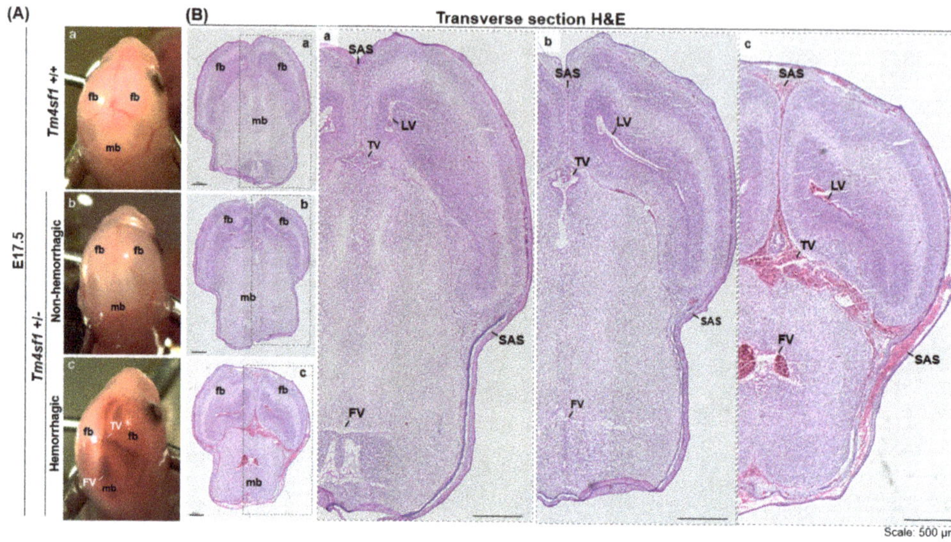

Figure 4. Brain hemorrhage in some *Tm4sf1* +/− embryos. Representative bright field head images of E17.5 *Tm4sf1* +/+ and +/− littermates are shown in (**A**) and their respective transverse H&E histological staining are shown in (**B**). (**A**) A lack of integrity in the fb (forebrain)-fb and fb-mb (midbrain) junctions and an accumulation of blood around the TV (third ventricle) and FV (fourth ventricle) were seen in the hemorrhagic *Tm4sf1* +/− embryo (**c**) but not in the non-hemorrhagic *Tm4sf1* +/− (**b**) and the *Tm4sf1* +/+ (**a**) embryos. (**B**) All four ventricles, TV, FV, and left and right LV (lateral ventricle), and the SAS (subarachnoid spaces) were filled with blood in the hemorrhagic *Tm4sf1* +/− embryo (**c**) whereas all structures appeared normal in both the non-hemorrhagic *Tm4sf1* +/− (**b**) and the *Tm4sf1* +/+ (**a**) embryos.

Whether or not brain hemorrhage was experienced, *Tm4sf1*-heterozygous embryos were smaller in body size at E15.5, but progressively caught up with wild type embryos at later ages and showed minimal difference in body size at the time of birth (Figure 3B(b); Supplementary Figure S5A). Tracking postnatal growth via body weight also demonstrated that *Tm4sf1*-heterozygous and wild type had a similar growth rate (Supplementary Figure S5B), and *Tm4sf1*-heterozygous mice also experienced normal fertility and lifespan. The ratio of live born males to females in the wild type and in *Tm4sf1*-heterozygous litters were 0.95 and 0.9, respectively (Supplementary Figure S5C). Overall, the results suggest that the challenging period for mice with deficient TM4SF1 protein expression is during embryonic development, and lethal embryonic brain hemorrhages are equally likely to occur in male and female *Tm4sf1*-heterozygous embryos.

4. Discussion

Through genetic manipulation approaches, we demonstrate here that *Tm4sf1*-null mice experience embryonic lethality by E9.5 due to a failure to form blood vessels, and that the suboptimal level of TM4SF1 in *Tm4sf1*-heterozygotes causes a smaller body size during early embryonic development with almost half of the embryos experiencing a lethal vascular defect in the head that led to intraventricular and subarachnoid hemorrhage.

Differential expression of TM4SF1 in different vascular beds was observed in vivo, with arteries having the highest expression, veins intermediate, and capillaries the lowest (Figure 1). This result was unanticipated since cultured endothelial cells, regardless of their in vivo bed of origin, showed uniformly high and similar levels of TM4SF1 expression in vitro (Supplementary Figure S1). We presume that this differential expression in vivo is the result of endothelial cell interactions with accessory cells to generate distinct vascular beds [16,17], a process supported by mesenchymal stem cell differentiation to smooth muscle cells for the formation of larger vessels like arteries or to pericytes for the support of veins and capillaries [18,19]. We exclude the potential involvement of shear stresses as our studies applying laminar flow (steady, non-turbulent) to cultured endothelial cells via an orbital shaker did not affect TM4SF1 level in our current ongoing studies.

VEGF-A plays a critical role in endothelial cell proliferation for vasculogenesis and angiogenesis through the activation of its receptors expressed on the cell surface [20–22]. Our prior studies demonstrated that, although VEGF-A was supplied in abundance, the depletion of TM4SF1 through knockdown [6] resulted in (i) an inability of endothelial cells to perform cytokinesis for proliferation, polarization for movement, or intercellular interactions in vitro; and (ii) the inhibition of blood vessel maturation in a mouse model of VEGF-A provoked pathological angiogenesis in vivo. In a similar manner, high VEGF-A expression failed to avert the avascular nature of E9.5 *Tm4sf1*-null embryos (Figure 2). The high VEGF-A expression seen in E9.5 *Tm4sf1*-heterozygous embryos, perhaps caused by hypoxia [23] for urging blood vessel development and conceivably originated from the choroid plexus, ventricular neuroectoderm cells, head mesenchyme, and between the somites [24–26], may have contributed to the ultimate recovery of normal body size in the heterozygous embryos over the course of developmental time (Figure 3B; Supplementary Figure S5). The smaller embryo body size observed during the *Tm4sf1*-heterozygotes' development is not uncommon among the knockout mice in others studies, in which the target gene is involved in blood vessel development; examples include *integrin-αvβ8* [27], *Erg* transcription factor [13], and proteins in the VEGF signaling cascade [25,28].

The circulatory system is the first and most essential organ system to begin functioning during embryonic development and is assembled through vasculogenesis and angiogenesis followed by vascular remodeling via the recruitment of vascular supporting accessory cells [29,30]. In mouse embryos, the initial establishment of vasculogenesis that gives rise to the first primitive vessels in the yolk sac occurs at E6.5; establishment of the cardiovascular system commences at E7, starting with the development of the endocardium [31]. The umbilical artery, the site of hematopoetic stem cell development [32], is initiated in the allantois at E7.5 and fused to the chorion at E8.5 to support the morphogenetic process of vascular remodeling and the circulation to and from the yolk sac and heart in the embryo proper [31,33]. Impairments in any of those steps, for example in mice with genetically manipulated VEGF [25,26] or VEGF receptor signaling [34–36], will lead to embryonic lethality by E9.5.

Given that TM4SF1 protein expression is specific to the vascular endothelium and the mesenchyme in vivo (Figure 1; Supplementary Figure S4) and cultured cells in vitro (Supplementary Figure S1), TM4SF1 is likely to influence embryonic blood vessel development through both the endothelial cells and the mesenchymal stem cells. Endothelial cells express high levels of VEGFR2/Flk-1 and PDGF-β, and mesenchymal stem cells express high levels of the corresponding ligand VEGF-A and receptor PDGFRβ [9,37,38], an indication that intimate interactions between the two cell types facilitate blood vessel formation [16]. As unpublished studies with cultured endothelial cells show that PLCγ-1, the immediate downstream mediator of VEGFR2/Flk-1 signaling, which activates ERK upon Flk-1 activation to produce an embryonic lethal phenotype, similar to that of Flk-1 null mice [39], is recruited to TMED. PLCγ-1 is also a downstream mediator of PDGFR in mesenchymal stem cells [38]. We anticipate that a defect in PLCγ-1 functional activities may be one of the molecular mechanisms underlying the *Tm4sf1*-null phenotype. Furthermore, the insufficient TMED-mediated intracellular transport of recruited molecules and/or intercellular interactions in both endothelial cells and mesenchymal stem cells may contribute to the impairment of vasculogenesis, and to the impaired remodeling seen in *Tm4sf1*-heterozygous embryos. A detailed characterization of TMED complexes to identify their recruited proteins via mass spectrometry is currently underway.

A number of questions remain unanswered: why was hemorrhage largely limited to the head but no other parts of the *Tm4sf1*-heterozygous embryos? Also, why did hemorrhage happen in only about half of the *Tm4sf1*-heterozygotes, and not all? We noted that the TM4SF1 protein is highly expressed in the cephalic mesenchyme of E10.5 mouse embryos, with no noticeable expression in the cortical neuroectoderm (Supplementary Figure S3); TM4SF1 protein expression is also seen in the choroid plexus epithelium of the lateral ventricle and the subependymal cell layer in adult mouse brains (Figure 1). However, we currently do not have data to elucidate how TM4SF1 expression in these locations influenced the phenotypic development of *Tm4sf1*-heterozygotes. Specifically, we do not know whether the expression in cephalic mesenchyme is transitory, and if differences in TM4SF1 expression regulation in the mesenchyme contributed to the occurrence of vascular hemorrhage. We anticipate, based on our occasional observations of the end-stage necrosis of *Tm4sf1*-heterozygous embryos as early as E14.5 (Figure 4A), that some *Tm4sf1*-heterozygous embryos may experience lethality before they have developed the apparent intraventricular and subarachnoid hemorrhage seen near birth (Figure 4B). This implies considerable variability among *Tm4sf1*-heterozygous embryos, with some encountering more severe vascular defects with early lethality, others intraventricular and subarachnoid hemorrhage, or normal development.

E14.5 is the transition stage from the embryonic to fetal period and is considered to be the most important time point for developmental disorder analyses [40]. In many other perinatally or prenatally lethal mutant mouse lines, half of the embryos survive until E14.5, including some in which the cause of death is intracerebral hemorrhage by mutations that affect endothelial cell function such as *Fli1* [41] and *Erg* [13] transcription factors, and *integrin-αvβ8* adhesion molecules [27,42]. Nonetheless, embryonic intraventricular and subarachnoid hemorrhage has not been reported in studies using genetically manipulated mice, including manipulations through the VEGF signaling cascade [25,26,34,35]. Ventricles are a communicating network of chambers filled with cerebrospinal fluid that in mouse embryos, is predominantly produced in the lateral ventricles, transported through the ventricular system, and enters the subarachnoid space through the foramina at the fourth ventricle for final absorbance [43]. In mouse embryos, the lateral (I and II), third, and fourth ventricles develop around E10 and the cerebral aqueduct connecting the third and fourth ventricle is formed and becomes the narrowest part of the cerebrospinal fluid system during late embryonic brain development [43]. We infer that the vascular integrity required to support the ventricles for cerebrospinal fluid clearance is especially challenged by the suboptimal level of TM4SF1 expression.

Future studies, combining proteomic analysis for protein identification and tracing the fate of *Tm4sf1*-null cells via lineage tracing with spatial biology imaging techniques, possibly aided by the knock-in of a reporter gene like lacZ to the *Tm4sf1* locus or conditional *Tm4sf1*-knockout, will address where and when TM4SF1 is expressed during embryo development, and how it works with recruited proteins for the regulation of blood vessel formation and maturation.

5. Conclusions

We report that *Tm4sf1*-null embryos fail to form blood vessels and experience embryonic lethality at E9.5; *Tm4sf1*-heterozygotes are smaller in size than wild type embryos during early embryonic growth, and half die in utero, often with intraventricular and subarachnoid hemorrhage. The half of *Tm4sf1*-heterozygous pups that are born alive appear normal, fertile, and have no apparent postnatal developmental defects. As noted in our prior reports, TMED form on the cell surface when TM4SF1 is highly expressed and serve as critical signaling mediators through internalization via microtubules to the nucleus and through nanopodia formation for intercellular interactions. This study displays that the timeliness of an optimal level of TM4SF1 expression to support TMED formation for endothelial cell and mesenchymal stem cell signaling and intercellular interactions are of critical importance during embryonic blood vessel development.

Supplementary Materials: The following supporting information can be downloaded at: https://www.mdpi.com/article/10.3390/cimb46110781/s1. Figure S1: TM4SF1 expression in cultured cells *in vitro*; Figure S2: Genomic structure of *Tm4sf1* and targeting strategy to generate *Tm4sf1*-knockout mice; Figure S3: Defective in yolk sac vascular development in *Tm4sf1* knockout embryo; Figure S4: TM4SF1 protein expression in E10.5 wild type embryos; Figure S5: Growth of *Tm4sf1* +/+ and non-hemorrhagic +/− mice during prenatal and postnatal stages; Table S1: PCR primer sequences used in generating *Tm4sf1*-knockout mice; Table S2: Real-time PCR primer sequences used in MGTP approach to quantitative real-time PCR; Movie S1: TM4SF1 wt; Movie S2: TM4SF1 ko.

Author Contributions: Conceptualization: S.-C.S.J.; methodology: C.-I.L., A.M., J.Z. and S.-C.S.J.; validation: C.-I.L., A.M., J.Z. and S.-C.S.J.; formal analysis: C.-I.L., A.M., J.Z., H.W. and S.-C.S.J.; investigation: C.-I.L., A.M., J.Z., H.W. and S.-C.S.J.; resources: C.-I.L., A.M., J.Z. and S.-C.S.J.; data curation: C.-I.L., A.M., J.Z. and S.-C.S.J.; writing—original draft: S.-C.S.J.; writing—review and editing: C.-I.L., A.M., H.W., J.Z. and S.-C.S.J.; visualization: C.-I.L., A.M., H.W. and S.-C.S.J.; supervision: S.-C.S.J.; project administration: S.-C.S.J.; funding acquisition: S.-C.S.J. All authors have read and agreed to the published version of the manuscript.

Funding: This work was supported by the BIDMC Chief Academic Officer Pilot Grants 2015 (S.C.S.J.), and an NIH grant P01 CA92644 (S.-C.S.J.: co-PI; H.F.D.: PI).

Institutional Review Board Statement: Approval for all animal experiments was granted by the Beth Israel Deaconess Medical Center Institutional Animal Care and Use Committee (protocol number 100-2011, approved on 19 December 2011; protocol number 111-2014, approved on 8 December 2014).

Informed Consent Statement: Not applicable.

Data Availability Statement: The datasets used and/or analyzed during the current study are available from the corresponding author on reasonable request.

Acknowledgments: We thank Harold F. Dvorak (HFD) for the assistance of capturing immunohistochemical images of mouse and human organs, the discussions of the manuscript, and the support of the funding. We thank William C. Aird and his group for their assistance strategizing the making of *Tm4sf1*-knockout mice, and especially David Beeler who supervised gene cloning, and Katherine Spoke who assisted with ES cell transfection. We thank Andrew Zukauskas for participation in the initial stages of cloning, Dan Li of MGTP core for the assistance with quantitative real-time PCR, and BIDMC's transgenic mice core for the generation of *Tm4sf1*-chimeric mice.

Conflicts of Interest: This paper has been published as a preprint before: https://www.researchsquare.com/article/rs-3245895/v1 (accessed on 17 August 2023). All authors confirm that the article has not been formally published elsewhere. Author Shou-Ching S. Jaminet was employed by the company Angiex Inc., Cambridge. The remaining authors declare that research was conducted in the absence of any commercial or financial relationships that could be construed as a potential conflicts of interest.

References

1. Wright, M.D.; Ni, J.; Rudy, G.B. The L6 membrane proteins--a new four-transmembrane superfamily. *Protein Sci.* **2000**, *9*, 1594–1600. [CrossRef] [PubMed]
2. Stipp, C.S.; Kolesnikova, T.V.; Hemler, M.E. Functional domains in tetraspanin proteins. *Trends Biochem. Sci.* **2003**, *28*, 106–112. [CrossRef] [PubMed]
3. Hellstrom, I.; Beaumier, P.L.; Hellstrom, K.E. Antitumor effects of L6, an IgG2a antibody that reacts with most human carcinomas. *Proc. Natl. Acad. Sci. USA* **1986**, *83*, 7059–7063. [CrossRef] [PubMed]
4. Marken, J.S.; Schieven, G.L.; Hellstrom, I.; Hellstrom, K.E.; Aruffo, A. Cloning and expression of the tumor-associated antigen L6. *Proc. Natl. Acad. Sci. USA* **1992**, *89*, 3503–3507. [CrossRef] [PubMed]
5. DeNardo, S.J.; O'Grady, L.F.; Macey, D.J.; Kroger, L.A.; DeNardo, G.L.; Lamborn, K.R.; Levy, N.B.; Mills, S.L.; Hellstrom, I.; Hellstrom, K.E. Quantitative imaging of mouse L-6 monoclonal antibody in breast cancer patients to develop a therapeutic strategy. *Int. J. Rad. Appl. Instrum. B* **1991**, *18*, 621–631. [CrossRef]
6. Shih, S.C.; Zukauskas, A.; Li, D.; Liu, G.; Ang, L.H.; Nagy, J.A.; Brown, L.F.; Dvorak, H.F. The L6 protein TM4SF1 is critical for endothelial cell function and tumor angiogenesis. *Cancer Res.* **2009**, *69*, 3272–3277. [CrossRef]
7. Sciuto, T.E.; Merley, A.; Lin, C.I.; Richardson, D.; Liu, Y.; Li, D.; Dvorak, A.M.; Dvorak, H.F.; Jaminet, S.C. Intracellular distribution of TM4SF1 and internalization of TM4SF1-antibody complex in vascular endothelial cells. *Biochem. Biophys. Res. Commun.* **2015**, *465*, 338–343. [CrossRef]

8. Zukauskas, A.; Merley, A.; Li, D.; Ang, L.H.; Sciuto, T.E.; Salman, S.; Dvorak, A.M.; Dvorak, H.F.; Jaminet, S.C. TM4SF1: A tetraspanin-like protein necessary for nanopodia formation and endothelial cell migration. *Angiogenesis* **2011**, *14*, 345–354. [CrossRef]
9. Lin, C.I.; Merley, A.; Sciuto, T.E.; Li, D.; Dvorak, A.M.; Melero-Martin, J.M.; Dvorak, H.F.; Jaminet, S.C. TM4SF1: A new vascular therapeutic target in cancer. *Angiogenesis* **2014**, *17*, 897–907. [CrossRef]
10. Lin, C.I.; Lau, C.Y.; Li, D.; Jaminet, S.C. Nanopodia-Thin, fragile membrane projections with roles in cell movement and intercellular interactions. *J. Vis. Exp. JoVE* **2014**, *86*, 51320. [CrossRef]
11. Visintin, A.; Knowlton, K.; Tyminski, E.; Lin, C.I.; Zheng, X.; Marquette, K.; Jain, S.; Tchistiakova, L.; Li, D.; O'Donnell, C.J.; et al. Novel Anti-TM4SF1 Antibody-Drug Conjugates with Activity against Tumor Cells and Tumor Vasculature. *Mol. Cancer Ther.* **2015**, *14*, 1868–1876. [CrossRef] [PubMed]
12. Yuan, L.; Janes, L.; Beeler, D.; Spokes, K.C.; Smith, J.; Li, D.; Jaminet, S.C.; Oettgen, P.; Aird, W.C. Role of RNA splicing in mediating lineage-specific expression of the von Willebrand factor gene in the endothelium. *Blood* **2013**, *121*, 4404–4412. [CrossRef] [PubMed]
13. Vijayaraj, P.; Le Bras, A.; Mitchell, N.; Kondo, M.; Juliao, S.; Wasserman, M.; Beeler, D.; Spokes, K.; Aird, W.C.; Baldwin, H.S.; et al. Erg is a crucial regulator of endocardial-mesenchymal transformation during cardiac valve morphogenesis. *Development* **2012**, *139*, 3973–3985. [CrossRef] [PubMed]
14. Wada, Y.; Li, D.; Merley, A.; Zukauskas, A.; Aird, W.C.; Dvorak, H.F.; Shih, S.C. A multi-gene transcriptional profiling approach to the discovery of cell signature markers. *Cytotechnology* **2010**, *63*, 25–33. [CrossRef] [PubMed]
15. Brown, L.F.; Dezube, B.J.; Tognazzi, K.; Dvorak, H.F.; Yancopoulos, G.D. Expression of Tie1, Tie2, and angiopoietins 1, 2, and 4 in Kaposi's sarcoma and cutaneous angiosarcoma. *Am. J. Pathol.* **2000**, *156*, 2179–2183. [CrossRef]
16. Swift, M.R.; Weinstein, B.M. Arterial-venous specification during development. *Circ. Res.* **2009**, *104*, 576–588. [CrossRef]
17. Beck, L., Jr.; D'Amore, P.A. Vascular development: Cellular and molecular regulation. *FASEB J.* **1997**, *11*, 365–373. [CrossRef]
18. Ding, R.; Darland, D.C.; Parmacek, M.S.; D'Amore, P.A. Endothelial-mesenchymal interactions in vitro reveal molecular mechanisms of smooth muscle/pericyte differentiation. *Stem Cells Dev.* **2004**, *13*, 509–520. [CrossRef]
19. Gu, W.; Hong, X.; Le Bras, A.; Nowak, W.N.; Issa Bhaloo, S.; Deng, J.; Xie, Y.; Hu, Y.; Ruan, X.Z.; Xu, Q. Smooth muscle cells differentiated from mesenchymal stem cells are regulated by microRNAs and suitable for vascular tissue grafts. *J. Biol. Chem.* **2018**, *293*, 8089–8102. [CrossRef]
20. Pettersson, A.; Nagy, J.A.; Brown, L.F.; Sundberg, C.; Morgan, E.; Jungles, S.; Carter, R.; Krieger, J.E.; Manseau, E.J.; Harvey, V.S.; et al. Heterogeneity of the angiogenic response induced in different normal adult tissues by vascular permeability factor/vascular endothelial growth factor. *Lab. Investig.* **2000**, *80*, 99–115. [CrossRef]
21. Conway, E.M.; Collen, D.; Carmeliet, P. Molecular mechanisms of blood vessel growth. *Cardiovasc. Res.* **2001**, *49*, 507–521. [CrossRef]
22. Nagy, J.A.; Vasile, E.; Feng, D.; Sundberg, C.; Brown, L.F.; Detmar, M.J.; Lawitts, J.A.; Benjamin, L.; Tan, X.; Manseau, E.J.; et al. Vascular permeability factor/vascular endothelial growth factor induces lymphangiogenesis as well as angiogenesis. *J. Exp. Med.* **2002**, *196*, 1497–1506. [CrossRef] [PubMed]
23. Liu, Y.; Cox, S.R.; Morita, T.; Kourembanas, S. Hypoxia regulates vascular endothelial growth factor gene expression in endothelial cells. Identification of a 5' enhancer. *Circ. Res.* **1995**, *77*, 638–643. [CrossRef] [PubMed]
24. Breier, G.; Albrecht, U.; Sterrer, S.; Risau, W. Expression of vascular endothelial growth factor during embryonic angiogenesis and endothelial cell differentiation. *Development* **1992**, *114*, 521–532. [CrossRef] [PubMed]
25. Ferrara, N.; Carver-Moore, K.; Chen, H.; Dowd, M.; Lu, L.; O'Shea, K.S.; Powell-Braxton, L.; Hillan, K.J.; Moore, M.W. Heterozygous embryonic lethality induced by targeted inactivation of the VEGF gene. *Nature* **1996**, *380*, 439–442. [CrossRef]
26. Carmeliet, P.; Ferreira, V.; Breier, G.; Pollefeyt, S.; Kieckens, L.; Gertsenstein, M.; Fahrig, M.; Vandenhoeck, A.; Harpal, K.; Eberhardt, C.; et al. Abnormal blood vessel development and lethality in embryos lacking a single VEGF allele. *Nature* **1996**, *380*, 435–439. [CrossRef]
27. Zhu, J.; Motejlek, K.; Wang, D.; Zang, K.; Schmidt, A.; Reichardt, L.F. beta8 integrins are required for vascular morphogenesis in mouse embryos. *Development* **2002**, *129*, 2891–2903. [CrossRef]
28. Shalaby, F.; Ho, J.; Stanford, W.L.; Fischer, K.D.; Schuh, A.C.; Schwartz, L.; Bernstein, A.; Rossant, J. A requirement for Flk1 in primitive and definitive hematopoiesis and vasculogenesis. *Cell* **1997**, *89*, 981–990. [CrossRef]
29. Risau, W. Mechanisms of angiogenesis. *Nature* **1997**, *386*, 671–674. [CrossRef]
30. Udan, R.S.; Culver, J.C.; Dickinson, M.E. Understanding vascular development. *Wiley Interdiscip. Rev. Dev. Biol.* **2013**, *2*, 327–346. [CrossRef]
31. Drake, C.J.; Fleming, P.A. Vasculogenesis in the day 6.5 to 9.5 mouse embryo. *Blood* **2000**, *95*, 1671–1679. [CrossRef] [PubMed]
32. de Bruijn, M.F.; Speck, N.A.; Peeters, M.C.; Dzierzak, E. Definitive hematopoietic stem cells first develop within the major arterial regions of the mouse embryo. *EMBO J.* **2000**, *19*, 2465–2474. [CrossRef] [PubMed]
33. Inman, K.E.; Downs, K.M. The murine allantois: Emerging paradigms in development of the mammalian umbilical cord and its relation to the fetus. *Genesis* **2007**, *45*, 237–258. [CrossRef] [PubMed]
34. Fong, G.H.; Rossant, J.; Gertsenstein, M.; Breitman, M.L. Role of the Flt-1 receptor tyrosine kinase in regulating the assembly of vascular endothelium. *Nature* **1995**, *376*, 66–70. [CrossRef]

35. Shalaby, F.; Rossant, J.; Yamaguchi, T.P.; Gertsenstein, M.; Wu, X.F.; Breitman, M.L.; Schuh, A.C. Failure of blood-island formation and vasculogenesis in Flk-1-deficient mice. *Nature* **1995**, *376*, 62–66. [CrossRef] [PubMed]
36. Koch, S.; Claesson-Welsh, L. Signal transduction by vascular endothelial growth factor receptors. *Cold Spring Harb. Perspect. Med.* **2012**, *2*, a006502. [CrossRef]
37. Hellstrom, M.; Kalen, M.; Lindahl, P.; Abramsson, A.; Betsholtz, C. Role of PDGF-B and PDGFR-beta in recruitment of vascular smooth muscle cells and pericytes during embryonic blood vessel formation in the mouse. *Development* **1999**, *126*, 3047–3055. [CrossRef]
38. Heldin, C.H.; Wasteson, A.; Westermark, B. Platelet-derived growth factor. *Mol. Cell Endocrinol.* **1985**, *39*, 169–187. [CrossRef]
39. Liao, H.J.; Kume, T.; McKay, C.; Xu, M.J.; Ihle, J.N.; Carpenter, G. Absence of erythrogenesis and vasculogenesis in Plcg1-deficient mice. *J. Biol. Chem.* **2002**, *277*, 9335–9341. [CrossRef]
40. Wilson, R.; McGuire, C.; Mohun, T.; Project, D. Deciphering the mechanisms of developmental disorders: Phenotype analysis of embryos from mutant mouse lines. *Nucleic Acids Res.* **2016**, *44*, D855–D861. [CrossRef]
41. Spyropoulos, D.D.; Pharr, P.N.; Lavenburg, K.R.; Jackers, P.; Papas, T.S.; Ogawa, M.; Watson, D.K. Hemorrhage, impaired hematopoiesis, and lethality in mouse embryos carrying a targeted disruption of the Fli1 transcription factor. *Mol. Cell Biol.* **2000**, *20*, 5643–5652. [CrossRef] [PubMed]
42. McCarty, J.H.; Monahan-Earley, R.A.; Brown, L.F.; Keller, M.; Gerhardt, H.; Rubin, K.; Shani, M.; Dvorak, H.F.; Wolburg, H.; Bader, B.L.; et al. Defective associations between blood vessels and brain parenchyma lead to cerebral hemorrhage in mice lacking alphav integrins. *Mol. Cell Biol.* **2002**, *22*, 7667–7677. [CrossRef] [PubMed]
43. Gupta, A.; Rarick, K.R.; Ramchandran, R. Established, New and Emerging Concepts in Brain Vascular Development. *Front. Physiol.* **2021**, *12*, 636736. [CrossRef] [PubMed]

current issues in molecular biology

MDPI

Article

Association of Glycoprotein IIIa PlA1/A2 Polymorphism with Risk of Stroke: Updated Meta-Analysis

Camelia Alexandra Coadă [1], Mihai Lupu [2,*], Iulia Florea [1], Stella Di Constanzo [3], Sara Coluccelli [4] and Ioan Şimon [5]

1 Faculty of Medicine, "Iuliu Haţieganu" University of Medicine and Pharmacy Cluj-Napoca, 400012 Cluj-Napoca, Romania; coada_camelia_alexandra@elearn.umfcluj.ro (C.A.C.); iulia.florea@elearn.umfcluj.ro (I.F.)

2 Morpho-Functional Sciences Department, "Iuliu Haţieganu" University of Medicine and Pharmacy Cluj-Napoca, 400012 Cluj-Napoca, Romania

3 Division of Oncologic Gynecology, IRCCS Azienda Ospedaliero-Universitaria di Bologna, 40138 Bologna, Italy; stella.dicostanzo@unibo.it

4 Solid Tumor Molecular Pathology Laboratory, IRCCS Azienda Ospedaliero-Universitaria di Bologna, 40138 Bologna, Italy; sara.coluccelli@aosp.bo.it

5 Department of Surgery, Faculty of Medicine, "Iuliu Haţieganu" University of Medicine and Pharmacy Cluj-Napoca, 400012 Cluj-Napoca, Romania; ioan.simon@umfcluj.ro

* Correspondence: lupu.mihai@umfcluj.ro

Abstract: Cardiovascular diseases are the main cause of death in the world, with ischemic heart disease (i.e., myocardial infarction) and cerebrovascular disease (i.e., stroke) taking the highest toll. Advances in diagnosis and treatment have led to a significant alleviation of ischemic complications, specifically in the realm of pharmacotherapy and interventional devices, while pharmacogenomics has yet to be fully leveraged to improve the burden of disease. Atherothrombotic events might occur earlier or respond worse to treatment in patients with genetic variants of GP IIb/IIIa. Therefore, we aimed to quantitate the involvement of the PlA2 variant in the risk of cerebral stroke events. A systematic search and meta-analysis were performed by pooling the risks of individual studies. A total of 31 studies comprising 5985 stroke patients and 7886 controls were analyzed. A meta-analysis of four studies on hemorrhagic stroke patients showed no association with the PIA2 rs5918(C) polymorphism in both fixed-effect (OR = 0.90 95%CI [0.71; 1.14]; p = 0.398) and random-effect models (OR = 0.86 95%CI [0.62; 1.20]; p-value = 0.386). The power of this analysis was below <30%, indicating a limited ability to detect a true effect. An analysis of the 28 studies on ischemic stroke revealed a significant association with the PIA2 rs5918(C) allele in both fixed-effect (OR = 1.16 95%CI [1.06; 1.27]; p = 0.001) and random-effect models (OR = 1.20 95%CI [1.04; 1.38]; p-value = 0.012), with a power of >80%. The PIA2 allele was associated with an increased risk of ischemic stroke. No association was found with hemorrhagic stroke, most likely due to the small number of available studies, which resulted in a lack of power.

Keywords: stroke; cerebrovascular disease; genetic variant; PIA2 polymorphism; platelet glycoprotein

1. Introduction

Ischemic disease is a leading cause of death worldwide, with the heart and brain being the predilect targets of atherosclerosis [1]. While mortality is higher in the case of cardiac events [1], brain infarction leads to a more prominent burden on quality of life, as neurological cell death affects functional independence and multiple aspects of life, more so than cardiac failure [2].

Antithrombotic treatment is the mainstay therapy in the secondary prevention of is-chemic conditions, whereas interventional procedures lead to a decrease in the complications of acute events. Furthermore, advances in genome sequencing and the advent of personalized pharmacogenomics have shown discrepancies in treatment response in patients with different rheologic phenotypes, especially in the case of genes involved in hemostasis [3]. Because arterial clots tend to be platelet-rich (as opposed to venous' clotting factor preponderance), clinical research has focused on primary hemostasis as a means to prevent arterial occlusions [3].

In the setting of endothelial disruption (atherosclerotic or otherwise) and shear stress, the hemostatic process is initiated by platelet adhesion and mediated by the von Willebrand factor (vWF) and glycoprotein (GP) Ib/Ia, which, in turn, leads to platelet aggregation and white thrombus formation, facilitated by GP IIb/IIIa. GP IIb/IIIa (also known as integrin αIIbβ3) is a platelet fibrinogen receptor with the form of a heterodimer (IIb and IIIa) that is assembled and stored in alpha granules inside platelets. It provides a key point between primary hemostasis and the coagulation cascade, as proven by its implication in both hemorrhagic and thrombotic events [4–7]. While patients with Glanzmann thrombasthenia exhibit higher rates of bleeding, single-nucleotide polymorphisms (SNP) of the IIIa subunit have been associated with an increased rate of platelet activation, earlier debut of atherosclerotic disease [4], and resistance to commonly used antiplatelet drugs (aspirin and clopidogrel) [8].

The most frequent SNP is rs5918 PlA1/A2 of the IIIa moiety and causes the replacement of leucine (PlA1) with proline (PlA2 variant), which has been evaluated in various studies regarding cardiac and brain ischemia, as well as other diseases [4–7]. Both ischemic and hemorrhagic stroke might be affected by GP IIb/IIIa variants, particularly in young patients, where the risk was evaluated to be higher in Caucasian women <45 years old, without reaching clear-cut statistical significance [9]. A decade-old meta-analysis appreciated that PlA2 homozygosity is positively associated with ischemic stroke, but not with hemorrhagic stroke [10]. Of note, all variants were observed to be clinically significant, mainly when associated with conventional cardiovascular risk factors such as diabetes, hypertension, or elevated homocysteine levels. This, together with the wide usage of GP IIb/IIIa inhibitors in cardiovascular disease [11], highlights the importance of a thorough investigation of the potential risks of its portents.

Over the years, the field of medical genetic testing has seen exponential growth as a result of ever-expanding knowledge, as well as increasingly affordable costs. Currently, there is a market for direct-to-consumer genetic panels, which are available to patients without the need for a medical prescription [12,13]. While this has exciting potential by enabling efficient screening for well-known cancer-associated germline mutations such as *BRCA1/2* mutations [14], panels aimed at multigenic and multifactorial traits may lack a robust foundation in studies, and consequently, their clinical utility raises concerns. Moreover, many such panels investigate polymorphisms rather than mutations, and are often based on small studies limited to specific populations without specifying their potential rarity in other populations [15].

Currently, the rs5918 PlA1/A2 SNP is included in various genetic testing panels dedicated to the exploration of cardiovascular risks. Therefore, the aim of this study was to conduct a meta-analysis of existing research to establish both the risk and magnitude of impact on the probability of developing a cerebral stroke. Moreover, we aimed to investigate whether the SNP is associated with a specific type of stroke.

2. Materials and Methods

2.1. Search Strategy and Article Identification

Public research databases were systematically queried for articles studying the relationship between the *GPIIIa* rs5918 and the risk of stroke. No limitations were imposed as to the type of stroke or the source of the population analyzed. The databases searched were PubMed, Scopus, Web of Science, and Cochrane Library. The GeneCards Human Gene Database [16] was used to identify all *GPIIIa* gene IDs' synonyms. Synonyms of rs5918 were identified through Google Scholar search and SNP databases (dbSNP and SNPedia). Advanced search strategies were employed by combining relevant key terms: "stroke", "cerebrovascular", "brain", "infarction", "ischemia", "hemorrhagic", "polymorphism", "BDPLT2", "integrin beta3", "ITGB3", "CD61", "GPIIIa", "GP3A", "BDPLT2", "GT2", "BDPLT16", "BDPLT24", "Human Platelet Antigen 1", "HPA-1", "rs5918", "PIA1/A2", "SNP", "genotype", "allele", "mutation", and "variant". Given that not all authors followed the HGVS [17] recommendations to use the SNP annotation on the primary protein sequence, further searches were conducted by including both the L33P (mature protein) and canonical L59P (primary protein). No language restrictions were in place during this search phase. The search results were exported and then loaded in Rayyan [18] for duplicate removal and article screening and selection.

References from the identified relevant articles were also considered and manually searched for additional data. The last updated research was conducted on the 5 April 2024.

2.2. Article Selection and Inclusion

The articles were screened according to the Preferred Reporting Items for Systematic Reviews and Meta-Analyses (PRISMA) guidelines and included if they met specific criteria. Eligible studies were those that analyzed the presence of the rs5918 SNP in patients diagnosed with stroke, had a case–control design and reported either the genotypes and/or allele frequencies in both cases and controls, or provided the computed odds ratio (OR) for the stroke group. Papers were excluded if they investigated outcomes other than stroke such as different cardiovascular pathologies, were of the wrong study type (case reports, conference abstracts, letters, or opinions), and were written in languages other than English. If a study analyzed multiple outcomes including stroke, only the analysis considering the relationship between rs5918 and stroke was considered for inclusion in the meta-analysis. Additionally, when multiple control populations were present within a study, the one that was the most similar (i.e., similar background risk factors) to the stroke group was considered. After the initial screening, the full texts of the identified articles were read to select the manuscripts for final inclusion in the meta-analysis.

In the case of studies showing the presence of overlap between patients, either the most recent publication was considered or the one comprising a larger number of patients.

2.3. Data Extraction

Relevant data were extracted from each article and stored in a database. Information regarding the number of cases and controls, country of origin, ethnic background, age, sex, distribution of genotype, and allele frequencies between groups and ORs (odds ratios) (if available) was collected.

2.4. Quality Assessment

A quality assessment of the included studies was conducted using the Newcastle–Ottawa Scale (NOS). Two authors independently reviewed the articles and assigned points to each item according to the NOS manual, for a maximum of 9 points. Discrepancies were discussed with a third author when necessary. The studies were considered to be of a high quality—"good"—if their cumulative score was above 6, "fair" if the score was between 4 and 6, and "poor" if the score was between 0 and 3.

2.5. Statistical Analysis

Analysis was conducted in R version 4.3.2 (31 March 2023 ucrt)—"Eye Holes" (R Foundation, Vienna, Austria) [19]. Meta-analysis was conducted using the meta package [20], and Forest plots were used to present the results. The association between the presence of the rs5918 polymorphism and the risk of stroke was assessed by pooling the ORs and their corresponding 95% confidence intervals (CI). Both the fixed-effects and random-effects models were used to compute the pooled ORs. The heterogeneity of the included studies was evaluated using Tau^2 for between-study variance and Higgin's and Thompson's I^2 for the percentage of variability in the effect sizes not due to sampling error [21,22]. The Restricted Maximum-Likelihood estimator was used to calculate Tau^2.

Publication bias was assessed using Egger's regression and visualized using Funnel plots [23]. A power-enhanced sunset Funnel plot was graphed to examine the power of individual studies [24]. The overall power of the pooled results was determined using the dmetar R package [25].

3. Results

The process and results of the identification, screening, and selection of the articles are presented in the PRISMA flowchart (Figure 1). Out of the 440 potentially relevant studies identified, 109 were duplicates and were eliminated. After title and abstract screening, 49 studies were selected for full text reading. Next, 17 studies were further discarded due to having a wrong outcome or design, overlapped patients, or no usable data. The study of Chen et al. [26] was eliminated from the analysis because they found no patient or control carriers of the SNP of interest. Finally, 31 studies were included in the meta-analysis, for a total number of 5985 patients diagnosed with stroke and a total of 7886 controls (Table 1).

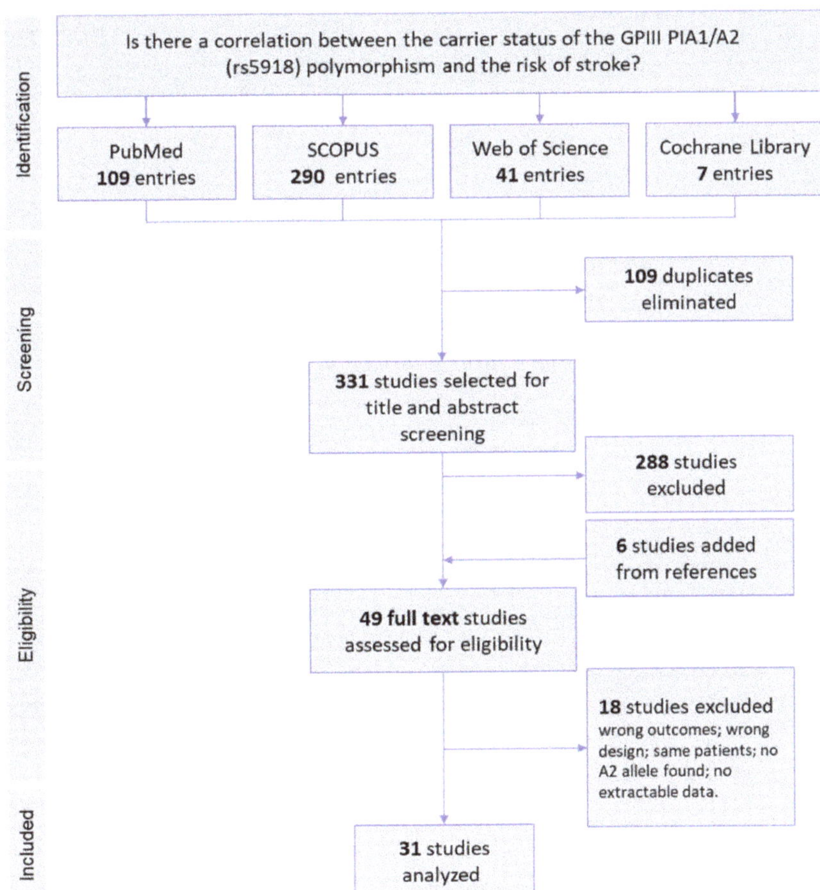

Figure 1. PRISMA flowchart illustrating the workflow for study identification, screening, eligibility evaluation, and inclusion in the meta-analysis. A total of 31 studies met the inclusion criteria and were pooled in the analysis.

Table 1. Description of the studies included in the meta-analysis. N: number.

Author, Year	Country/Race	Type of stroke	No. of Stroke Cases/Controls	Selection Criteria of Stroke Cases/Controls
Adamski et al., 2009 [27]	Poland Caucasian	Hemorrhagic	288/457	White patients with aneurysmal SAH recruited between 2001 and 2007 at the Stroke Unit and the Department of Neurosurgery, Jagiellonian University, Krakow, Poland. Controls matched by gender, race, and age (±10 years) without any history of cerebrovascular disease.
Berge et al., 2007 [28]	Norway	Ischemic	363/482	Patients with acute ischemic stroke and atrial fibrillation. Unselected healthy blood donors recruited from the Blood Bank.
Carlsson et al., 1997 [29]	Germany	Ischemic	218/486	Patients from 1995–1996, admitted to the stroke service of the Department of Neurology. Subarachnoid hemorrhage or patients with a stroke due to vasculitis were excluded. Control group were neurological inpatients without signs for vascular pathology and healthy blood donors.
Carter et al., 1998 [30]	England Caucasian	Ischemic	505/402	Clinical diagnosis of acute stroke. Control subjects were healthy white Europeans free from vascular disease.
Corral et al., 1997 [31]	Spain	Ischemic	103/103	Unselected patients with cerebrovascular disease with age-, race-, and sex-matched controls with no history of cerebrovascular disease.
Derle et al., 2015 [32]	Türkiye	ischemic	42/133	Patients with acute ischemic stroke or TIA. Controls were patients at a high risk for vascular events but with a negative history.
Iniesta et al., 1999 [33]	Spain	Ischemic	124/236	Patients diagnosed with ischemic attack, minor stroke, and cerebral infarction. Controls were individuals with no documented history of migraine, venous, and/or arterial thrombosis and sex- and age-matched individuals with the same risk factors as the patients with ischemic CVD and migraine.
Iniesta et al., 2003 [34]	Spain Caucasian	Hemorrhagic	141/141	Adult patients with a first episode of PIH. Controls were matched by age, sex, race, and selected risk factors for PIH (smoking history, blood pressure, and alcohol consumption), without a documented history of vascular disease.
Iniesta et al., 2004 [35]	Spain Caucasian	Hemorrhagic	103/103	Patients diagnosed with aneurysmal and spontaneous nonaneurysmal SAH. Controls were subjects without a documented history of vascular disease nor a personal history of thromboembolic or hemorrhagic disease and were not undergoing antithrombotic therapy.
Jimenez-Gonzalez et al., 2021 [36]	Mexico	Ischemic	200/200	CT-confirmed diagnosis of ischemic CVD without risk factors for cardioembolic CVD. Controls were healthy potential blood bank donors with no relevant history of CVD and normal biochemical parameters.
Kamberi et al., 2016 [37]	Macedonia Caucasian	Ischemic	39/102	Consecutive patients admitted to the Neurology Department with confirmed diagnosis acute first IS, (age: 18–90 years). Controls were subjects free from CVD.
Kekomaki et al., 1999 [38]	Finland Caucasian	Ischemic	234/200	Survivors of IS or transient ischemic attack. Controls had no signs of CVD and also lived in North Karelia.
Konialis et al., 2016a [39]	Greece Caucasian	Ischemic	166/159	Adult patients newly diagnosed with IS. Control group: adult participants entering the hospital for a confirmed reason unrelated to cerebro- or cardiovascular disease.
Konialis et al., 2016b [39]	Greece Caucasian	Ischemic	34/165	Adult patients newly diagnosed with IS and CHD.

Table 1. *Cont.*

Author, Year	Country/Race	Type of stroke	No. of Stroke Cases/Controls	Selection Criteria of Stroke Cases/Controls
Lanni et al., 2007 [40]	Italy Caucasian	Ischemic	28/180	Hypertensive individuals with IS. Control group were hypertensive individuals without ischemic stroke.
Lu et al., 2014 [41]	China Asian	Ischemic	350/300	Included IS patients (males and females, mean age, 55 years), some with a smoking history and an alcohol consumption history. Controls were healthy people who received physical examination in hospital during the same time period.
Maguire et al., 2011 [42]	Australia Caucasian (94.3%) and non-Caucasian	Ischemic	609/627	Patients with IS, as defined by World Health Organization clinical criteria. Controls were random healthy community men and women (age 55–85) from the New South Wales Electoral Roll between 2004 and 2007.
Meiklejohn et al., 2001 [43]	Scotland Caucasian	Ischemic	150/150	Patients with a cerebrovascular accident without a history of atrial fibrillation, valvular heart disease, or connective tissue disease. Controls were patients born in same year as the subjects with no history of stroke, TIA, peripheral vascular, or ischemic heart disease.
Reiner et al., 2000 [9]	USA Caucasian	Ischemic and hemorrhagic	78 (42 + 36)/346	Caucasian women with nonfatal stroke with brain imaging signs from the Baltimore-Washington Cooperative Young Stroke Study.
Reuner et al., 1997 [44]	Germany	Ischemic	79/79	Patients were male and female with stroke or with TIA. Controls were age- and sex-matched healthy subjects without history of arterial vascular disease.
Ridker et al., 1997 [45]	USA	Ischemic	209/704	Subjects from the Physicians' Health Study randomized, double-blind, placebo-controlled trial. Men aged 40–84, with no history of myocardial infarction, stroke, TIA, or cancer who developed IS. Control group were subjects without signs of CVD.
Rivera-Garcia et al., 2013 [46]	Mexico/ Mexican	Ischemic	200/200	Consecutive unrelated patients <45 years old, with diagnosis of IS. The control group was constituted by individuals without history of CVD.
Roldan et al., 2009 [47]	Spain	Ischemic	96/119	Patients with atrial fibrillation with or without IS. Cases with recent venous thromboembolism, myocardial infarction, acute coronary syndrome, infection or inflammatory disease, surgery, malignancy, renal/liver impairment, and with hormone replacement therapy or oral anticoagulation were excluded.
Rubattu et al., 2005 [48]	Italy	Ischemic	115/180	First-ever IS patient. Patients with TIA were excluded. Controls were age-matched healthy unrelated subjects recruited from the same medical center, among a population of blood donors. They were free from any pharmacological treatment.
Saidi et al., 2008 [29]	Tunisia Arabs	Ischemic	329/444	Tunisian hospitalized patients with first IS from the Neurology service of CHU Sahloul, Sousse, Tunisia. Controls were unrelated healthy individuals from the same geographical area as the patients.
Slowik et al., 2004 [49]	Poland Caucasian	Ischemic	377/572	Unrelated patients with IS. Controls were unrelated individuals, free of clinically detectable CVD and without any stroke history.

Table 1. *Cont.*

Author, Year	Country/Race	Type of stroke	No. of Stroke Cases/Controls	Selection Criteria of Stroke Cases/Controls
Salvarani et al., 2007 [50]	Italy Caucasian	Ischemic	31/109	Patients with biopsy-proven giant cell arteritis who were residents of Reggio Emilia, Italy, with cranial ischemic complications. Controls were population-based healthy subjects from the same geographic area.
Szolnoki et al., 2003 [51]	Hungary Caucasian	Ischemic	545/158	Consecutive Caucasian Hungarian patients with a first acute IS. Controls were random neurological patients without stroke and with normal MRI.
van Goor et al., 2002 [52]	Netherlands	Ischemic	45/60	Patients with TIA and ischemic stroke. Controls were healthy blood donors. Risk factors and clinical data were not assessed.
Wagner et al., 1998 [53]	USA Caucasian Black	Ischemic	65 (33 + 32)/ 123 (76 + 47)	Female patients from 15 to 44 years of age with a first cerebral infarction, identified by discharge surveillance at 59 regional hospitals and through direct referral by regional neurologists. Controls were women without a history of stroke, frequency-matched by age and geographic region of residence to the cases.
Zhang et al., 2007 [54]	China Asian	Ischemic	119/166	Patients with a diagnosis of IS. Cardioembolic strokes, hemorrhagic stroke, TIA, cerebral venous thrombosis, and hemorrhagic transformation of an infarct were excluded. Controls were age- and sex-matched apparently healthy subjects without a previous history of stroke, myocardial infarction and peripheral arterial disease.

SAH: subarachnoid hemorrhage; CVD: cardiovascular disease; IS: ischemic stroke; CHD: coronary heart disease; and TIA: transient ischemic attack.

3.1. Quality Assessment of the Included Studies

The NOS tool for case–control studies was applied to all 31 studies selected for the meta-analysis. The criteria with the highest points across studies were those in the "Selection" section, suggesting strength in the methodology of participant selection. Conversely, the area with the lowest points was that of "Comparability". Nevertheless, the overall scores indicated that the studies were of a good quality. The assigned points for each item are reported in Supplementary Table S1.

3.2. Publication Bias and Power Estimation

Publication bias was inspected through the construction of a funnel plot (Figure 2). There was a slight asymmetrical pattern given by four studies with high effect sizes, namely Saidi et al. [29], Roldan et al. [47], Derle et al. [32], and Iniesta et al. [35] (Supplementary Figure S1). However, testing for the presence of publication bias using Eggers' test did not indicate a significant funnel plot asymmetry ($t = 0.85$, $df = 30$, p-value = 0.401). An examination of the power-enhanced funnel plot revealed that most of the published studies had a low power (Supplementary Figure S2). The overall power of this meta-analysis was calculated to be 95.59% for a fixed-effect model and 85% when assuming a random-effects model. When considering the meta-analysis conducted on studies with hemorrhagic stroke, the power was 20.14% for a fixed-effect model and 23.65% for a random-effects model. The meta-analysis of the ischemic stroke studies showed a power of 94.11% for the fixed-effect model and 82.15% for the random-effect model.

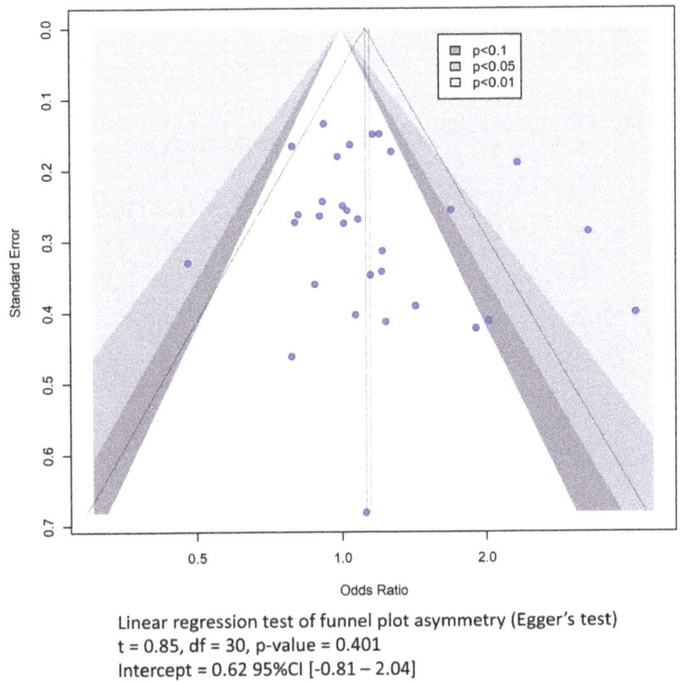

Figure 2. Funnel plot showing the publication bias. The vertical dashed line represents the average effect size while the diagonal dashed lines represent the idealized funnel shape that the studies are expected to follow. CI: confidence interval; df: degrees of freedom; t: t-statistic.

3.3. *Carrier Status of the PIA2 rs5918(C) Allele and Stroke Risk*

The selected studies comprised both hemorrhagic and ischemic stroke subtypes. Most of them focused on ischemic strokes (n = 28), while only four studies analyzed hemorrhagic strokes. Among these, one study included patients from both subtypes of stroke [9].

A meta-analysis of the four studies [9,27,34,35] including hemorrhagic stroke patients showed no association with the carrier status of the PIA2 rs5918(C) polymorphism in both fixed-effect (OR = 0.90 95%CI [0.71; 1.14]; $p = 0.398$) and random-effect models (OR = 0.86 95%CI [0.62; 1.20]; p-value = 0.386). The I^2 index showed a moderate but insignificant heterogeneity of the studies ($I^2 = 41\%$ [0%; 80%]), $Tau^2 = 0.05$ [0.00; 1.80]. Cochran's Q was 5.06 with a $p = 0.167$. The prediction interval was wide [0.26; 2.83] (Figure 3; Supplementary Figure S3).

An analysis of the 28 studies [9,28–33,36–54] focusing on ischemic stroke revealed a significant association with the carrier status of the PIA2 rs5918(C) allele in both the fixed-effect (OR = 1.16 95%CI [1.06; 1.27]; $p = 0.001$) and random-effect models (OR = 1.20 95%CI [1.04; 1.38]; p-value = 0.012). The I^2 index showed a moderate/substantial percentage of effect sizes variation due to heterogeneity ($I^2 = 54\%$ [29%, 70%]) and $Tau^2 = 0.07$ [0.01, 0.20]. Cochran's Q was 58.22 with a significant p-value < 0.001. The prediction interval was also wide [0.68; 2.09] (Figure 3; Supplementary Figure S4).

A total pooled analysis showed that carriers of the PIA2 rs5918(C) allele had a higher risk of stroke with an OR of 1.12 95%CI [1.03; 1.22], p-value = 0.006, when considering a fixed-effect model. In the random-effect model, the pooled OR was 1.15 95%CI [1.01; 1.31], p-value = 0.042. The heterogeneity index showed an $I^2 = 54\%$ [31%; 69%], corresponding to a moderate/substantial heterogeneity and inconsistency across studies. Tau^2 was 0.068 with its CI [0.02, 0.20], suggesting the presence of between-study heterogeneity and Q = 67.03 with a p-value < 0.001. Moreover, the prediction interval of the studies fell in the range of [0.66; 1.99], suggesting a high dispersion of the effects seen across individual studies (Figure 3).

Study or Subgroup	logOR	SE	Weight (fixed)	Weight (random)	Odds Ratio IV, Fixed + Random, 95% CI
`Type of stroke` = hemorrhagic					
Adamski 2009	0.042	0.164	6.7%	4.7%	1.04 [0.76, 1.44]
Iniesta 2003	−0.207	0.263	2.6%	3.3%	0.81 [0.49, 1.36]
Iniesta 2004	−0.734	0.330	1.7%	2.5%	0.48 [0.25, 0.92]
Reiner 2000h	0.138	0.348	1.5%	2.4%	1.15 [0.58, 2.27]
Total (fixed effect, 95% CI)			**12.4%**	.	**0.90 [0.71, 1.14]**
Total (random effect, 95% CI)			.	**12.9%**	**0.86 [0.62, 1.20]**
Heterogeneity: $Tau^2 = 0.047$; $Chi^2 = 5.07$, df = 3 (P = 0.167); $I^2 = 41\%$					
Test for overall effect (fixed effect): Z = -0.84 (P = 0.398)					
Test for overall effect (random effects): Z = -0.87 (P = 0.386)					
`Type of stroke` = ischemic					
Berge 2007	−0.236	0.166	6.5%	4.7%	0.79 [0.57, 1.09]
Carlsson 1997	0.239	0.174	6.0%	4.6%	1.27 [0.90, 1.79]
Carter 1998	0.184	0.149	8.1%	5.0%	1.20 [0.90, 1.61]
Corral 1997	0.196	0.314	1.8%	2.7%	1.22 [0.66, 2.25]
Derle 2015	1.406	0.400	1.1%	2.0%	4.08 [1.86, 8.94]
Iniesta 1999	0.028	0.257	2.7%	3.3%	1.03 [0.62, 1.70]
Jimenez-Gonzalez 2021	−0.105	0.265	2.6%	3.2%	0.90 [0.54, 1.51]
Kamberi 2016	0.211	0.413	1.1%	1.9%	1.23 [0.55, 2.77]
Kekomaki 1999	0.006	0.250	2.9%	3.4%	1.01 [0.62, 1.64]
Konialis 2016a	−0.091	0.244	3.0%	3.5%	0.91 [0.57, 1.47]
Konialis 2016b	0.067	0.403	1.1%	1.9%	1.07 [0.48, 2.36]
Lanni 2007	0.704	0.412	1.1%	1.9%	2.02 [0.90, 4.54]
Lu 2014	0.354	0.391	1.2%	2.0%	1.42 [0.66, 3.07]
Maguire 2011	−0.083	0.135	9.9%	5.2%	0.92 [0.71, 1.20]
Meiklejohn 2001	−0.224	0.274	2.4%	3.1%	0.80 [0.47, 1.37]
Reiner 2000i	0.079	0.269	2.5%	3.2%	1.08 [0.64, 1.84]
Reuner 1997	−0.130	0.361	1.4%	2.3%	0.88 [0.43, 1.78]
Ridker 1997	−0.018	0.181	5.5%	4.5%	0.98 [0.69, 1.40]
Rivera-Garcia 2013	−0.105	0.265	2.6%	3.2%	0.90 [0.54, 1.51]
Roldan 2009	1.180	0.287	2.2%	3.0%	3.25 [1.85, 5.71]
Rubattu 2005a	0.012	0.275	2.4%	3.1%	1.01 [0.59, 1.73]
Saidi 2008	0.844	0.190	5.0%	4.3%	2.33 [1.60, 3.37]
Salvarani 2007	0.643	0.422	1.0%	1.8%	1.90 [0.83, 4.35]
Slowik 2004	0.151	0.150	8.1%	5.0%	1.16 [0.87, 1.56]
Szolnoki 2003	0.525	0.257	2.7%	3.4%	1.69 [1.02, 2.79]
van Goor 2002	−0.241	0.462	0.8%	1.6%	0.79 [0.32, 1.94]
Wagner 1998	0.193	0.343	1.5%	2.4%	1.21 [0.62, 2.37]
Zhang 2007	0.113	0.682	0.4%	0.8%	1.12 [0.29, 4.26]
Total (fixed effect, 95% CI)			**87.6%**	.	**1.16 [1.06, 1.27]**
Total (random effect, 95% CI)			.	**87.1%**	**1.20 [1.04, 1.38]**
Heterogeneity: $Tau^2 = 0.068$; $Chi^2 = 58.22$, df = 27 (P < 0.001); $I^2 = 54\%$					
Test for overall effect (fixed effect): Z = 3.24 (P = 0.001)					
Test for overall effect (random effects): Z = 2.51 (P = 0.012)					
Total (fixed effect, 95% CI)			**100.0%**	.	**1.12 [1.03, 1.22]**
Total (random effect, 95% CI)			.	**100.0%**	**1.15 [1.01, 1.31]**
Prediction interval					**[0.66, 1.99]**
Heterogeneity: $Tau^2 = 0.068$; $Chi^2 = 67.03$, df = 31 (P < 0.001); $I^2 = 54\%$					
Test for overall effect (fixed effect): Z = 2.74 (P = 0.006)					
Test for overall effect (random effects): Z = 2.04 (P = 0.042)					
Test for subgroup differences (fixed effect): $Chi^2 = 3.74$, df = 1 (P = 0.053)					
Test for subgroup differences (random effects): $Chi^2 = 3.14$, df = 1 (P = 0.076)					

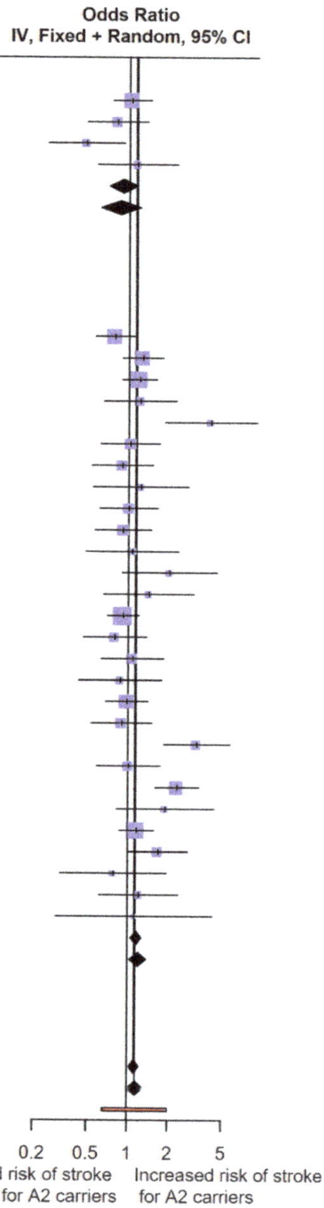

Figure 3. Forest plot showing the OR of individual studies [9,27–54] and the pooled analysis divided by type of stroke and overall effects. Both fixed effect and random effect models were used. OR: odds ratio; CI: confidence interval; SE: standard error.

A separate analysis was conducted on the studies reporting the ethnicity of the included subjects. A total of 16 studies included Caucasian subjects and two studies included

Asian subjects. The meta-analysis of the 16 studies revealed an OR of 1.17 95%CI [1.02, 1.35] for both fixed and random effects; p-value = 0.026 (Supplementary Figures S5 and S6).

4. Discussion

Stroke represents a major health problem, as it ranks among the topmost contributors to global mortality and long-term disability [55,56]. The underlying pathophysiology of stroke encompasses multiple risk factors, both genetic as well as environmental factors [57]. Amid the non-genetic modifiable risk factors, the most well-known are heart disease, blood hypertension, sedentary lifestyle, smoking and alcohol consumption, dyslipidemia, and obesity [58,59]. These factors interact between themselves as well as with genetic predisposition factors in a complex manner, collectively contributing to its onset. Among the explored genetic risk factors, the PIA2 allele of GPIIIa has gained considerable attention, so much so that it is part of various genetic panels for cardiovascular risk assessment. We conducted a meta-analysis of the available studies to assess the relationship of this SNP with the risk of developing a stroke and found a small but significant association with the ischemic type, but not with the hemorrhagic type.

One important aspect that stands out from the pooling of the individual studies is that the effect sizes varied substantially across them. The implications of this variation need to be considered and discussed. Indeed, with a high number of studies, heterogeneity is inevitable [21]. However, the high heterogeneity here could denote that the mean effect may not be representative for all populations. This was further highlighted by the large width of the prediction interval which included 1. The wide dispersion of the PIA2 effects suggests that, at one extreme, the studies point out a trivial or even nil risk of stroke, while the other extreme points out an elevated risk of stroke. Moreover, most studies were conducted on Caucasians, further limiting the generalizability of these results.

4.1. The PIA2 Allele in the Context of Different Populations

It is worth discussing that allele frequency distribution among different populations is highly diverse, thus, studies tackling this topic should assess their required sample size in a population-dependent manner [60,61]. None of the studies included in this meta-analysis conducted an a priori sample size calculation, making most of the studies underpowered. This was confirmed by the power-enhanced funnel plot showing that the studies had a power less than 40%.

rs5918 is a relatively uncommon SNP present with different frequencies among populations. The Genome Aggregation Database (gnomAD) [62], a platform encompassing data from more than 800,000 exomes and genomes from various human sequencing studies, revealed that the minor allele frequencies (MAFs) of the rs5918 SNP are highest in Europeans, with MAFs of around 0.15, while the number significantly drops for East Asian populations, who report a MAF of 0.004. This explains the inexistence of PIA2 carriers in the study of Chen et al. [26], who conducted their research on cases from Taiwan, where this SNP is rare. Therefore, their study would have required an infeasibly large population to test the hypothesis. Another study, that of Zhang et al. [54], included 285 patients from China, where the MAF was reported to be around 0.09, and, in fact, had the lowest power of all studies included in our meta-analysis.

4.2. Clinical Relevance of the Results in the Context of Direct-to-Consumer Genetic Testing Panels

The issue of underpowered studies is widely spread, even among Cochrane reviews, impacting the precision and heterogeneity of results. Moreover, many meta-analyses themselves do not estimate their power [63]. The present meta-analysis conducted on the 28 studies with ischemic stroke patients had an adequate power of more than 80%. The pooled ORs for ischemic stroke studies were 1.16 and 1.20 for fixed- and random-effect models, respectively. This suggests that the impact, albeit statistically significant, may not be sufficiently large to offer practical clinical utility, and that testing for this SNP for stroke risk assessment may be futile in certain cases [64]. On the other hand, with only four studies available for the meta-analysis of hemorrhagic strokes, the statistical power was slightly above 20%, considerably lower than the recommended threshold. This discrepancy in the available literature is not surprising, since the vast majority of strokes are of an ischemic nature [1]. Nevertheless, the relationship between PIA2 and hemorrhagic stroke seemed to have an opposite trend. While these results need to be taken cautiously due to the underpowered meta-analysis, the pooled ORs were 0.86 and 0.90. Given that the PIA2 SNP is associated with an increase in platelet aggregability [65], it is not surprising that carriers may have some level of protection from hemorrhagic events.

Overall, one could argue that a genetic variant with such a high variability and such negligible impact should have limited clinical relevance, at least when interpreted alone. Indeed, the presence of several genetic variants with minor impacts alone could result in a significant cumulative impact, should they be present simultaneously in a patient [66]. However, to the best of our knowledge, many current genetic testing panels for multifactorial diseases lack compelling evidence for the genetic variants they include. Moreover, the reports accompanying the results are often scarce, without offering further genetic counseling [12,67]. Nowadays, direct-to-consumer genetic testing allows patients to order such tests without specific medical indication, exposing them to unintended information and to false-positive or -negative results [13,68]. The discovery of a health risk, albeit sometimes with low probabilities, can lead to psychological distress, as many patients do not fully grasp the notion that a risk does not imply that they will actually develop the respective disease [13,69]. Furthermore, the rapid growth and high availability of these panels has long surpassed the training of many medical doctors, limiting their capacity to accurately interpret their patients' results [12,13,67]. Therefore, well-conducted systematic analyses such as the one presented here are much needed to fill in this knowledge gap.

4.3. Study Limitations and Future Directions

A relevant limitation of the analyzed studies is the heterogeneity of cases, and most importantly controls, which were taken both from the general population as well as from patients matched for comorbidities known to influence the risk of stroke, such as atrial fibrillation, body mass index, age, and other factors. This might also have contributed to the high variability and heterogeneity of the results. Despite this issue, strong predictors, whether genetic or not, should have a more pronounced association that is independent of other comorbidities.

Despite advancements in and the widespread availability of genetic testing, the evidence regarding the importance of the PIA1/A2 polymorphism in stroke risk remains modest. While a positive association with the risk of ischemic stroke was found, no progress in understanding its role in hemorrhagic stroke has been recorded. This issue was due to an insufficient number of adequately powered studies on this topic. Further research is needed to elucidate the association between PIA2 and this type of stroke.

5. Conclusions

The carrier status for the PIA2 polymorphism was significantly associated with a marginally elevated risk of ischemic stroke. No significant association could be found with hemorrhagic stroke, due to the small number of studies analyzing such patients. Care must be taken when evaluating the risk of stroke through genetic testing, due to its multifactorial nature.

Supplementary Materials: The following supporting information can be downloaded at: https://www.mdpi.com/article/10.3390/cimb46060321/s1. Table S1: Quality assessment of the studies included in the meta-analysis; Figure S1: Funnel plot with labels for the inspection of study bias showing the scatter of the studies effect sizes and the measure of their standard error; Figure S2: Sunset power enhanced funnel plot showing study-level statistical power; Figure S3: Forest plot for haemorrhagic stroke studies; Figure S4: Forest plot for ischemic stroke studies; Figure S5: Forest plot for stroke studies specifying the race: Caucasian; Figure S6: Forest plot for stroke studies specifying the race: Asian.

Author Contributions: Conceptualization: C.A.C. and M.L. Data curation: I.F., S.D.C. and S.C. Formal analysis: C.A.C. and M.L. Funding acquisition: I.Ş. Investigation: I.F., S.D.C. and S.C. Methodology: C.A.C. and M.L. Project administration: I.Ş. Resources: I.Ş. Software: C.A.C. Supervision: I.Ş. Validation: M.L. and S.D.C. Visualization: C.A.C. Writing—original draft preparation: C.A.C. and M.L. Writing—review and editing: I.Ş. All authors have read and agreed to the published version of the manuscript.

Funding: This research received no external funding.

Institutional Review Board Statement: Not applicable.

Informed Consent Statement: Not applicable.

Data Availability Statement: All data generated in this work is presented in the manuscript.

Acknowledgments: The graphical abstract was created with icons from Bioincons (https://bioicons.com/) and Servier Medical Art (https://smart.servier.com/).

Conflicts of Interest: The authors declare no conflicts of interest.

References

1. Roth, G.A.; Abate, D.; Abate, K.H.; Abay, S.M.; Abbafati, C.; Abbasi, N.; Abbastabar, H.; Abd-Allah, F.; Abdela, J.; Abdelalim, A.; et al. Global, Regional, and National Age-Sex-Specific Mortality for 282 Causes of Death in 195 Countries and Territories, 1980–2017: A Systematic Analysis for the Global Burden of Disease Study 2017. *Lancet* **2018**, *392*, 1736–1788. [CrossRef]
2. Lourenço, E.; Sampaio, M.R.d.M.; Sánchez-Ramos, J.L.; Costa, E.I.M.T.d. A Comparative Study on the Quality of Life of Survivors of Stroke and Acute Myocardial Infarction. *Healthcare* **2024**, *12*, 254. [CrossRef] [PubMed]
3. Lane, D.A.; Grant, P.J. Role of Hemostatic Gene Polymorphisms in Venous and Arterial Thrombotic Disease. *Blood* **2000**, *95*, 1517–1532. [CrossRef] [PubMed]
4. Floyd, C.N.; Mustafa, A.; Ferro, A. The PlA1/A2 Polymorphism of Glycoprotein IIIa as a Risk Factor for Myocardial Infarction: A Meta-Analysis. *PLoS ONE* **2014**, *9*, e101518. [CrossRef] [PubMed]
5. Wang, J.; Liu, J.; Zhou, Y.; Wang, F.; Xu, K.; Kong, D.; Bai, J.; Chen, J.; Gong, X.; Meng, H.; et al. Association among PlA1/A2 Gene Polymorphism, Laboratory Aspirin Resistance and Clinical Outcomes in Patients with Coronary Artery Disease: An Updated Meta-Analysis. *Sci. Rep.* **2019**, *9*, 13177. [CrossRef] [PubMed]
6. Khatami, M.; Heidari, M.M. Common Rs5918 (PlA1/A2) Polymorphism in the ITGB3 Gene and Risk of Coronary Artery Disease. *Arch. Med. Sci. Atheroscler. Dis.* **2016**, *1*, 9–15. [CrossRef] [PubMed]
7. Komsa-Penkova, R.; Golemanov, G.; Tsankov, B.; Ivanov, P.; Beshev, L.; Tonchev, P. Rs5918ITGB3 Polymorphism, Smoking, and BMI as Risk Factors for Early Onset and Recurrence of DVT in Young Women. *Clin. Appl. Thromb. Hemost.* **2017**, *23*, 585–595. [CrossRef] [PubMed]
8. Zuern, C.S.; Schwab, M.; Gawaz, M.; Geisler, T. Platelet Pharmacogenomics. *J. Thromb. Haemost.* **2010**, *8*, 1147–1158. [CrossRef] [PubMed]
9. Reiner, A.P.; Kumar, P.N.; Schwartz, S.M.; Longstreth, W.T.; Pearce, R.M.; Rosendaal, F.R.; Psaty, B.M.; Siscovick, D.S. Genetic Variants of Platelet Glycoprotein Receptors and Risk of Stroke in Young Women. *Stroke* **2000**, *31*, 1628–1633. [CrossRef] [PubMed]
10. Floyd, C.N.; Ellis, B.H.; Ferro, A. The PlA1/A2 Polymorphism of Glycoprotein IIIa as a Risk Factor for Stroke: A Systematic Review and Meta-Analysis. *PLoS ONE* **2014**, *9*, e100239. [CrossRef] [PubMed]
11. Tavenier, A.H.; Claassens, D.M.F.; Hermanides, R.S.; Vos, G.J.A.; Bergmeijer, T.O.; Kelder, J.C.; Deneer, V.H.M.; van 't Hof, A.W.J.; ten Berg, J.M. Efficacy and Safety of Glycoprotein IIb/IIIa Inhibitors in Addition to P2Y12 Inhibitors in ST-Segment Elevation Myocardial Infarction: A Subanalysis of the POPular Genetics Trial. *Catheter. Cardiovasc. Interv.* **2022**, *99*, 676–685. [CrossRef] [PubMed]
12. Malgorzata, M.; Maria, S.; Michał, W. Genetic Testing—Whether to Allow Complete Freedom? Direct to Consumer Tests versus Genetic Tests for Medical Purposes. *J. Appl. Genet.* **2022**, *63*, 119–126. [CrossRef]
13. Horton, R.; Crawford, G.; Freeman, L.; Fenwick, A.; Wright, C.F.; Lucassen, A. Direct-to-Consumer Genetic Testing. *BMJ* **2019**, *367*, l5688. [CrossRef] [PubMed]
14. Direct-to-Consumer Test for BRCA Mutations Authorized. *Cancer Discov.* **2018**, *8*, OF5. [CrossRef]
15. Sherkow, J.S.; Park, J.K.; Lu, C.Y. Regulating Direct-to-Consumer Polygenic Risk Scores. *JAMA* **2023**, *330*, 691–692. [CrossRef] [PubMed]
16. Stelzer, G.; Rosen, N.; Plaschkes, I.; Zimmerman, S.; Twik, M.; Fishilevich, S.; Stein, T.I.; Nudel, R.; Lieder, I.; Mazor, Y.; et al. The GeneCards Suite: From Gene Data Mining to Disease Genome Sequence Analyses. *Curr. Protoc. Bioinform.* **2016**, *54*, 1.30.1–1.30.33. [CrossRef]
17. den Dunnen, J.T.; Dalgleish, R.; Maglott, D.R.; Hart, R.K.; Greenblatt, M.S.; McGowan-Jordan, J.; Roux, A.-F.; Smith, T.; Antonarakis, S.E.; Taschner, P.E.M. HGVS Recommendations for the Description of Sequence Variants: 2016 Update. *Hum. Mutat.* **2016**, *37*, 564–569. [CrossRef]

18. Ouzzani, M.; Hammady, H.; Fedorowicz, Z.; Elmagarmid, A. Rayyan—A Web and Mobile App for Systematic Reviews. *Syst. Rev.* **2016**, *5*, 210. [CrossRef]
19. R Core Team. European Environment Agency. 2020. Available online: https://www.eea.europa.eu/data-and-maps/indicators/oxygen-consuming-substances-in-rivers/r-development-core-team-2006 (accessed on 1 May 2024).
20. Balduzzi, S.; Rücker, G.; Schwarzer, G. How to Perform a Meta-Analysis with R: A Practical Tutorial. *BMJ Ment Health* **2019**, *22*, 153–160. [CrossRef]
21. Higgins, J.P.T.; Thompson, S.G. Quantifying Heterogeneity in a Meta-Analysis. *Stat. Med.* **2002**, *21*, 1539–1558. [CrossRef]
22. Borenstein, M. How to Understand and Report Heterogeneity in a Meta-Analysis: The Difference between I-Squared and Prediction Intervals. *Integr. Med. Res.* **2023**, *12*, 101014. [CrossRef]
23. Egger, M.; Smith, G.D.; Schneider, M.; Minder, C. Bias in Meta-Analysis Detected by a Simple, Graphical Test. *BMJ* **1997**, *315*, 629–634. [CrossRef] [PubMed]
24. Kossmeier, M.; Tran, U.S.; Voracek, M. Metaviz: Forest Plots, Funnel Plots, and Visual Funnel Plot Inference for Meta-Analysis 2020. Available online: https://github.com/Mkossmeier/metaviz (accessed on 25 May 2024).
25. Harrer, M.; Cuijpers, P.; Furukawa, T.; Ebert, D.D. Dmetar: Companion R Package for the Guide 'Doing Meta-Analysis in R 2019. Available online: https://github.com/MathiasHarrer/dmetar (accessed on 25 May 2024).
26. Chen, C.-H.; Lo, Y.-K.; Ke, D.; Liu, C.-K.; Liou, C.-W.; Wu, H.-L.; Lai, M.-L. Platelet Glycoprotein Ia C807T, Ib C3550T, and IIIa PlA1/A2 Polymorphisms and Ischemic Stroke in Young Taiwanese. *J. Neurol. Sci.* **2004**, *227*, 1–5. [CrossRef] [PubMed]
27. Adamski, M.G.; Borratynska, A.; Krupa, M.; Wloch-Kopec, D.; Turaj, W.; Wolkow, P.; Wnuk, M.; Urbanik, A.; Moskala, M.; Szczudlik, A.; et al. A1/A2 Polymorphism of GpIIIa Gene and a Risk of Aneurysmal Subarachnoid Haemorrhage. *Biochem. Biophys. Res. Commun.* **2009**, *383*, 228–230. [CrossRef] [PubMed]
28. Berge, E.; Haug, K.B.F.; Charlotte Sandset, E.; Kristine Haugbro, K.; Turkovic, M.; Sandset, P.M. The Factor V Leiden, Prothrombin Gene 20210GA, Methylenetetrahydrofolate Reductase 677CT and Platelet Glycoprotein IIIa 1565TC Mutations in Patients With Acute Ischemic Stroke and Atrial Fibrillation. *Stroke* **2007**, *38*, 1069–1071. [CrossRef] [PubMed]
29. Saidi, S.; Mahjoub, T.; Slamia, L.B.; Ammou, S.B.; Al-Subaie, A.M.; Almawi, W.Y. Polymorphisms of the Human Platelet Alloantigens HPA-1, HPA-2, HPA-3, and HPA-4 in Ischemic Stroke. *Am. J. Hematol.* **2008**, *83*, 570–573. [CrossRef] [PubMed]
30. Carter, A.M.; Catto, A.J.; Bamford, J.M.; Grant, P.J. Platelet GP IIIa PlA and GP Ib Variable Number Tandem Repeat Polymorphisms and Markers of Platelet Activation in Acute Stroke. *Arterioscler. Thromb. Vasc. Biol.* **1998**, *18*, 1124–1131. [CrossRef] [PubMed]
31. Corral, J.; González-Conejero, R.; Rivera, J.; Iniesta, J.A.; Lozano, M.L.; Vicente, V. HPA-1 Genotype in Arterial Thrombosis—Role of HPA-1b Polymorphism in Platelet Function. *Blood Coagul. Fibrinolysis* **1997**, *8*, 284. [CrossRef]
32. Derle, E.; Öcal, R.; Kibaroglu, S.; Çelikkol, C.; Bayraktar, N.; Verdi, H.; Ataç, B.F.; Can, U. Aspirin Resistance in Cerebrovascular Disease and the Role of Glycoprotein IIIa Polymorphism in Turkish Stroke Patients. *Blood Coagul. Fibrinolysis* **2016**, *27*, 169. [CrossRef]
33. Iniesta, J.A.; Corral, J.; González-Conejero, R.; Rivera, J.; Vicente, V. Prothrombotic Genetic Risk Factors in Patients With Coexisting Migraine and Ischemic Cerebrovascular Disease. *Headache J. Head Face Pain* **1999**, *39*, 486–489. [CrossRef]
34. Iniesta, J.A.; Corral, J.; González-Conejero, R.; Piqueras, C.; Vicente, V. Polymorphisms of Platelet Adhesive Receptors: Do They Play a Role in Primary Intracerebral Hemorrhage? *Cerebrovasc. Dis.* **2003**, *15*, 51–55. [CrossRef] [PubMed]
35. Iniesta, J.A.; González-Conejero, R.; Piqueras, C.; Vicente, V.; Corral, J. Platelet GP IIIa Polymorphism HPA-1 (PlA) Protects Against Subarachnoid Hemorrhage. *Stroke* **2004**, *35*, 2282–2286. [CrossRef]
36. Jiménez-González, M.C.; Santiago-Germán, D.; Castillo-Henkel, E.F.; Alvarado-Moreno, J.A.; Hernández-Juárez, J.; Leaños-Miranda, A.; Majluf-Cruz, A.; Isordia-Salas, I. Identification of Genetic Risk Factors Associated with Ischaemic Stroke in Young Mexican Patients. *Neurol. (Engl. Ed.)* **2021**, *36*, 337–345. [CrossRef] [PubMed]
37. Kamberi, B.; Kamberi, F.; Spiroski, M. Vascular Genetic Variants and Ischemic Stroke Susceptibility in Albanians from the Republic of Macedonia. *Open Access Maced. J. Med. Sci.* **2016**, *4*, 556–564. [CrossRef]
38. Kekornaki, S.; Harnalainen, L.; Kauppinen-Makelin, R.; Palornaki, H.; Kaste, M.; Kontula, K. Genetic Polymorphism of Platelet Glycoprotein Lila in Patients with Acute Myocardial Infarction and Acute Ischaemic Stroke. *J. Cardiovasc. Risk* **1999**, *6*, 13–17. [CrossRef] [PubMed]
39. Konialis, C.; Spengos, K.; Iliopoulos, P.; Karapanou, S.; Gialatos, E.; Hagnefelt, B.; Vemmos, K.; Zakopoulos, N.; Pangalos, C. The APOE E4 Allele Confers Increased Risk of Ischemic Stroke Among Greek Carriers. *Adv. Clin. Exp. Med.* **2016**, *25*, 471–478. [CrossRef]
40. Lanni, F.; Santulli, G.; Izzo, R.; Rubattu, S.; Zanda, B.; Volpe, M.; Iaccarino, G.; Trimarco, B. The PlA1/A2 Polymorphism of Glycoprotein IIIa and Cerebrovascular Events in Hypertension: Increased Risk of Ischemic Stroke in High-Risk Patients. *J. Hypertens.* **2007**, *25*, 551. [CrossRef] [PubMed]
41. Lu, J.X.; Lu, Z.Q.; Zhang, S.L.; Zhi, J.; Chen, Z.P.; Wang, W.X. Polymorphism in Integrin ITGA2 Is Associated with Ischemic Stroke and Altered Serum Cholesterol in Chinese Individuals. *Balk. Med. J.* **2014**, *33*, 55–59. [CrossRef] [PubMed]
42. Maguire, J.; Thakkinstian, A.; Levi, C.; Lincz, L.; Bisset, L.; Sturm, J.; Scott, R.; Whyte, S.; Attia, J. Impact of *COX-2 Rs5275* and *Rs20417* and *GPIIIa Rs5918* Polymorphisms on 90-Day Ischemic Stroke Functional Outcome: A Novel Finding. *J. Stroke Cerebrovasc. Dis.* **2011**, *20*, 134–144. [CrossRef] [PubMed]

43. Meiklejohn, D.J.; Vickers, M.A.; Morrison, E.R.; Dijkhuisen, R.; Moore, I.; Urbaniak, S.J.; Greaves, M. In Vivo Platelet Activation in Atherothrombotic Stroke Is Not Determined by Polymorphisms of Human Platelet Glycoprotein IIIa or Ib. *Br. J. Haematol.* **2001**, *112*, 621–631. [CrossRef] [PubMed]
44. Reuner, K.H.; Elgas, M.; Kaps, M.; Ruf, A.; Patscheke, H. The Human Platelet Antigen HPA-La/Lb (PIA1/PIA2)Polymorphism and Cerebral Ischaemia. *Thromb. Haemost.* **1997**, *78*, 964–965. [CrossRef] [PubMed]
45. Ridker, P.M.; Hennekens, C.H.; Schmitz, C.; Stampfer, M.J.; Lindpaintner, K. PIA1/A2 Polymorphism of Platelet Glycoprotein IIIa and Risks of Myocardial Infarction, Stroke, and Venous Thrombosis. *Lancet* **1997**, *349*, 385–388. [CrossRef]
46. Rivera-García, B.E.; Esparza-García, J.C.; Aceves-Chimal, J.L.; Leaños-Miranda, A.; Majluf-Cruz, A.; Isordia-Salas, I. Platelet Glycoprotein IIIA PIA1/A2 Polymorphism in Young Patients with ST Elevation Myocardial Infarction and Idiopathic Ischemic Stroke. *Mol. Cell Biochem.* **2013**, *384*, 163–171. [CrossRef] [PubMed]
47. Roldán, V.; Marín, F.; González-Conejero, R.; García-Honrubia, A.; Martí, S.; Alfaro, A.; Valdés, M.; Corral, J.; Lip, G.Y.H.; Vicente, V. Factor VII –323 Decanucleotide D/I Polymorphism in Atrial Fibrillation: Implications for the Prothrombotic State and Stroke Risk. *Ann. Med.* **2008**, *40*, 553–559. [CrossRef] [PubMed]
48. Rubattu, S.; Speranza, R.; Ferrari, M.; Evangelista, A.; Beccia, M.; Stanzione, R.; Assenza, G.E.; Volpe, M.; Rasura, M. A Role of TNF-α Gene Variant on Juvenile Ischemic Stroke: A Case–Control Study. *Eur. J. Neurol.* **2005**, *12*, 989–993. [CrossRef] [PubMed]
49. Slowik, A.; Dziedzic, T.; Turaj, W.; Pera, J.; Glodzik-Sobanska, L.; Szermer, P.; Malecki, M.T.; Figlewicz, D.A.; Szczudlik, A. A2 Alelle of GpIIIa Gene Is a Risk Factor for Stroke Caused by Large-Vessel Disease in Males. *Stroke* **2004**, *35*, 1589–1593. [CrossRef] [PubMed]
50. Salvarani, C.; Casali, B.; Farnetti, E.; Pipitone, N.; Formisano, D.; Nicoli, D.; Macchioni, P.; Cimino, L.; Bajocchi, G.; Grazia Catanoso, M.; et al. PlA1/A2 Polymorphism of the Platelet Glycoprotein Receptor IIIA and Risk of Cranial Ischemic Complications in Giant Cell Arteritis. *Arthritis Rheum.* **2007**, *56*, 3502–3508. [CrossRef] [PubMed]
51. Szolnoki, Z.; Somogyvári, F.; Kondacs, A.; Szabó, M.; Bene, J.; Havasi, V.; Komlósi, K.; Melegh, B. Increased Prevalence of Platelet Glycoprotein IIb/IIIa PLA2 Allele in Ischaemic Stroke Associated with Large Vessel Pathology. *Thromb. Res.* **2003**, *109*, 265–269. [CrossRef] [PubMed]
52. van Goor, M.-L.P.J.; García, E.G.; Brouwers, G.-J.; Leebeek, F.W.G.; Koudstaal, P.J.; Dippel, D.W.J. PLA1/A2 Polymorphism of the Platelet Glycoprotein Receptor IIb/IIIa in Young Patients with Cryptogenic TIA or Ischemic Stroke. *Thromb. Res.* **2002**, *108*, 63–65. [CrossRef]
53. Wagner, K.R.; Giles, W.H.; Johnson, C.J.; Ou, C.-Y.; Bray, P.F.; Goldschmidt-Clermont, P.J.; Croft, J.B.; Brown, V.K.; Stern, B.J.; Feeser, B.R.; et al. Platelet Glycoprotein Receptor IIIa Polymorphism P1A2 and Ischemic Stroke Risk. *Stroke* **1998**, *29*, 581–585. [CrossRef]
54. Zhang, Y.; Wang, Y.; Wang, Y.; Cui, C.; Huang, P.; Li, X.; Liu, S.; Lendon, C.; Guo, N. Platelet Glycoprotein Polymorphisms: Risk, in Vivo Expression and Severity of Atherothrombotic Stroke in Chinese. *Clin. Chim. Acta* **2007**, *378*, 99–104. [CrossRef] [PubMed]
55. Feigin, V.L.; Brainin, M.; Norrving, B.; Martins, S.; Sacco, R.L.; Hacke, W.; Fisher, M.; Pandian, J.; Lindsay, P. World Stroke Organization (WSO): Global Stroke Fact Sheet 2022. *Int. J. Stroke* **2022**, *17*, 18–29. [CrossRef] [PubMed]
56. Katan, M.; Luft, A. Global Burden of Stroke. *Semin. Neurol.* **2018**, *38*, 208–211. [CrossRef] [PubMed]
57. Barthels, D.; Das, H. Current Advances in Ischemic Stroke Research and Therapies. *Biochim. Biophys. Acta (BBA)-Mol. Basis Dis.* **2020**, *1866*, 165260. [CrossRef]
58. Zhang, H.; Xu, Y.; Xu, Y. The Association of the Platelet/High-Density Lipoprotein Cholesterol Ratio with Self-Reported Stroke and Cardiovascular Mortality: A Population-Based Observational Study. *Lipids Health Dis.* **2024**, *23*, 121. [CrossRef] [PubMed]
59. Severino, P.; D'Amato, A.; Pucci, M.; Infusino, F.; Adamo, F.; Birtolo, L.I.; Netti, L.; Montefusco, G.; Chimenti, C.; Lavalle, C.; et al. Ischemic Heart Disease Pathophysiology Paradigms Overview: From Plaque Activation to Microvascular Dysfunction. *Int. J. Mol. Sci.* **2020**, *21*, 8118. [CrossRef] [PubMed]
60. Politi, C.; Roumeliotis, S.; Tripepi, G.; Spoto, B. Sample Size Calculation in Genetic Association Studies: A Practical Approach. *Life* **2023**, *13*, 235. [CrossRef] [PubMed]
61. Goddard, K.A.B.; Hopkins, P.J.; Hall, J.M.; Witte, J.S. Linkage Disequilibrium and Allele-Frequency Distributions for 114 Single-Nucleotide Polymorphisms in Five Populations. *Am. J. Hum. Genet.* **2000**, *66*, 216–234. [CrossRef] [PubMed]
62. Karczewski, K.J.; Francioli, L.C.; Tiao, G.; Cummings, B.B.; Alföldi, J.; Wang, Q.; Collins, R.L.; Laricchia, K.M.; Ganna, A.; Birnbaum, D.P.; et al. The Mutational Constraint Spectrum Quantified from Variation in 141,456 Humans. *Nature* **2020**, *581*, 434–443. [CrossRef] [PubMed]
63. Turner, R.M.; Bird, S.M.; Higgins, J.P.T. The Impact of Study Size on Meta-Analyses: Examination of Underpowered Studies in Cochrane Reviews. *PLoS ONE* **2013**, *8*, e59202. [CrossRef] [PubMed]
64. Chen, H.; Cohen, P.; Chen, S. How Big Is a Big Odds Ratio? Interpreting the Magnitudes of Odds Ratios in Epidemiological Studies. *Commun. Stat.—Simul. Comput.®* **2010**, *39*, 860–864. [CrossRef]
65. Feng, D.; Lindpaintner, K.; Larson, M.G.; Rao, V.S.; O'Donnell, C.J.; Lipinska, I.; Schmitz, C.; Sutherland, P.A.; Silbershatz, H.; D'Agostino, R.B.; et al. Increased Platelet Aggregability Associated With Platelet GPIIIa PlA2 Polymorphism. *Arterioscler. Thromb. Vasc. Biol.* **1999**, *19*, 1142–1147. [CrossRef]
66. Ekkert, A.; Šliachtenko, A.; Grigaitė, J.; Burnytė, B.; Utkus, A.; Jatužis, D. Ischemic Stroke Genetics: What Is New and How to Apply It in Clinical Practice? *Genes* **2022**, *13*, 48. [CrossRef] [PubMed]

67. Martins, M.F.; Murry, L.T.; Telford, L.; Moriarty, F. Direct-to-Consumer Genetic Testing: An Updated Systematic Review of Healthcare Professionals' Knowledge and Views, and Ethical and Legal Concerns. *Eur. J. Hum. Genet.* **2022**, *30*, 1331–1343. [CrossRef] [PubMed]
68. Tandy-Connor, S.; Guiltinan, J.; Krempely, K.; LaDuca, H.; Reineke, P.; Gutierrez, S.; Gray, P.; Davis, B.T. False-Positive Results Released by Direct-to-Consumer Genetic Tests Highlight the Importance of Clinical Confirmation Testing for Appropriate Patient Care. *Genet. Med.* **2018**, *20*, 1515–1521. [CrossRef] [PubMed]
69. Peck, L.; Borle, K.; Folkersen, L.; Austin, J. Why Do People Seek out Polygenic Risk Scores for Complex Disorders, and How Do They Understand and React to Results? *Eur. J. Hum. Genet.* **2022**, *30*, 81–87. [CrossRef] [PubMed]

current issues in molecular biology

MDPI

Review

Neurological Sequelae of Post-COVID-19 Fatigue: A Narrative Review of Dipeptidyl Peptidase IV-Mediated Cerebrovascular Complications

Che Mohd Nasril Che Mohd Nassir [1,*], Muhammad Danial Che Ramli [2,*], Usman Jaffer [3], Hafizah Abdul Hamid [4], Muhammad Zulfadli Mehat [4], Mazira Mohamad Ghazali [1,5] and Ebrahim Nangarath Kottakal Cheriya [6]

1 Department of Anatomy and Physiology, Faculty of Medicine, School of Basic Medical Sciences, Universiti Sultan Zainal Abidin, Kuala Terengganu 20400, Terengganu, Malaysia; mazira.mohamadghazali@gmail.com
2 Faculty of Health and Life Sciences, Management and Science University, Shah Alam 40150, Selangor, Malaysia
3 Kulliyyah of Islamic Revealed Knowledge and Human Sciences, International Islamic University Malaysia, Kuala Lumpur 50728, Malaysia; jafferu@iium.edu.my
4 Department of Human Anatomy, Faculty of Medicine and Health Sciences, Universiti Putra Malaysia (UPM), Serdang 43400, Selangor, Malaysia; a_hafizah@upm.edu.my (H.A.H.); m_zulfadli@upm.edu.my (M.Z.M.)
5 Department of Neurosciences, School of Medical Sciences, Universiti Sains Malaysia, Kubang Kerian, Kota Bharu 16150, Kelantan, Malaysia
6 Department of Physiology, International Medical School (IMS), Management and Science University, Shah Alam 40150, Selangor, Malaysia; ebrahim_nangarath@msu.edu.my
* Correspondence: nasrilnassir@unisza.edu.my (C.M.N.C.M.N.); muhddanial_cheramli@msu.edu.my (M.D.C.R.)

Abstract: Coronavirus disease 2019 (COVID-19) has been a global pandemic affecting millions of people's lives, which has led to 'post-COVID-19 fatigue'. Alarmingly, severe acute respiratory syndrome-coronavirus 2 (SARS-CoV-2) not only infects the lungs but also influences the heart and brain. Endothelial cell dysfunction and hypercoagulation, which we know occur with this infection, lead to thrombo-inflammation that can manifest as many myriad cardio-cerebrovascular disorders, such as brain fog, fatigue, cognitive dysfunction, etc. Additionally, SARS-CoV-2 has been associated with oxidative stress, protein aggregation, cytokine storm, and mitochondrial dysfunction in neurodegenerative diseases. Accordingly, the identification of molecular targets involved in these actions could provide strategies for preventing and treating this disease. In particular, the very common enzyme dipeptidyl peptidase IV (DPPIV) has recently been identified as a candidate co-receptor for the cell entry of the SARS-CoV-2 virus with its involvement in infection. In addition, DPPIV has been reported as a co-receptor for some viruses such as Middle East respiratory syndrome-coronavirus (MERS-CoV). It mediates immunologic reactions and diseases such as type 2 diabetes mellitus, obesity, and hypertension, which have been considered the prime risk factors for stroke among other types of cardio-cerebrovascular diseases. Unlike angiotensin-converting enzyme 2 (ACE2), DPPIV has been implicated in aggravating the course of infection due to its disruptive effect on inflammatory signaling networks and the neuro–glia–vascular unit. Regarding the neurological, physiological, and molecular grounds governing post-COVID-19 fatigue, this review focuses on DPPIV as one of such reasons that progressively establishes cerebrovascular grievances following SARS-CoV infection.

Keywords: post-COVID-19 fatigue; neurodegenerative diseases; dipeptidyl peptidase IV; cerebrovascular; SARS-CoV-2

1. Introduction

Public worry is still raised by the long-lasting physical and mental health problems that have resulted from severe acute respiratory syndrome-coronavirus 2 (SARS-CoV-2) infection.

The COVID-19 pandemic's worldwide effects include aspects related to sociopsychology, health, and the economy [1–3]. Many of the patients who have recovered from coronavirus disease 2019 (COVID-19) have been described as having symptoms that continue for several months after they first became ill. Long COVID, post-COVID-19 syndrome, chronic COVID-19 syndrome, post-acute sequelae of COVID-19 (PASC), or post-COVID-19 fatigue are some names for this illness [4–6]. However, this manuscript will adopt the term "post-COVID-19 fatigue".

Alarmingly, post-COVID-19 fatigue is not limited to elderly patients with severe illness; younger generations with mild-to-moderate symptoms are also affected. Moreover, more than half of young adults have reported symptoms that last longer than six months, suggesting that long-term physical and mental repercussions might occur even in cases of mild-to-moderate illness [7–9]. Many theories concerning the duration of symptoms after SARS-CoV-2 infection are being offered by new research. Additionally, post-COVID-19 fatigue is thought to be caused by the immune system imbalance and residual virus particles in the body. Results showing that SARS-CoV-2 ribonucleic acid (RNA) was still present in COVID-19-infected people's excretions several months after infection lend support to this. Moreover, studies have shown evidence of a viral presence in several tissues, including the brain [10,11]. Moreover, patients with chronic autoimmune diseases also experience persistent symptoms associated with post-COVID-19 fatigue [12]. The probability or intensity of long-term symptoms does not seem to be correlated with the severity of the initial illness [13].

In addition to the pathological consequences, lockdowns, and the increasing use of remote work have had a profound impact on socio-economic dynamics and face-to-face contact. These changes have introduced temporal distortions, likely contributing to prolonged psychological and physiological disruptions [14,15]. Nutritional modifications and a reduction in daily physical activity may potentially have a substantial impact on the results of post-COVID-19 fatigue [16]. It is becoming more and more clear that post-COVID-19 fatigue may trigger a secondary public health emergency, especially when it comes to cerebrovascular illnesses including depression, stroke, and cognitive decline [17,18].

Moreover, multiple reports have shown that the COVID-19 pandemic's isolation and lockdown protocols have harmed children's emotional and physical health. While many young people and children infected with SARS-CoV-2 have moderate disease or show no symptoms at all, new research suggests that, in the months after infection, 5–10% of all children may show new symptoms, including neurological and neurocognitive sequelae [19,20]. Owing to the wide range of symptoms linked to COVID-19, there is no specific treatment option available at this time. Although this topic needs further research through additional studies, it is fortunate that most children with post-COVID-19 fatigue tend to recover faster than most adults. There is currently a great deal of clinical and basic research being carried out to try and understand the processes behind post-COVID-19 fatigue produced by SARS-CoV-2, even though the possible mechanisms of vascular and neurological involvement during SARS and MERS infections have been largely understood. Further research is needed to improve our capacity to predict and mitigate the impact of post-infectious sequelae, specifically on cerebrovascular outcomes. The fact that post-COVID-19 fatigue is a complex illness involving multiple organs, cell types, and chemicals highlights the significance of thorough research in this field.

Thus, highlighting the neurovascular, cellular, and molecular mechanisms linked to post-COVID-19 fatigue, this narrative review focuses on the molecular consequences of dipeptidyl peptidase IV (DPPIV). DPPIV's putative role as a co-receptor for SARS-CoV-2 has attracted a lot of attention. Specifically, DPPIV is an important regulator of diabetes, obesity, hypertension, inflammation, and glucose metabolism—all of which are linked to the chronic cerebrovascular dysfunctions that follow COVID-19.

2. SARS-CoV-2, Post-COVID-19 Fatigue, and Multi-Organ Injury

The coronavirus is described as a glycoprotein-based surface spiked positive-stranded RNA virus enclosed in an envelope. It is a member of a large family of viruses that have long been present in both humans and animals around the globe. These viruses can cause everything from minor respiratory ailments to serious respiratory diseases. Of these, the following seven distinct coronavirus types have been found in humans: Middle East respiratory syndrome coronavirus (MERS-CoV), severe acute respiratory syndrome coronavirus (SARS-CoV), and human coronaviruses (HCoV) including HCoV-OC43, HCoV-229E, HKU1, and HCoV-NL63 [21]. These variants stand out due to their high pathogenicity and transmissibility. Moreover, SARS-CoV-2 is the sixth coronavirus reported to infect people. It is genetically identical to SARS-CoV and MERS-CoV, which are also present in bats. The neurological symptoms of the three major SARS-CoV, MERS-CoV, and SARS-CoV-2 outbreaks that resulted in worldwide pandemics show both variances and commonalities [21–23]. Moreover, SARS-CoV, which was discovered in 2002, and MERS-CoV, which was discovered in 2011, share sequence similarities according to recent research using the next-generation sequencing of the SARS-CoV-2 genome [24,25].

Even though most people infected with COVID-19 eventually recover, a consider-able percentage of people in different age groups experience several symptoms, including post-exertional malaise, for a lengthy time [1–9]. Recent findings indicate that an unbalanced autoimmune reaction and persistent virus particles in the body could cause these post-COVID-19 fatigue symptoms. Studies have shown that the virus is present in several tissues, including the brain [10,11]. Moreover, post-COVID-19 fatigue symptoms are similarly persistent in those with chronic autoimmune diseases [10,26].

Thus, the goal of the ongoing study is to understand the origins and consequences of SARS-CoV-2 infection, with a focus on the long-term effects on various organs and tissues. COVID-19 was first identified as a respiratory disease, but it has now been linked to multiple organ failures and neurological disorders [27–29]. In this case, the viral infection in the lungs triggers responses that can eventually develop into a systemic event impacting several organs, such as the heart and brain, and different hematologic abnormalities, regardless of the disease's original severity. A slow and progressive failure of the gastrointestinal, endocrine, hepatic, renal, and circulatory systems may result from this trend [27–30]. The multi-organ injury and sequelae symptoms in post-COVID-19 fatigue are summarized in Table 1.

Table 1. Multi-organ injury and sequelae symptoms in post-COVID-19 fatigue.

Organ/System.	Sequel Symptoms	References
Nervous System	Brain fog; cognitive and memory Dizziness Headache Stroke Joint pain EC dysfunction—BBB damage Microthromboembolism Anxiety, stress, and depression Anosmia (loss of sense of smell) Ageusia (loss of taste) Insomnia (sleep disturbance) Guillain–Barré syndrome Chronic fatigue Encephalopathy	[31–37]

Table 1. *Cont.*

Organ/System.	Sequel Symptoms	References
Respiratory	Cough Sore throat Dyspnea (difficulty of breathing) Modified diffusion capacity restricting pattern, and obstructive pattern	[4,5]
Cardiovascular	Myocardial damage Palpitations An irregular pulse Chest pain Myocarditis	[38]
Gastrointestinal	Diarrhea Vomiting/nausea	[39–41]

EC, endothelial cell; BBB, blood–brain barrier.

Heart and Brain Crosstalk: Roles of Thrombo-Inflammation and Post-COVID-19 Fatigue

As COVID-19 symptoms persist, manifestations affecting the cardio-cerebrovascular systems are increasingly apparent. COVID-19 affects both large and small blood vasculature, raising the risk of thrombosis, and has a major effect on the coagulation cascade, hence leading to thrombo-inflammation. Over time, these problems may result in increased rates of morbidity and mortality from cardio-cerebrovascular diseases [30,42]. According to a recent study by the National Institute of Health and other studies, it was reported that the coronavirus can travel through systemic microcirculation to the brain and heart within a few days of infection and can stay there for a long time. Furthermore, research has shown that micro-clots can form in smaller blood vessels, which can obstruct cerebral blood flow. About 20% to 30% of COVID-19 patients who are sick have these clots [43,44].

There is evidence that SARS-CoV-2 is present in the brain's hypothalamus and cerebrospinal fluid of COVID-19 patients. It is thought that the remaining virus population causes vascular inflammation, which results in symptoms related to cerebrovascular accidents. Reports of ischemic strokes, intracranial bleeding, and cerebral venous thrombosis have all been connected to the COVID-19-related clotting condition [31,45]. Uncontrolled immunological and inflammatory responses, which mediate hypercoagulability, endothelial dysfunction, and thrombosis, most likely start the clotting process. These processes may result in a rise in heart attacks, strokes, and deaths [32,46]. It is interesting to note that neuro-invasion and cerebrovascular symptoms are present in about two-thirds of post-COVID-19 fatigue patients. Headache, taste and smell abnormalities, encephalopathy, ischemic stroke, Guillain–Barré syndrome, cognitive impairment, dizziness, fast heartbeat, post-traumatic stress disorder, mini-strokes, anxiety, depression, and other mental health conditions are some of these symptoms [31–37]. Therefore, it is conceivable that the harm caused by COVID-19 to the central nervous system (CNS) and various organs may be progressive, necessitating extended periods for the recovery and restoration of normal activities [40,41].

3. Current Updates on Molecular and Cellular Mechanism of Post-COVID-19 Fatigue

Although the involvement of multiple microstructural (i.e., cellular) and molecular components in SARS-CoV-2 infection is well-known, the exact processes underlying post-COVID-19 fatigue are still vague. Extensive studies have been conducted to identify the complex pathways associated with post-COVID-19 fatigue. It is well-reported that the coronavirus attaches itself to the cell surface during the first stage of infection, acting as a portal for further infection. The vulnerability of a cell to viral infection is determined by the presence of receptors and cofactors on the cell surface. On the other hand, COVID-19 primarily targets cells in the kidneys, heart, lungs, immune system, and brain's neuronal and glial cells. However, the digestive epithelium offers a different entrance point for SARS-CoV-2, which results in gastrointestinal symptoms. This may lead to the disruption of the intestinal barrier and systemic inflammation, which could impact the function of the CNS and the enteric nervous system [47,48]. Hence, post-COVID-19 fatigue may be linked to a variety of microstructural alterations or determinants, molecular effects, meso-structural reactions, and symptoms.

3.1. Microstructural Changes—Endothelial Cells (ECs)

Endothelial cells (ECs) are essential for maintaining barrier integrity, restricting the entry of circulating inflammatory cells and chemicals and regulating vascular tone under normal physiological conditions. The lining of blood vessels, such as arteries, veins, capillaries, and lymphatic vessels, is made up of ECs, which come into direct touch with virus particles and blood in circulation. ECs also produce several factors and enzymes that control blood coagulation, platelet adhesion, immunological response, and vascular relaxation and contraction [49]. ECs in blood arteries play a critical role in SARS-CoV-2 infection, causing damage, advancing COVID-19 illness, and perhaps spreading the infection to other organs. Increased thrombotic and inflammatory processes are intimately associated with microvascular EC dysfunction and have been connected to the symptoms of post-COVID-19 fatigue [50].

Angiotensin-converting enzyme 2 (ACE2) receptors are expressed on the surface of ECs, which allows SARS-CoV-2 to enter through its high-affinity spike protein. Endothelial injury, malfunction, and blood vessel damage result from this interaction. As a result, cytokines, chemokines, P-selectin, E-selectin, and intercellular adhesion molecule 1 (ICAM-1) are created, and these adhesion molecules draw different cells, including platelets, to the artery walls. Aggregation, coagulopathy, and a higher frequency of thrombotic events are all encouraged by this. Moreover, vasculitis is triggered by SARS-CoV-2 infection in the heart, which also causes dysfunction and chronic inflammation. Following EC dysfunction, SARS-CoV-2 virus particles in the brain may lead to blood–brain barrier (BBB) damage, triggering an innate immune reaction and escalating inflammation within brain microvascular ECs [50,51].

3.2. Microstructural Changes—Neuro–Glia–Vascular Unit (NGVU)

Recently, it has been well-documented that CNS cells are involved in COVID-19 infection. It has been discovered that SARS-CoV-2 directly infects human peripheral sensory neurons through interaction with the ACE2 receptor, leading to deficits in chemosensory perception [47]. Furthermore, COVID-19 has been linked, particularly in the hippocampus region of the brain, to neuronal degeneration and a reduction in neurogenesis. These results raise the possibility of memory impairment and long-term neurological problems, possibly through oxidative stress and neuro-inflammation processes [52]. Additionally, there is a recognized regulatory role that this connection between ECs and the NGVU plays in cerebrovascular disorders. Research has indicated that SARS-CoV-2 causes hypoxic brain alterations that impact the cerebellum and cerebrum, among other brain regions. As a result, this causes the death of neurons in important regions such as the hippocampus, cerebral cortex, and the cerebellum's Purkinje cell layer [53].

The systemic inflammatory reactions that begin in the peripheral areas may be the cause of the CNS issues seen in post-COVID-19 fatigue patients. Furthermore, viral infection activates glial cells, including microglia and astrocytes, which may increase the vulnerability to neurodegenerative illnesses [54]. According to recent studies, the SARS-CoV-2 infection causes an inflammatory response that activates both astrocytic and microglial processes [55]. These processes are linked to aberrant phenotypes in the interactions between astrocytes and the vasculature, which are essential for preserving the BBB integrity. Peripheral cell infiltration into the brain is facilitated in COVID-19 patients by an impaired BBB. Moreover, even in the absence of overt clinical neurological symptoms, increased levels of secreted neural and glial proteins, such as serum neurofilament (sNfL) and serum glial fibrillary acidic protein (sGFAP), suggest neuro–glial degeneration in people with COVID-19 [56,57].

3.3. Meso-Structural Responses—Pro-Inflammatory Cells

It has been shown that the SARS-CoV-2-infection-induced dysregulated host inflammatory response exacerbates hypercoagulability, EC dysfunction, and thrombosis. These variables exacerbate cerebrovascular disease, which raises the risk of stroke and its associated mortality [58]. Previous studies have extensively described the complex and distinct cellular immune response during acute, resolving, or progressive COVID-19. Activated immune cells have been found in the blood of COVID-19 patients for a significant amount of time, from one to eight months after infection [59]. Research has demonstrated the involvement of monocytes and different subtypes of macrophages in the lung, which is correlated with outcomes in COVID-19-affected people [60]. Furthermore, lymphocytes, which are made up of B cells, T cells, and natural killer cells, are other vital immune components. Studies have also shown that those with post-COVID-19 fatigue had different T cell characteristics, such as worn-out T cells and fewer CD4+ and CD8+ effector memory cells [61,62].

3.4. Molecular Mechanism of Post-COVID-19 Fatigue

Understanding the molecular processes and feedback loops that underlie COVID-19 is essential for developing treatment strategies that will lessen the disease's long-term effects. Through its interactions with the genome, the virus modifies the intricate structures of proteins and DNA, changing gene expression and causing immunological dysregulation. Since these molecular changes may have long-term effects, it is critical to comprehend and target them in therapeutic interventions [63].

An important biological mechanism linked to increased illness severity and persistence is the overproduction of pro-inflammatory cytokines (or cytokine storms) caused by the virus. These include interferon (IFN)-γ, IFN-γ-inducible protein 10, monocyte chemoattractant protein-1 (MCP-1), granulocyte colony-stimulating factor (GCSF), macrophage inflammatory protein-1 alpha (MIP-1α), and tumor necrotic factor alpha (TNF-α), in addition to interleukins such IL-1β, IL-2, IL-6, IL-7, IL-8, IL-10, and IL-17 [64,65]. When the transcription factor such as nuclear factor kappa B (NF-κB) is activated, cytokines and chemokines are produced, which attract and activate immune cells from the circulation. An uncontrolled cytokine storm worsens organ failure, vascular permeability, adhesion characteristics, pro-coagulation of blood, and COVID-19-mediated inflammation and oxidative stress, along with long-lasting tissue damage. Higher blood biomarkers of the vascular system in post-COVID-19 fatigue patients suggest that angiogenesis-regulating molecules are important components of this pathophysiological mechanism [66,67].

Additional molecular biomarkers linked to COVID-19 comprise pro-calcitonin, phospholipids, D-dimer, serum ferritin, and C-reactive protein. These biomarkers function as indications of severe organ failure and are implicated in the production of micro-clots. Furthermore, the increased expression of surface proteins that indicate endothelial dysfunction includes E-selectin, vascular cell adhesion molecule 1 (VCAM-1), and ICAM-1. These molecular factors have the potential to worsen issues in the neurological, cardiovascular, gastrointestinal, liver, and kidney systems [68,69]. Thus, the identification of the post-COVID-19 fatigue prevalence and the molecular factors that underlie it has great potential for identifying therapeutic targets that will either stop the illness from starting or lessen its symptoms. Through a comprehensive comprehension of the distinct biochemical pathways implicated in post-COVID-19 fatigue, scholars might identify possible therapies aimed at attenuating its impact, thereby providing alleviation to individuals enduring symptoms.

On the other hand, the first phase of the infectious viral life cycle is represented by the recognition of and interactions with cellular receptors. It is a well-known fact that the SARS-CoV-2 virus binds to the ACE2 receptor, which is located on the surface of cells in the digestive and respiratory systems, to begin the infection process [70]. However, DPPIV, neuropilin-1 (NRP1), and transmembrane serine protease type 2 (TMPRSS2) are other molecules that have been discovered as possible targets for the virus [71]. The expression of several molecular receptors on the cell surface could therefore be connected to different routes of viral entry for SARS-CoV-2 [71]. Moreover, DPPIV offers unique insights that distinguish it from the ACE2 pathway, particularly in its broader involvement in metabolic and inflammatory pathways. Unlike ACE2, which is mainly linked to direct viral entry, DPPIV is deeply involved in regulating glucose metabolism, immune responses, and chronic inflammation, all of which are key in the progression of post-COVID-19 fatigue and cerebrovascular complications [72]. The most common comorbidities seen in COVID-19 patients were immune system, cardiovascular, neurological, and diabetic diseases, as well as hypertension, atherosclerosis, and other conditions. Surprisingly, type 2 diabetes mellitus (T2DM) is linked to DPPIV in these illnesses. Further investigation into how DPPIV affects COVID-19 intensity and the length of persistent symptoms is, therefore, necessary [73–75].

Furthermore, DPPIV's enzymatic role in degrading glucagon-like peptide-1 (GLP-1) links it to hyperglycemia, a significant risk factor for severe COVID-19 outcomes [74]. This hyperglycemic state, often exacerbated in patients with metabolic disorders such as diabetes, fuels systemic inflammation and endothelial dysfunction, which increases the risk of microvascular damage in the brain [76]. Unlike ACE2, DPPIV not only facilitates viral entry, but also amplifies the metabolic and inflammatory disruptions that contribute to the long-term neurovascular damage seen in post-COVID-19 patients.

Thus, while ACE2 primarily acts as the gateway for SARS-CoV-2 infection, DPPIV may putatively serve as both a co-receptor and a key mediator of the metabolic and inflammatory cascades that aggravate the course of infection [77,78]. This dual role makes DPPIV a potential therapeutic target not just for limiting viral entry, but for managing the chronic complications, such as stroke and cognitive decline, that are linked to metabolic and cerebrovascular dysregulation in COVID-19 survivors.

4. Novel Insights into DPPIV's Role in Metabolic Syndrome and COVID-19

The course and results of COVID-19 are considerably impacted by metabolic diseases such as T2DM [74]. Chemokines and their receptors have been identified as a shared pathway between T2DM and COVID-19 by an investigation of pathway connections. By controlling the functions of chemokines, glucose homeostasis, and metabolism, DPPIV is essential in T2DM [74]. Notably, there is a substantial correlation between hyperglycemia and diabetes and elevated DPPIV activity. It is well-known that DPPIV causes disruptions in glucose metabolism by degrading active GLP-1, which raises blood glucose levels [76]. As illustrated in Figure 1, DPPIV's enzymatic activity reduces GLP-1 levels, exacerbating hyperglycemia, which, alongside the SARS-CoV-2-induced cytokine storm, leads to endothelial dysfunction, increased microthrombi, and heightened cerebrovascular risks.

It is conceivable that metabolic syndrome associated with comorbidities including obesity, diabetes, and cerebrovascular disorders is related to hyperglycemia, which may promote viral shedding and reproduction. This could, therefore, worsen COVID-19-related inflammation and cause endothelial dysfunction. Moreover, a hyperglycemic state is brought on by the multi-systemic inflammation associated with SARS-CoV-2 infection; on the other hand, this hyperglycemic state may facilitate viral replication [79]. The uncontrolled glycosylation of SARS-CoV-2 modifies viral epitopes, facilitating the evasion of immune system detection. This evasion leads to a more severe and prolonged COVID-19 infection in hyperglycemic diabetic individuals [80].

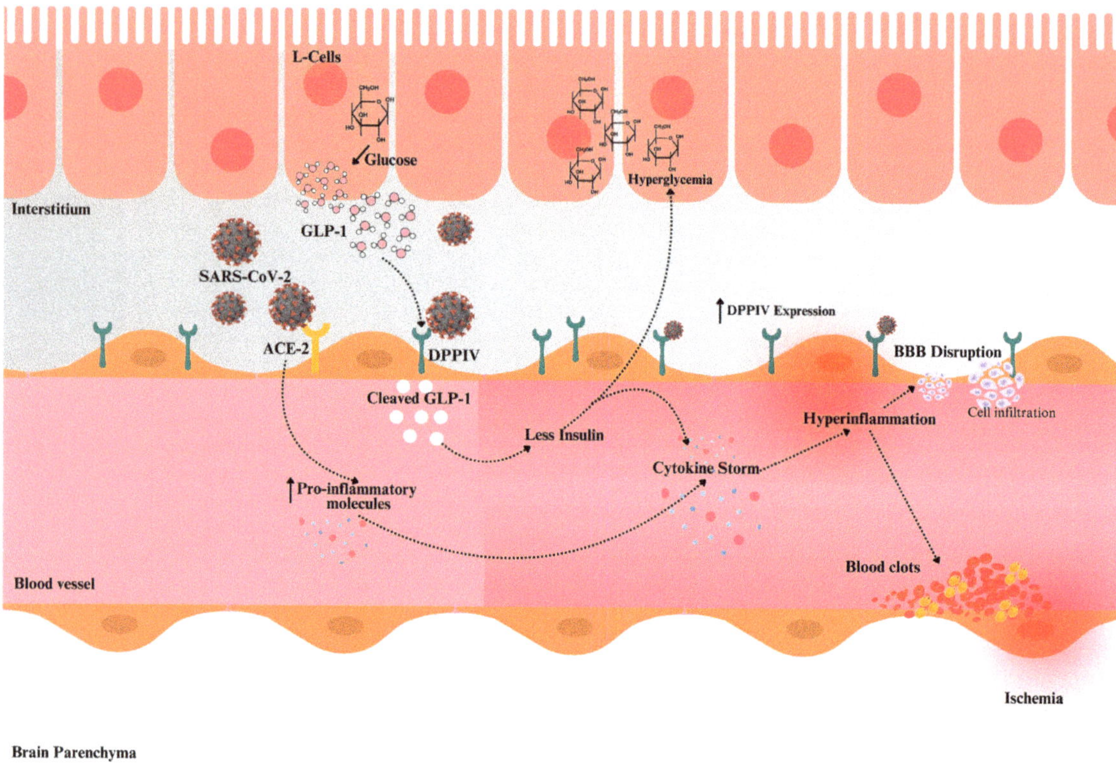

Figure 1. Proposed mechanism illustrating DPPIV-mediated SARS-CoV-2 infection and its impact on glucose metabolism and cerebrovascular health. DPPIV degrades GLP-1 in the bloodstream, leading to reduced insulin secretion and hyperglycemia. This hyperglycemic state, along with the cytokine storm triggered by SARS-CoV-2, exacerbates endothelial dysfunction and inflammation, resulting in blood–brain barrier (BBB) disruption, cell infiltration, blood clot and microthrombus formation, and increased risk of cerebrovascular complications, including stroke. Darker red in the blood vessel structure indicates an inflamed vessel. The upward arrow indicates "increase".

Moreover, many physiological systems, such as the metabolism, obesity, autoimmunity, the immune system, endocrine functions, inflammation, cellular activities, and thrombotic cerebrovascular disorders like stroke, are known to be regulated by the adaptable transmembrane and circulating protein DPPIV [81]. Numerous cell types, including T cells, B cells, NK cells, certain macrophage subsets, hematopoietic stem cells, and hematopoietic progenitor cells, have a broad expression of this protein. It can also be found on the surface of acinar, endothelial, and bone marrow cells in a variety of organs, including the kidney, liver, intestines, lung, spleen, and pancreas [81,82]. Moreover, when DPPIV is enzymatically active, it behaves as a homodimer, whereby numerous cytokines, including interferons, hepatocyte NF-1α, and hypoxia-inducible factor-1α, can strongly stimulate its expression. Pro-inflammatory diseases like obesity, diabetes, and atherosclerosis, and cerebral and cardiovascular disorders like stroke are associated with markedly elevated DPPIV levels [81–83]. Through its effects on inflammatory signaling pathways, stimulation of vascular smooth cell proliferation, and induction of oxidative stress in various cell types, imbalanced levels of DPPIV may make these pathological states worse. Alarmingly, DPPIV's immunomodulatory effects can cause inflammation on both a local and systemic level. It has been shown that DPPIV may promote T cell activation, proliferation, and signal transduction, among other aspects of lymphocyte function. Moreover, after an acute infection, DPPIV is likely involved in a new regulatory mechanism that suppresses the immune response against the SARS-CoV-2 virus [84].

4.1. Potential Role of DPPIV as Receptor for SARS-CoV-2 Infection

In addition to controlling inflammation and hyperglycemia, DPPIV has recently attracted attention concerning SARS-CoV-2 infection because it may act as a co-receptor for the virus. Remarkably, DPPIV has been found to function as a co-receptor for the entry of some viruses, such as the coronavirus MERS-CoV and the human immunodeficiency virus (HIV) [85]. Studies reveal that DPPIV's critical binding sites required for this interaction coincide with those that MERS-CoV-S binds to. Additionally, it has been established that DPPIV serves as a useful receptor for cellular invasion when it comes to Human Coronavirus-Erasmus Medical Center (hCoV-EMC). Antibodies directed against DPP4 did, in fact, successfully prevent hCoV-EMC infection in Huh-7 and human bronchial epithelial cells [85–87].

Moreover, protein–protein docking experiments using crystal structures have demonstrated a strong affinity between DPPIV and the spike-receptor-binding domain of SARS-CoV-2, which is particularly noteworthy [88]. Furthermore, silico computational studies have indicated DPPIV as a likely SARS-CoV-2 receptor. Building on these discoveries and hypotheses, a growing body of research suggests that DPPIV might serve as a functional co-receptor for human coronaviruses, hence aiding in the transmission of SARS-CoV-2 [88–91].

4.2. Emerging Research on DPPIV and COVID-19

In this context, DPPIV has been analyzed in COVID-19 studies ranging from preclinical, including animal models and in vitro studies, to clinical and observational ones. These collectively provide important insights into how DPPIV influences disease progression, with particular emphasis on inflammation, the glucose metabolism, and cerebrovascular outcomes.

Pre-clinical animal studies, including its involvement in COVID-19, have investigated the involvement of DPPIV during viral infections. Evidence from recent animal models indicated that DPPIV inhibition led to lower inflammation and neurovascular damage attributed to COVID-19-like symptoms [92]. These findings point out the enzyme's contribution to the inflammation process, particularly in exacerbating the cytokine storm leading to endothelial dysfunction and cerebrovascular complications.

For instance, recent findings have identified that the inhibition of DPPIV in animal models could mitigate SARS-CoV-2-induced endothelial injury in the brain, reducing the incidence of microthrombi [93,94]. Such a model has, until now, helped establish the role of the enzyme in increasing COVID-19-related cerebrovascular risks, especially in aged and diabetic animal models. DPPIV inhibitors such as gliptins (i.e., saxagliptin, alogliptin sitagliptin, and linagliptin) have also been a focus of various clinical studies concerning COVID-19 patients, specifically those with T2DM. They are said to decrease the severity of COVID-19 through the limited viral entry and modulation of the immune response [95,96]. Additionally, diabetic patients receiving DPPIV inhibitors had lower viral loads, reduced levels of inflammatory markers, and a lower incidence of cerebrovascular complications compared to controls [97].

Observational studies further stress the high activity of DPPIV with the poor outcomes in COVID-19 disease. For example, elderly COVID-19 patients reported that higher levels of DPPIV were related to increased cerebrovascular complications such as stroke and cognitive decline [98,99]. Such observations have shown that DPPIV acts as a potential biomarker for the anticipation of cerebrovascular risks among post-COVID-19 patients, particularly among patients with comorbid conditions like diabetes and hypertension.

All in all, the studies presented highlight DPPIV's central role in modulating COVID-19 severity, particularly through its effects on inflammation, immune response, and glucose metabolism. In animal models, DPPIV inhibition has shown promise in reducing the impact of COVID-19 on the brain and cardiovascular system. Meanwhile, clinical trials suggest that DPPIV inhibitors may offer therapeutic potential, especially for diabetic patients who are more prone to severe outcomes. Observational studies further support the enzyme's involvement in post-COVID-19 cerebrovascular complications.

5. DDPIV and Post-COVID-19 Fatigue: Impact on Cerebrovascular Disease

Vascular disorders, including embolisms and thrombotic cerebrovascular diseases, as well as cardiovascular diseases, hypertension, diabetes, and obesity, are the main comorbidities linked to post-COVID-19 fatigue. Severe COVID-19 cases can cause vascular integrity to be disrupted, coagulation and inflammatory responses to be triggered, and atherosclerosis to be encouraged, all of which increase the risk of cerebrovascular disorders. The virus can also harm the structure and function of the brain, which can lead to long-term cognitive impairment.

Alarmingly, those with COVID-19 who have abnormalities in their glucose metabolism are more likely to die from vascular–neurological issues and to experience other COVID-19-related complications down the road. This could put them at risk for COVID-19's long-term repercussions. Thus, DPPIV, i.e., a vascular system component that affects glucose metabolism, is essential in controlling these disorders. A recent study has shown that DPPIV is dysregulated in the brain after a stroke. Through an in vitro functional study, recent reports have discovered that DPPIV inhibits the migration of neural progenitor cells and decreases the angiogenic capacity of brain ECs mediated by stromal cell-derived factor 1 (SDF1). The processes of tissue remodeling and repair in the brain following a stroke may be hampered by the overexpression of DPPIV in response to ischemia/reperfusion injury [100]. Table 2 summarizes the role of DPPIV in post-COVID-19 cerebrovascular complications.

Table 2. Summary of DPPIV's role in post-COVID-19 cerebrovascular complication.

Mechanism	DPPIV Role	COVID-19-Related Cerebrovascular Complications
Endothelial dysfunction	DPPIV degrades GLP-1, leading to endothelial damage and impaired vascular function	Increases risk of stroke, microthrombi, and blood–brain barrier disruption
Inflammatory pathways	DPPIV activates pro-inflammatory cytokines (TNF-α, IL-6, and IL-1β), enhancing immune response	Exacerbates vascular inflammation and thrombotic events
Glucose metabolism	DPPIV's regulation of glucose homeostasis is disrupted, particularly in diabetic patients	Hyperglycemia worsens inflammation and thrombosis, leading to stroke
Therapeutic inhibition	DPPIV inhibitors (gliptins) reduce inflammation, endothelial dysfunction, and hyperglycemia	Potential to lower cerebrovascular risks in COVID-19 patients

DDPIV, dipeptidyl peptidase IV; GLP-1, glucagon-like peptide-1; ILs, interleukins; TNF-α, tumor necrosis factor-alpha.

DPPIV plays a multifaceted role in the pathophysiology of COVID-19, particularly concerning cerebrovascular complications. Its ability to exacerbate endothelial dysfunction, promote inflammation, and disrupt glucose metabolism highlights the enzyme's critical role in increasing the risk of post-COVID-19 cerebrovascular events. As such, DPPIV may represent both a key biomarker for predicting cerebrovascular complications and a potential therapeutic target for mitigating these risks in high-risk COVID-19 patients, especially those with metabolic comorbidities. While the exact mechanisms of DPPIV's involvement in SARS-CoV-2 infection are still being explored, the existing evidence suggests that targeting DPPIV may reduce the risk of long-term cerebrovascular complications, including stroke and cognitive decline, in post-COVID-19 patients. Future research should continue to explore the therapeutic potential of DPPIV inhibitors in reducing both acute and chronic COVID-19 outcomes.

Studies have shown that DPPIV inhibitors may protect against stroke and vascular illnesses. Research on DPPIV, both scientific and clinical, has drawn interest not just in diabetes but also in cerebrovascular disorders. Thus, it is possible that increased DPPIV levels, which support hyperglycemia and increased inflammation brought on by cytokine storms, are connected to the long-term neurological problems and vascular cognitive impairment linked to COVID-19. DPPIV has been identified as a potential factor aggravating inflammation and the long-term effects of viral infections, according to the research evaluated in this context. Consequently, using DPPIV inhibitors in COVID-19 patients may offer a simple method of reducing viral entrance and replication, even if the patient has diabetes. This strategy may provide efficient therapeutic approaches to counteract inflammation and metabolic disruptions brought on by COVID-19. However, a more thorough investigation is required to clarify DPPIV's role in post-COVID-19 fatigue.

5.1. Confounding Factors Influencing DPPIV-Mediated Post-COVID-19 Fatigue and Cerebrovascular Complications

Age, genetic predisposition, and the presence of comorbidities, including diabetes, hypertension, and obesity, may further modulate the severity of fatigue and cerebrovascular complications post-COVID-19 through the interaction with DPPIV activity [101]. Indeed, it has been documented that these comorbid conditions modulate the levels and activities of DPPIV, leading to exacerbated inflammation and metabolic pathways that drive severe outcomes in COVID-19 patients.

Aging is associated with a natural increase in DPPIV activity [102], which, in the cases of elderly patients affected by COVID-19, may worsen the outcome. According to the obtained data, elderly patients exhibit increased circulating levels of DPPIV, which are associated with an increased inflammatory response and a higher risk of developing cerebrovascular complications [74]. For example, one multicenter cohort study reported that, among elderly COVID-19 patients, high levels of DPPIV were associated with a higher incidence of stroke and cognitive decline, probably as a result of increased endothelial dysfunction and microthrombus formation [103]. This age-dependent increase in the enzymatic activity of DPPIV may further potentiate the cytokine storm observed in severe COVID-19 and thus contribute to post-COVID-19 fatigue and neurovascular damage. Thus, the increased inflammation burden in elderly subjects, along with high levels of DPPIV, places them at greater risk for neurological complications long after SARS-CoV-2 infection.

Moreover, metabolic disorders such as T2DM, obesity, and hypertension are directly associated with increased DPPIV activity, which, in turn, is associated with the severity of COVID-19 and its complications [74]. DPPIV has an important role in glucose metabolism; patients with T2DM have increased levels of DPPIV due to GLP-1 degradation [74]. This leads to hyperglycemia, considered a risk factor for poor outcomes in COVID-19. Moreover, in diabetic or obese patients, the hyperglycemic environment mediates viral replication and systemic inflammation, increasing the risks of cerebrovascular complications such as stroke and cognitive impairment [104,105]. Very recent studies also reported that COVID-19 diabetic patients with high DPPIV levels had more severe cerebrovascular events compared to non-diabetic COVID-19 patients, further underlining the pivotal role this enzyme plays in bridging metabolic dysfunction to COVID-19 severity [106].

This was also true for hypertension, another frequent comorbidity in COVID-19 patients, which presented higher levels of DPPIV [74]. Perhaps through the exacerbation of endothelial dysfunction and inflammation, known comorbid conditions that may be particularly exacerbated by the presence of DPPIV, individuals with hypertension would indeed suffer worse cerebrovascular outcomes. Moreover, genetic predispositions related to single nucleotide polymorphisms in the gene encoding DPPIV or related pathways could be relevant in determining the poor outcomes of some individuals affected by COVID-19 [107]. Although direct studies relating genetic predisposition to DPPIV-mediated COVID-19 complications are still emerging, there is evidence to date that polymorphisms in the gene encoding DPPIV influence its expression and activity [76]. These genetic variations may modulate the predisposition of an individual to the neurovascular and metabolic complications associated with COVID-19.

Moreover, genetic factors related to immune response and inflammation may interact with DPPIV activity. For example, individuals with a predisposition to heightened inflammatory responses may experience a more severe cytokine storm when combined with elevated DPPIV activity, leading to prolonged post-COVID-19 fatigue and cerebrovascular dysfunction. Table 3 summarizes the confounding factors influencing DPPIV-mediated COVID-19 outcomes.

Table 3. Summary of confounding factors influencing DPPIV-mediated COVID-19 outcomes.

Confounding Factor	Impact on DPPIV Activity	Influence on Post-COVID-19 Complications
Age	Increases with age, leading to elevated inflammation and endothelial dysfunction	Heightened risk of stroke, cognitive decline, and post-COVID-19 fatigue
Diabetes and obesity	Elevated DPPIV levels due to glucose metabolism dysregulation	Worsens hyperglycemia and inflammation, and increases cerebrovascular risks
Hypertension	Increased DPPIV linked to worsened endothelial dysfunction and inflammation	Greater risk of cerebrovascular complications and neurovascular damage
Genetic predisposition	Genetic variations may influence DPPIV expression and activity	May heighten susceptibility to cerebrovascular and inflammatory complications

These confounding factors highlight the multifaceted nature of DPPIV's influence on COVID-19 severity, particularly in post-COVID-19 fatigue and cerebrovascular complications. Age-related increases in DPPIV activity, combined with comorbidities such as diabetes and hypertension, significantly exacerbate the inflammatory and metabolic disruptions caused by COVID-19. Future research should focus on understanding the genetic predispositions that may further modulate DPPIV activity and how these factors collectively contribute to long-term COVID-19 sequelae.

5.2. Future Recommendations

Post-COVID-19 fatigue presents a multifaceted challenge, characterized by a diverse array of symptoms that can manifest regardless of the initial severity of the disease. Since these features of post-COVID-19 fatigue are complex and multifactorial, research and clinical practices should aim at the identification of specific biomarkers of early diagnosis, which may allow clinicians to better predict symptom trajectories and timely management. The primary focus in addressing post-COVID-19 fatigue lies in comprehending its progression and identifying effective treatment strategies, i.e., in this case, focusing on DPPIV. Researchers are beginning to delineate the specific clinical, cellular, and molecular mechanisms underlying post-COVID-19 fatigue symptoms and DPPIV. Neurovascular disorders may play a role in driving the progression of post-COVID-19 fatigue, with SARS-CoV-2 infection and the role of DPPIV as a co-receptor potentially causing tissue and cellular damage, as well as exacerbating inflammatory or autoimmune responses due to disruptions in the proteolytic enzyme activities, catalysis, and molecular signaling pathways.

While existing research has contributed to our understanding of the mechanisms underlying post-COVID-19 fatigue, numerous unanswered questions persist and warrant urgent attention to mitigate its effects. Identifying additional molecular determinants and target proteins, i.e., DPPIV, is crucial for alleviating the substantial burden of post-COVID-19 fatigue. Applying advanced molecular profiling in searching for the singular biomarker of post-COVID-19 fatigue and neurovascular dysfunction, specifically those related to DPPIV and other proteolytic enzyme pathways by using techniques such as single-cell RNA sequencing, metabolomics, proteomics, whole-genome sequencing, and gut microbiome studies offer promising avenues to unravel the pathophysiological processes underlying post-COVID-19 fatigue.

Further extension of pharmacological research into DPPIV inhibitors and modulators of neuroinflammatory responses might allow for targeted treatments that could alleviate symptoms directly related to immune and inflammatory dysregulation. This may minimize neuroinflammatory effects promoting fatigue and possibly reduce the risk of chronic neurovascular complications in these patients. In addition, such a study on the interplay between the gut microbiome and neurovascular health in post-COVID-19 fatigue is clinically important because there is emerging evidence that dysbiosis of the microbiome might be associated with immune and neurological dysfunction.

Longitudinal clinical trials should also be a focus of investigation in the assessment of the long-term effects of SARS-CoV-2 infection on neurovascular and cognitive health. Such trials would allow the identification of at-risk individuals, helping clinicians tailor preventative and therapeutic interventions, especially in those likely to develop chronic cerebrovascular conditions. Integration of assessment for cognitive function, neuroimaging biomarkers, and vascular health over extended follow-ups can give robust insights into disease progression and intervention efficacy.

The integration of data from basic science and clinical research is essential for gaining insight into the viral invasion into various tissues, particularly the brain, where it leads to neurological dysfunctions. Developing improved therapies to mitigate long-term cerebrovascular disease incidence and progression resulting from SARS-CoV-2 infection hinges on these investigations. Finally, further extensive research is undoubtedly necessary to evaluate and devise targeted treatment strategies for combating the persistent effects of post-COVID-19 fatigue.

Limitations of Current Narrative Review

We acknowledge that such a narrative approach has its limitations, including selection bias and the lack of quantitative synthesis. These limitations are partly offset by the comprehensiveness and richness of our expert analysis, which provides a very strong foundation for understanding neurological sequelae following post-COVID-19 fatigue. When the body of research builds up, a future systematic review could analyze the well-defined studies to quantify the impact and underlying mechanisms of DPPIV-mediated cerebrovascular complications. A review might consider extensive inclusion criteria and data analysis to define more precise relationships. We strongly believe that our narrative review epitomizes the current knowledge on the subject and thus lays a foundation for further systematic studies in the future as the body of research grows.

6. Conclusions

In conclusion, from this review, the immense potential of DPPIV has become clear not only as a co-receptor of SARS-CoV-2 but also as a main mediator in metabolic–inflammatory cascades leading to the aggravation of post-COVID-19 fatigue and cerebrovascular complications. Its participation in glucose metabolism, immune regulation, and endothelial dysfunction, therefore, places it at the heart of the chronic cerebrovascular outcomes that accompany COVID-19 survivors, such as neurovascular damage, stroke, and cognitive decline. Targeting DPPIV with therapeutic inhibitors holds immense promise toward the mitigation of long-term cardiovascular and cerebrovascular risks, particularly in post-

COVID-19 fatigue patients with comorbidities that include diabetes and hypertension. Further studies are necessary to demonstrate the exact molecular mechanisms and establish the therapeutic value of DPPIV inhibitors in the management of post-COVID-19 fatigue as an effective way to reduce the public health burden of this debilitating condition.

Author Contributions: Conceptualization, C.M.N.C.M.N. and M.D.C.R.; resources, C.M.N.C.M.N. and M.D.C.R.; data curation, C.M.N.C.M.N.; writing—original draft preparation, C.M.N.C.M.N.; writing—review and editing, C.M.N.C.M.N., M.D.C.R., U.J., M.Z.M., H.A.H., M.M.G., and E.N.K.C.; funding acquisition, M.D.C.R. All authors have read and agreed to the published version of the manuscript.

Funding: This study received the MSU Publication Grant (MPG-004-022024-FHLS).

Acknowledgments: We would like to express our sincere gratitude to all the author members for their invaluable guidance, support, and encouragement throughout the preparation of this manuscript. We are also deeply appreciative of Management and Science University (MSU) for providing the MSU Publication Grant (MPG-004-022024-FHLS). Special thanks go to colleagues, friends, and family members for their understanding and continuous support.

Conflicts of Interest: The authors declare no conflicts of interest.

References

1. Centers for Disease Control and Prevention. COVID Data Tracker. 2022. Available online: https://covid.cdc.gov/covid-data-tracker/#datatracker-home (accessed on 8 November 2023).
2. Hu, B.; Guo, H.; Zhou, P.; Shi, Z.L. Characteristics of SARS-CoV-2 and COVID-19. *Nat. Rev. Microbiol.* **2021**, *19*, 141–154. [CrossRef] [PubMed]
3. Thompson, M.; Ferrando, S.J.; Dornbush, R.; Lynch, S.; Shahar, S.; Klepacz, L.; Smiley, A. Impact of COVID-19 on employment: Sociodemographic, medical, psychiatric and neuropsychological correlates. *Front. Rehabil. Sci.* **2023**, *4*, 1150734. [CrossRef] [PubMed]
4. Davis, H.E.; McCorkell, L.; Vogel, J.M.; Topol, E.J. Long COVID: Major findings, mechanisms and recommendations. *Nat. Rev. Microbiol.* **2023**, *21*, 133–146. [CrossRef] [PubMed]
5. Kuznik, B.I.; Shapovalov, K.G.; Chalisova, N.I. Changes in the state of vital systems with long COVID-19. *Biol. Bull. Rev.* **2023**, *13*, 112–123. [CrossRef]
6. Becerra-García, J.A.; Sánchez-Gutiérrez, T. Long-COVID psychological symptoms in child and adolescent population: A standardized proposal for its exploration. *Enfermedades Infecc. Microbiol. Clínica* **2023**, *41*, 384–385. [CrossRef]
7. Gupta, M.; Gupta, N.; Esang, M. Long COVID in children and adolescents. *Prim. Care Companion CNS Disord.* **2022**, *24*, 21r03218. [CrossRef]
8. Guido, C.A.; Lucidi, F.; Midulla, F.; Zicari, A.M.; Bove, E.; Avenoso, F.; Amedeo, I.; Mancino, E.; Nenna, R.; De Castro, G.; et al. Neurological and psychological effects of long COVID in a young population: A cross-sectional study. *Front. Neurol.* **2022**, *13*, 925144. [CrossRef]
9. Batra, A.; Clark, J.R.; Kang, A.K.; Ali, S.; Patel, T.R.; Shlobin, N.A.; Hoffman, S.C.; Lim, P.H.; Orban, Z.S.; Visvabharathy, L.; et al. Persistent viral RNA shedding of SARS-CoV-2 is associated with delirium incidence and six-month mortality in hospitalized COVID-19 patients. *Geroscience* **2022**, *44*, 1241–1254. [CrossRef]
10. Natarajan, A.; Zlitni, S.; Brooks, E.F.; Vance, S.E.; Dahlen, A.; Hedlin, H.; Park, R.M.; Han, A.; Schmidtke, D.T.; Verma, R.; et al. Gastrointestinal symptoms and fecal shedding of SARS-CoV-2 RNA suggest prolonged gastrointestinal infection. *Med* **2022**, *3*, 371–387. [CrossRef]
11. Herman, J.D.; Atyeo, C.; Zur, Y.; Cook, C.E.; Patel, N.J.; Vanni, K.M.; Kowalski, E.N.; Qian, G.; Srivatsan, S.; Shadick, N.A.; et al. Humoral immunity to an endemic coronavirus is associated with postacute sequelae of COVID-19 in individuals with rheumatic diseases. *Sci. Transl. Med.* **2023**, *6*, 15. [CrossRef]
12. van Kessel, S.A.M.; Olde Hartman, T.C.; Lucassen, P.L.B.J.; van Jaarsveld, C.H.M. Post-acute and long-COVID-19 symptoms in patients with mild diseases: A systematic review. *Fam. Pract.* **2022**, *39*, 159–167. [CrossRef] [PubMed]
13. Rioux, P.; Chaumon, M.; Demers, A.; Fitzback-Fortin, H.; Kübel, S.L.; Lebrun, C.; Mendoza-Duran, E.; Micillo, L.; Racine, C.; Thibault, N.; et al. Psychological time during the COVID-19 lockdown: Canadian data. *Timing Time Percept.* **2022**, *10*, 326–343. [CrossRef]
14. Medline, A.; Hayes, L.; Valdez, K.; Hayashi, A.; Vahedi, F.; Capell, W.; Sonnenberg, J.; Glick, Z.; Klausner, J.D. Evaluating the impact of stay-at-home orders on the time to reach the peak burden of COVID-19 cases and deaths: Does timing matter? *BMC Public Health* **2020**, *20*, 1750. [CrossRef] [PubMed]
15. Elisabeth, A.L.; Karlen, S.B.; Magkos, F. The Effect of COVID-19-related Lockdowns on Diet and Physical Activity in Older Adults: A Systematic Review. *Aging Dis.* **2021**, *12*, 1935–1947. [CrossRef] [PubMed]

16. Holdsworth, D.A.; Chamley, R.; Barker-Davies, R.; O'Sullivan, O.; Ladlow, P.; Mitchell, J.L.; Dewson, D.; Mills, D.; May SL, J.; Cranley, M.; et al. Comprehensive clinical assessment identifies specific neurocognitive deficits in working-age patients with long-COVID. *PLoS ONE* **2022**, *17*, e0267392. [CrossRef]
17. Leng, A.; Shah, M.; Ahmad, S.A.; Premraj, L.; Wildi, K.; Li Bassi, G.; Pardo, C.A.; Choi, A.; Cho, S.M. Pathogenesis underlying neurological manifestations of long COVID syndrome and potential therapeutics. *Cells* **2023**, *12*, 816. [CrossRef]
18. Avittan, H.; Kustovs, D. Cognition and mental health in pediatric patients following COVID-19. *Int. J. Environ. Res. Public Health* **2023**, *20*, 5061. [CrossRef]
19. Govil-Dalela, T.; Sivaswamy, L. Neurological effects of COVID-19 in children. *Pediatr. Clin. N. Am.* **2021**, *68*, 1081–1091. [CrossRef]
20. Bradley, B.T.; Bryan, A. Emerging respiratory infections: The infectious disease pathology of SARS, MERS, pandemic influenza, and Legionella. *Semin. Diagn. Pathol.* **2019**, *36*, 152–159. [CrossRef]
21. Rees, A.R. Immunological challenges of the "new" infections: Corona viruses. In *A New History of Vaccines for Infectious Diseases*; Elsevier: Amsterdam, The Netherlands, 2022; pp. 395–450.
22. Desforges, M.; Le Coupanec, A.; Dubeau, P.; Bourgouin, A.; Lajoie, L.; Dubé, M.; Talbot, P.J. Human coronaviruses and other respiratory viruses: Underestimated opportunistic pathogens of the central nervous system? *Viruses* **2019**, *12*, 14. [CrossRef]
23. Chiara, M.; D'Erchia, A.M.; Gissi, C.; Manzari, C.; Parisi, A.; Resta, N.; Zambelli, F.; Picardi, E.; Pavesi, G.; Horner, D.S.; et al. Next generation sequencing of SARS-CoV-2 genomes: Challenges, applications and opportunities. *Brief. Bioinform.* **2021**, *22*, 616–630. [CrossRef] [PubMed]
24. John, G.; Sahajpal, N.S.; Mondal, A.K.; Ananth, S.; Williams, C.; Chaubey, A.; Rojiani, A.M.; Kolhe, R. Next-generation sequencing (NGS) in COVID-19: A tool for SARS-CoV-2 diagnosis, monitoring new strains and phylodynamic modeling in molecular epidemiology. *Curr. Issues Mol. Biol.* **2021**, *43*, 845–867. [CrossRef] [PubMed]
25. Galeotti, C.; Bayry, J. Autoimmune and inflammatory diseases following COVID-19. *Nat. Rev. Rheumatol.* **2020**, *16*, 413–414. [CrossRef] [PubMed]
26. Stein, S.R.; Ramelli, S.C.; Grazioli, A.; Chung, J.Y.; Singh, M.; Yinda, C.K.; Winkler, C.W.; Sun, J.; Dickey, J.M.; Ylaya, K.; et al. SARS-CoV-2 infection and persistence throughout the human body and brain. *Nature* **2022**, *612*, 758–763. [CrossRef]
27. Dennis, A.; Wamil, M.; Alberts, J.; Oben, J.; Cuthbertson, D.J.; Wootton, D.; Crooks, M.; Gabbay, M.; Brady, M.; Hishmeh, L.; et al. Multiorgan impairment in low-risk individuals with post-COVID-19 syndrome: A prospective, community-based study. *BMJ Open* **2021**, *11*, e048391. [CrossRef]
28. Conway, E.M.; Mackman, N.; Warren, R.Q.; Wolberg, A.S.; Mosnier, L.O.; Campbell, R.A.; Gralinski, L.E.; Rondina, M.T.; van de Veerdonk, F.L.; Hoffmeister, K.M.; et al. Understanding COVID-19-associated coagulopathy. *Nat. Rev. Immunol.* **2022**, *22*, 639–649. [CrossRef]
29. Kansagra, A.P.; Goyal, M.S.; Hamilton, S.; Albers, G.W. Collateral effect of COVID-19 on stroke evaluation in the United States. *N. Engl. J. Med.* **2020**, *383*, 400–401. [CrossRef]
30. Newell, K.L.; Waickman, A.T. Inflammation, immunity, and antigen persistence in post-acute sequelae of SARS-CoV-2 infection Immunity and inflammation in post-acute sequelae of SARS-CoV-2 infection. *Curr. Opin. Immunol.* **2022**, *77*, 102228. [CrossRef]
31. Toscano, G.; Palmerini, F.; Ravaglia, S.; Ruiz, L.; Invernizzi, P.; Cuzzoni, M.G.; Franciotta, D.; Baldanti, F.; Daturi, R.; Postorino, P.; et al. Guillain-Barré syndrome associated with SARS-CoV-2. *N. Engl. J. Med.* **2020**, *382*, 2574–2576. [CrossRef]
32. Lao, W.P.; Imam, S.A.; Nguyen, S.A. Anosmia, hyposmia, and dysgeusia as indicators for positive SARS-CoV-2 infection. *World J. Otolaryngol. Head Neck Surg.* **2020**, *6*, S22–S25. [CrossRef]
33. Spudich, S.; Nath, A. Nervous system consequences of COVID-19. *Science* **2022**, *375*, 267–269. [CrossRef] [PubMed]
34. Akimoto, T.; Hara, M.; Tasaki, K.; Kurosawa, Y.; Nakamoto, T.; Hirose, S.; Mizoguchi, T.; Yokota, Y.; Ninomiya, S.; Nakajima, H. Delayed encephalopathy after COVID-19: A case series of six patients. *Medicine* **2022**, *101*, e31029. [CrossRef] [PubMed]
35. Taube, M. Depression and brain fog as long-COVID mental health consequences: Difficult, complex and partially successful treatment of a 72-year-old patient-A case report. *Front. Psychiatry* **2023**, *14*, 1153512. [CrossRef] [PubMed]
36. Savardashtaki, A.; Johnston, T.P.; Sahebkar, A. COVID-19 and cardiac injury: Clinical manifestations, biomarkers, mechanisms, diagnosis, treatment, and follow up. *Expert. Rev. Anti Infect. Ther.* **2021**, *19*, 345–357.
37. Bowe, B.; Xie, Y.; Xu, E.; Al-Aly, Z. Kidney outcomes in Long COVID. *J. Am. Soc. Nephrol.* **2021**, *32*, 2851–2862. [CrossRef]
38. Wang, S.; Farland, L.V.; Gaskins, A.J.; Mortazavi, J.; Wang, Y.X.; Tamimi, R.M.; Rich-Edwards, J.W.; Zhang, D.; Terry, K.L.; Chavarro, J.E.; et al. Association of laparoscopically confirmed endometriosis with long COVID-19: A prospective cohort study. *Am. J. Obstet. Gynecol.* **2023**, *228*, 714.e1–714.e13. [CrossRef]
39. Zhang, D.; Weng, S.; Xia, C.; Ren, Y.; Liu, Z.; Xu, Y.; Yang, X.; Wu, R.; Peng, L.; Sun, L.; et al. Gastrointestinal symptoms of long COVID-19 related to the ectopic colonization of specific bacteria that move between the upper and lower alimentary tract and alterations in serum metabolites. *BMC Med.* **2023**, *21*, 264. [CrossRef]
40. Levi, M.; Thachil, J.; Iba, T.; Levy, J.H. Coagulation abnormalities and thrombosis in patients with COVID-19. *Lancet Haematol.* **2020**, *7*, e438–e440. [CrossRef]
41. Lukiw, W.J.; Pogue, A.; Hill, J.M. SARS-CoV-2 Infectivity and Neurological Targets in the Brain. *Cell Mol. Neurobiol.* **2022**, *42*, 217–224. [CrossRef]
42. Douaud, G.; Lee, S.; Alfaro-Almagro, F.; Arthofer, C.; Wang, C.; McCarthy, P.; Lange, F.; Andersson JL, R.; Griffanti, L.; Duff, E.; et al. SARS-CoV-2 is associated with changes in brain structure in UK biobank. *Nature* **2022**, *604*, 697–707. [CrossRef]

43. Ntaios, G.; Michel, P.; Georgiopoulos, G.; Guo, Y.; Li, W.; Xiong, J.; Calleja, P.; Ostos, F.; González-Ortega, G.; Fuentes, B.; et al. Characteristics and outcomes in patients with COVID-19 and acute ischemic stroke: The global COVID-19 stroke registry. *Stroke* **2020**, *51*, e254–e258. [CrossRef] [PubMed]
44. Karnik, M.; Beeraka, N.M.; Uthaiah, C.A.; Nataraj, S.M.; Bettadapura AD, S.; Aliev, G.; Madhunapantula, S.V. A Review on SARS-CoV-2-induced neuroinflammation, neurodevelopmental complications, and recent updates on the vaccine development. *Mol. Neurobiol.* **2021**, *58*, 4535–4563. [CrossRef] [PubMed]
45. Lyoo, K.S.; Kim, H.M.; Lee, B.; Che, Y.H.; Kim, S.J.; Song, D.; Hwang, W.; Lee, S.; Park, J.H.; Na, W.; et al. Direct neuronal infection of SARS-CoV-2 reveals cellular and molecular pathology of chemosensory impairment of COVID-19 patients. *Emerg. Microbes Infect.* **2022**, *11*, 406–411. [CrossRef] [PubMed]
46. Fahlberg, M.D.; Blair, R.V.; Doyle-Meyers, L.A.; Midkiff, C.C.; Zenere, G.; Russell-Lodrigue, K.E.; Monjure, C.J.; Haupt, E.H.; Penney, T.P.; Lehmicke, G.; et al. Cellular events of acute, resolving or progressive COVID-19 in SARS-CoV-2 infected non-human primates. *Nat. Commun.* **2020**, *11*, 6078. [CrossRef] [PubMed]
47. Michiels, C. Endothelial cell functions. *J. Cell Physiol.* **2003**, *196*, 430–443. [CrossRef]
48. Gavriilaki, E.; Eftychidis, I.; Papassotiriou, I. Update on endothelial dysfunction in COVID-19: Severe disease, long COVID-19 and pediatric characteristics. *J. Lab. Med.* **2021**, *45*, 293–302. [CrossRef]
49. Tarnawski, A.S.; Ahluwalia, A. Endothelial cells and blood vessels are major targets for COVID-19-induced tissue injury and spreading to various organs. *World J. Gastroenterol.* **2022**, *28*, 275–289. [CrossRef]
50. Bayat, A.H.; Azimi, H.; Hassani Moghaddam, M.; Ebrahimi, V.; Fathi, M.; Vakili, K.; Mahmoudiasl, G.R.; Forouzesh, M.; Boroujeni, M.E.; Nariman, Z.; et al. COVID-19 causes neuronal degeneration and reduces neurogenesis in human hippocampus. *Apoptosis* **2022**, *27*, 852–868. [CrossRef]
51. Plantone, D.; Locci, S.; Bergantini, L.; Manco, C.; Cortese, R.; Meocci, M.; Cavallaro, D.; d'Alessandro, M.; Bargagli, E.; De Stefano, N. Brain neuronal and glial damage during acute COVID-19 infection in absence of clinical neurological manifestations. *J. Neurol. Neurosurg. Psychiatry* **2022**, *93*, 1343–1348. [CrossRef]
52. Klein, R.S.; Garber, C.; Howard, N. Infectious immunity in the central nervous system and brain function. *Nat. Immunol.* **2017**, *18*, 132–141. [CrossRef]
53. Wongchitrat, P.; Chanmee, T.; Govitrapong, P. Molecular mechanisms associated with neurodegeneration of neurotropic viral infection. *Mol. Neurobiol.* **2023**, *61*, 2881–2903. [CrossRef] [PubMed]
54. Beckman, D.; Bonillas, A.; Diniz, G.B.; Ott, S.; Roh, J.W.; Elizaldi, S.R.; Schmidt, B.A.; Sammak, R.L.; Van Rompay KK, A.; Iyer, S.S.; et al. SARS-CoV-2 infects neurons and induces neuroinflammation in a non-human primate model of COVID-19. *Cell Rep.* **2022**, *41*, 111573. [CrossRef]
55. Zorzo, C.; Solares, L.; Mendez, M.; Mendez-Lopez, M. Hippocampal alterations after SARS-CoV-2 infection: A systematic review. *Behav. Brain Res.* **2023**, *455*, 114662. [CrossRef]
56. Klein, J.; Wood, J.; Jaycox, J.R.; Dhodapkar, R.M.; Lu, P.; Gehlhausen, J.R.; Tabachnikova, A.; Greene, K.; Tabacof, L.; Malik, A.A.; et al. Distinguishing features of Long COVID identified through immune profiling. *Nature* **2023**, *623*, 139–148. [CrossRef] [PubMed]
57. Glynne, P.; Tahmasebi, N.; Gant, V.; Gupta, R. Long COVID following mild SARS-CoV-2 infection: Characteristic T cell alterations and response to antihistamines. *J. Investig. Med.* **2022**, *70*, 61–67. [CrossRef]
58. Arun, S.; Storan, A.; Myers, B. Mast cell activation syndrome and the link with long COVID. *Br. J. Hosp. Med.* **2022**, *83*, 1–10. [CrossRef]
59. Peluso, M.J.; Lu, S.; Tang, A.F.; Durstenfeld, M.S.; Ho, H.E.; Goldberg, S.A.; Forman, C.A.; Munter, S.E.; Hoh, R.; Tai, V.; et al. Markers of immune activation and inflammation in individuals with postacute sequelae of severe acute respiratory syndrome coronavirus 2 infection. *J. Infect. Dis.* **2021**, *224*, 1839–1848. [CrossRef] [PubMed]
60. Barnes, B.J.; Adrover, J.M.; Baxter-Stoltzfus, A.; Borczuk, A.; Cools-Lartigue, J.; Crawford, J.M.; Daßler-Plenker, J.; Guerci, P.; Huynh, C.; Knight, J.S.; et al. Targeting potential drivers of COVID-19: Neutrophil extracellular traps. *J. Exp. Med.* **2020**, *217*, e20200652. [CrossRef]
61. Wang, R.; Lee, J.H.; Kim, J.; Xiong, F.; Hasani, L.A.; Shi, Y.; Simpson, E.N.; Zhu, X.; Chen, Y.T.; Shivshankar, P.; et al. SARS-CoV-2 restructures host chromatin architecture. *Nat. Microbiol.* **2023**, *8*, 679–694. [CrossRef]
62. Mohamed, M.S.; Johansson, A.; Jonsson, J.; Schiöth, H.B. Review dissecting the molecular mechanisms surrounding post-COVID-19 syndrome and neurological features. *Int. J. Mol. Sci.* **2022**, *23*, 4275. [CrossRef]
63. Ragab, D.; Salah Eldin, H.; Taeimah, M.; Khattab, R.; Salem, R. The COVID-19 cytokine storm; what we know so far. *Front. Immunol.* **2020**, *11*, 1446. [CrossRef] [PubMed]
64. Scurati, R.; Papini, N.; Giussani, P.; Alberti, G.; Tringali, C. The challenge of Long COVID-19 management: From disease molecular hallmarks to the proposal of exercise as therapy. *Int. J. Mol. Sci.* **2022**, *23*, 12311. [CrossRef] [PubMed]
65. Patel, M.A.; Knauer, M.J.; Nicholson, M.; Daley, M.; Van Nynatten, L.R.; Martin, C.; Patterson, E.K.; Cepinskas, G.; Seney, S.L.; Dobretzberger, V.; et al. Elevated vascular transformation blood biomarkers in Long-COVID indicate angiogenesis as a key pathophysiological mechanism. *Mol. Med.* **2022**, *28*, 122. [CrossRef]
66. Gomes, S.M.R.; Brito, A.C.S.; Manfro, W.F.P.; Ribeiro-Alves, M.; Ribeiro, R.S.A.; da Cal, M.S.; Lisboa, V.D.C.; Abreu, D.P.B.; Castilho, L.D.R.; Porto, L.C.M.S.; et al. High levels of pro-inflammatory SARS-CoV-2-specific biomarkers revealed by in vitro whole blood cytokine release assay (CRA) in recovered and long-COVID-19 patients. *PLoS ONE* **2023**, *18*, e0283983. [CrossRef]

67. Petrella, C.; Nenna, R.; Petrarca, L.; Tarani, F.; Paparella, R.; Mancino, E.; Di Mattia, G.; Conti, M.G.; Matera, L.; Bonci, E.; et al. Serum NGF and BDNF in Long-COVID-19 adolescents: A pilot study. *Diagnostics* **2022**, *12*, 1162. [CrossRef]
68. Hoffmann, M.; Kleine-Weber, H.; Schroeder, S.; Krüger, N.; Herrler, T.; Erichsen, S.; Schiergens, T.S.; Herrler, G.; Wu, N.H.; Nitsche, A.; et al. SARS-CoV-2 cell entry depends on ACE2 and TMPRSS2 and is blocked by a clinically proven protease inhibitor. *Cell* **2020**, *181*, 271–280. [CrossRef]
69. Adimulam, T.; Arumugam, T.; Gokul, A.; Ramsuran, V. Genetic Variants within SARS-CoV-2 Human Receptor Genes May Contribute to Variable Disease Outcomes in Different Ethnicities. *Int. J. Mol. Sci.* **2023**, *24*, 8711. [CrossRef] [PubMed]
70. Li, Y.; Zhang, Z.; Yang, L.; Lian, X.; Xie, Y.; Li, S.; Xin, S.; Cao, P.; Lu, J. The MERS-CoV receptor DPP4 as a candidate binding target of the SARS-CoV-2 spike. *iScience* **2020**, *23*, 101160. [CrossRef] [PubMed]
71. Nag, S.; Mandal, S.; Mukherjee, O.; Mukherjee, S.; Kundu, R. DPP-4 Inhibitors as a savior for COVID-19 patients with diabetes. *Future Virol.* **2023**, *18*, 321–333. [CrossRef]
72. Bassendine, M.F.; Bridge, S.H.; McCaughan, G.W.; Gorrell, M.D. COVID-19 and comorbidities: A role for dipeptidyl peptidase 4 (DPP4) in disease severity? *J. Diabetes* **2020**, *12*, 649–658. [CrossRef]
73. Wu, K.C.H.; He, Q.; Bennett, A.N.; Li, J.; Chan, K.H.K. Shared genetic mechanism between type 2 diabetes and COVID-19 using pathway-based association analysis. *Front. Genet.* **2022**, *13*, 1063519. [CrossRef] [PubMed]
74. Deacon, C.F. Dipeptidyl peptidase 4 inhibition with sitagliptin: A new therapy for type 2 diabetes. *Expert. Opin. Investig. Drugs* **2007**, *16*, 533–545. [CrossRef] [PubMed]
75. Roy, A.N.; Gupta, A.M.; Banerjee, D.; Chakrabarti, J.; Raghavendra, P.B. Unraveling DPP4 receptor interactions with SARS-CoV-2 variants and MERS-CoV: Insights into pulmonary disorders via immunoinformatics and molecular dynamics. *Viruses* **2023**, *15*, 2056. [CrossRef]
76. Sebastián-Martín, A.; Sánchez, B.G.; Mora-Rodríguez, J.M.; Bort, A.; Díaz-Laviada, I. Role of dipeptidyl peptidase-4 (DPP4) on COVID-19 physiopathology. *Biomedicines* **2022**, *10*, 2026. [CrossRef]
77. Icard, P.; Lincet, H.; Wu, Z.; Coquerel, A.; Forgez, P.; Alifano, M.; Fournel, L. The key role of Warburg effect in SARS-CoV-2 replication and associated inflammatory response. *Biochimie* **2021**, *180*, 169–177. [CrossRef] [PubMed]
78. Mehdipour, A.R.; Hummer, G. Dual nature of human ACE2 glycosylation in binding to SARS-CoV-2 spike. *Proc. Natl. Acad. Sci. USA* **2021**, *118*, e2100425118. [CrossRef]
79. Lambeir, A.M.; Durinx, C.; Scharpé, S.; De Meester, I. Dipeptidyl-peptidase IV from bench to bedside: An update on structural properties functions clinical aspects of the enzyme, DPP IV. *Crit. Rev. Clin. Lab. Sci.* **2003**, *40*, 209–294. [CrossRef]
80. Arscott, W.T.; LaBauve, A.E.; May, V.; Wesley, U.V. Suppression of neuroblastoma growth by dipeptidyl peptidase IV: Relevance of chemokine regulation and caspase activation. *Oncogene* **2009**, *28*, 479–491. [CrossRef] [PubMed]
81. Gorrell, M.D. Dipeptidyl peptidase IV and related enzymes in cell biology and liver disorders. *Clin. Sci.* **2005**, *108*, 277–292. [CrossRef]
82. Klemann, C.; Wagner, L.; Stephan, M.; von Hörsten, S. Cut to the chase: A review of CD26/dipeptidyl peptidase-4's (DPP4) entanglement in the immune system. *Clin. Exp. Immunol.* **2016**, *185*, 1–21. [CrossRef]
83. Lu, G.; Hu, Y.; Wang, Q.; Qi, J.; Gao, F.; Li, Y.; Zhang, Y.; Zhang, W.; Yuan, Y.; Bao, J.; et al. Molecular Basis of Binding between Novel Human Coronavirus MERS-CoV and Its Receptor CD26. *Nature* **2013**, *500*, 227–231. [CrossRef] [PubMed]
84. Callebaut, C.; Krust, B.; Jacotot, E.; Hovanessian, A.G. T cell activation antigen, CD26, as a cofactor for entry of HIV in CD4+ cells. *Science* **1993**, *262*, 2045–2050. [CrossRef] [PubMed]
85. Raj, V.S.; Mou, H.; Smits, S.L.; Dekkers, D.H.; Müller, M.A.; Dijkman, R.; Muth, D.; Demmers, J.A.; Zaki, A.; Fouchier, R.A.; et al. Dipeptidyl peptidase 4 is a functional receptor for the emerging human coronavirus-EMC. *Nature* **2013**, *495*, 251–254. [CrossRef] [PubMed]
86. Brooke, G.N.; Prischi, F. Structural and functional modelling of SARS-CoV-2 entry in animal models. *Sci. Rep.* **2020**, *10*, 15917. [CrossRef]
87. Vankadari, N.; Wilce, J.A. Emerging WuHan (COVID-19) coronavirus: Glycan shield and structure prediction of spike glycoprotein and its interaction with human CD26. *Emerg. Microbes Infect.* **2020**, *9*, 601–604. [CrossRef]
88. Radzikowska, U.; Ding, M.; Tan, G.; Zhakparov, D.; Peng, Y.; Wawrzyniak, P.; Wang, M.; Li, S.; Morita, H.; Altunbulakli, C.; et al. Distribution of ACE2, CD147, CD26, and other SARS-CoV-2 associated molecules in tissues and immune cells in health and in asthma, COPD, obesity, hypertension, and COVID-19 risk factors. *Allergy* **2020**, *75*, 2829–2845. [CrossRef]
89. Gao, Q.; Zhang, W.; Li, T.; Yang, G.; Zhu, W.; Chen, N.; Jin, H. Interrelationship between 2019-nCov receptor DPP4 and diabetes mellitus targets based on protein interaction network. *Sci. Rep.* **2022**, *12*, 188. [CrossRef]
90. Singh, A.; Singh, R.S.; Sarma, P.; Batra, G.; Joshi, R.; Kaur, H.; Sharma, A.R.; Prakash, A.; Medhi, B. A Comprehensive Review of Animal Models for Coronaviruses: SARS-CoV-2, SARS-CoV, and MERS-CoV. *Virol. Sin.* **2020**, *35*, 290–304. [CrossRef]
91. Pitocco, D.; Tartaglione, L.; Viti, L.; Di Leo, M.; Pontecorvi, A.; Caputo, S. SARS-CoV-2 and DPP4 inhibition: Is it time to pray for Janus Bifrons? *Diabetes Res. Clin. Pract.* **2020**, *163*, 108162. [CrossRef]
92. Groppa, S.A.; Ciolac, D.; Duarte, C.; Garcia, C.; Gasnaș, D.; Leahu, P.; Efremova, D.; Gasnaș, A.; Bălănuță, T.; Mîrzac, D.; et al. Molecular mechanisms of SARS-CoV-2/COVID-19 pathogenicity on the central nervous system: Bridging experimental probes to clinical evidence and therapeutic interventions. *Adv. Exp. Med. Biol.* **2022**, *1376*, 1–27.

93. Mahgoub, S.; Fatahala, S.S.; Sayed, A.I.; Atya, H.B.; El-Shehry, M.F.; Afifi, H.; Awad, S.M.; El-Hameed, R.H.A.; Taha, H. Novel hit of DPP-4Is as promising antihyperglycemic agents with dual antioxidant/anti-inflammatory effects for type 2 diabetes with/without COVID-19. *Bioorg. Chem.* **2022**, *128*, 106092. [CrossRef] [PubMed]
94. Chen, Z.; Su, X.; Cao, W.; Tan, M.; Zhu, G.; Gao, J.; Zhou, L. The discovery and characterization of a potent DPP-IV inhibitory peptide from oysters for the treatment of type 2 diabetes based on computational and experimental studies. *Mar. Drugs* **2024**, *22*, 361. [CrossRef] [PubMed]
95. Kang, S.M.; Park, J.H. Pleiotropic benefits of DPP-4 inhibitors beyond glycemic control. *Clin. Med. Insights Endocrinol. Diabetes* **2021**, *14*, 11795514211051698. [CrossRef]
96. Rajpal, A.; Sayyed Kassem, L.; Aron, D.C. Management of diabetes in elderly patients during the COVID-19 pandemic: Current and future perspectives. *Expert. Rev. Endocrinol. Metab.* **2021**, *16*, 181–189. [CrossRef] [PubMed]
97. Nádasdi, Á.; Sinkovits, G.; Bobek, I.; Lakatos, B.; Förhécz, Z.; Prohászka, Z.Z.; Réti, M.; Arató, M.; Cseh, G.; Masszi, T.; et al. Decreased circulating dipeptidyl peptidase-4 enzyme activity is prognostic for severe outcomes in COVID-19 inpatients. *Biomark. Med.* **2022**, *16*, 317–330. [CrossRef]
98. Wesley, U.V.; Hatcher, J.F.; Ayvaci, E.R.; Klemp, A.; Dempsey, R.J. Regulation of Dipeptidyl Peptidase IV in the post-stroke rat brain and in vitro ischemia: Implications for chemokine-mediated neural progenitor cell migration and angiogenesis. *Mol. Neurobiol.* **2017**, *54*, 4973–4985. [CrossRef]
99. Rudroff, T.; Workman, C.D.; Bryant, A.D. Potential factors that contribute to post-COVID-19 fatigue in women. *Brain Sci.* **2022**, *12*, 556. [CrossRef]
100. Hong, S.; Jung, C.H.; Han, S.; Park, C.Y. Increasing age associated with higher dipeptidyl peptidase-4 inhibition rate is a predictive factor for efficacy of dipeptidyl peptidase-4 inhibitors. *Diabetes Metab. J.* **2022**, *46*, 63–70. [CrossRef]
101. Akinosoglou, K.; Schinas, G.; Bletsa, E.; Bristianou, M.; Lanaras, L.; Michailides, C.; Katsikas, T.; Barkas, F.; Liberopoulos, E.; Kotsis, V.; et al. COVID-19 outcomes and diabetes mellitus: A comprehensive multicenter prospective cohort study. *Microorganisms* **2023**, *11*, 1416. [CrossRef]
102. Park, M.J.; Hwang, J.; Ahn, J.; Park, S.J.; Song, E.; Jang, A.; Choi, K.M.; Baik, S.H.; Yoo, H.J. Ischaemic stroke in patients with diabetes requiring urgent procedures during the COVID-19 pandemic in South Korea: A retrospective, nationwide, population-based cohort study using data from the National Emergency Department Information System. *BMJ Open* **2023**, *13*, e074381. [CrossRef]
103. Malempati, M.; Patel, M.; Patel, J. Ischemic stroke and COVID-19 infection—A review of clinical case reports. *Egypt J. Intern. Med.* **2024**, *36*, 50. [CrossRef]
104. Narayanan, N.; Naik, D.; Sahoo, J.; Kamalanathan, S. Dipeptidyl peptidase 4 inhibitors in COVID-19: Beyond glycemic control. *World J. Virol.* **2022**, *11*, 399–410. [CrossRef] [PubMed]
105. Posadas-Sánchez, R.; Sánchez-Muñoz, F.; Guzmán-Martín, C.A.; Hernández-Díaz Couder, A.; Rojas-Velasco, G.; Fragoso, J.M.; Vargas-Alarcón, G. Dipeptidylpeptidase-4 levels and DPP4 gene polymorphisms in patients with COVID-19. Association with disease and with severity. *Life Sci.* **2021**, *276*, 119410. [CrossRef] [PubMed]
106. Bhargave, A.; Devi, K.; Ahmad, I.; Yadav, A.; Gupta, R. Genetic variation in DPP-IV gene linked to predisposition of T2DM: A case control study. *J. Diabetes Metab. Disord.* **2022**, *21*, 1709–1716. [CrossRef]
107. Adli , A.; Rahimi, M.; Khodaie, R.; Hashemzaei, N.; Hosseini, S.M. Role of genetic variants and host polymorphisms on COVID-19: From viral entrance mechanisms to immunological reactions. *J. Med. Virol.* **2022**, *94*, 1846–1865. [CrossRef]

current issues in molecular biology

MDPI

Article

Brain Tumor-Induced Changes in Routine Parameters of the Lipid Spectrum of Blood Plasma and Its Short-Chain Fatty Acids

Larisa Obukhova *, Natalia Shchelchkova, Igor Medyanik, Konstantin Yashin, Artem Grishin, Oksana Bezvuglyak and Ilkhom Abdullaev

Federal State Budgetary Educational Institution of Higher Education, The "Privolzhsky Research Medical University" of the Ministry of Health of the Russian Federation, 603005 Nizhny Novgorod, Russia; n.shchelchkova@mail.ru (N.S.); med_neuro@inbox.ru (I.M.); yashinmed@yandex.ru (K.Y.); zhest8242@mail.ru (A.G.); bezvugliak.oksana@gmail.com (O.B.); ilhomabdullaev@yahoo.com (I.A.)

* Correspondence: obuhovalm@yandex.ru; Tel.: +7-951-914-55-45

Abstract: The aim of this research was to provide a comparative analysis of the major parameters of the blood lipid spectrum found both in the case of brain tumors and in atherosclerosis, as well as to assess the correlation of these indicators with the proliferative activity index Ki-67 in cells. Blood analyses were conducted on samples from 50 patients with brain tumors and 50 patients with cerebral atherosclerosis. Blood plasma from 50 essentially healthy people was used for controls. Significant differences were found in the parameter values between the atherosclerosis sufferers and the control group only for their ratios of neutral lipids to cholesterol. Of the short-chain fatty acids, butyric acid is of greatest interest due to the significant differences of its levels from the control group in the blood of both patients with meningiomas and of those with gliomas. Statistically significant correlation coefficients between the levels of the Ki-67 cell proliferation marker and, in particular, butyric acid were found when compared with the neutral lipids to cholesterol ratios. These identified parameters of the blood plasma lipid spectrum can be used for preoperative diagnostics of brain tumors. However, these ratios cannot be used as preoperative noninvasive predictors of the level of the Ki-67 mitotic index, as no significant differences corresponding to this were found for low-grade or for high-grade anaplasia of brain tumors.

Keywords: blood plasma; lipids; cholesterol; neutral lipid/phospholipid ratio; short-chain fatty acids; butyric acid; molecular genetic markers of gliomas; Ki-67

1. Introduction

The investigation of aberrations in the metabolome, in particular, of levels of those molecules involved in lipid metabolism is a promising research area in tumor diagnostics [1]. Compared to non-malignant tissues, glial tumors, for example, are characterized by overall increases in lipid levels and by high levels of lipids within these specific tumors [2]. However, lipid metabolism disorders are also typical of other diseases, primarily atherosclerosis [3].

Therefore, a new approach is needed to understand lipid metabolism in the pathogenesis of brain tumors, including the investigation of the levels of short-chain fatty acids (SCFAs). These are monocarboxylic acids with a chain length of up to six carbon atoms and are the main products of the anaerobic fermentation of indigestible polysaccharides, such as dietary fiber and resistant starch, being carried out by the microbiota in the colon [4]. As well as having an impact on the immune system, such SCFAs are involved in the regulation of neutrophil chemotaxis, the induction of regulatory T-cell development, IL-10 secretion, NF-kB inhibition, and the suppression of cytokine production by myeloid cells. Changes in the microbial flora can therefore alter not only the local immune environment but also the central nervous system (CNS). SCFAs are thought to have a key role in the bi-directional metabolic pathways between the microbiota, gut, and brain [5]. SCFAs can penetrate the blood–brain barrier using monocarboxylate transporters (MCTs) located on the endothelial cells and then increase the expression of tight junction proteins.

It has now been realized that it is not possible to predict tumor behavior without considering both the immunohistochemical and molecular genetic aspects. While the WHO classification of central nervous system tumors (2016) was the first step in the development of a new approach, the WHO classification (2021) consolidated it [6]. However, the level of the Ki-67 mitotic index remains an important factor in glioma diagnosis [7] as it allows for the differentiation between different degrees of anaplasia in brain tumors. The Ki-67 cell marker, encoded by the *MKI67* gene, is a non-histone protein associated with cell proliferation, and it is actively expressed in the S, G2, late G1, and mitotic phases of cell division. In the morphological diagnosis of central nervous system tumors, the level of this marker is particularly useful for reflecting the proliferative activity of the tumor and distinguishing between low-grade and high-grade gliomas [8,9].

The purpose of this study was to provide a comparative analysis of the major parameters of the blood lipid spectrum in brain tumors and in atherosclerosis as well as to assess any correlation of these indicators with the Ki-67 proliferative activity index of the cells involved.

2. Materials and Methods

2.1. Materials

Blood plasma was analyzed from 50 patients with brain tumors (25 men aged 32–74 years and 25 women aged 22–81) and from 50 patients with cerebral atherosclerosis (38 men aged 44–79 and 12 women aged 55–77). The cerebral atherosclerosis condition includes chronic elastic and muscular-elastic disease of the arteries of the brain, arising as a result of violations of lipid and protein metabolism, accompanied by the deposition of cholesterol and some fractions of lipoproteins in the intima of the blood vessels with the formation of atheromatous plaques, leading to a slowly progressive disruption of the blood supply to the brain. The brain tumor group had the following tumors predominating: meningiomas (21 patients, 42%), glioblastomas (16 patients, 32%), and astrocytomas (6 patients, 12%); it also included oligodendrogliomas (5 patients, 10%), ependymoma (1 patient, 2%), and gliosarcoma (1 patient, 2%) of varying grades of anaplasia: Grade I—2%, Grade II—8%, Grade III—26%, and Grade IV—64%. Blood plasma from 50 essentially healthy people, in whom brain MRIs revealed no brain tumors, was used to provide comparisons (controls) (16 men aged 24–64 and 34 women aged 24–65). Cerebral atherosclerosis was not detected in either the patient groups with brain tumors or in the healthy individuals (controls). With informed consent, tumor tissues were sampled as postoperative fragments at the Federal State Budgetary Educational Institution of Higher Education, the "Privolzhsky Research Medical University" of the Ministry of Health of the Russian Federation prior to anticancer therapy. Histological diagnoses were made using the WHO classification of CNS tumors [6]. Detailed information regarding the patients is listed in File S1: Supplementary Table S1. The study was approved by the Ethics Committee of the Federal State Budgetary Educational Institution of Higher Education, the "Privolzhsky Research Medical University" of the Ministry of Health of the Russian Federation (No. 6 of 17.04.2019). Biological objects were obtained during routine treatment procedures with informed consent of patients (File S2).

2.2. Measurement of Tumor Markers

The postoperative sample material was fixed in 10% formalin for at least 24 h. Basic histological processing of the sampled material was conducted using an Excelsior ES (Thermo Scientific, Waltham, MA, USA), and embedding was performed using a HistoStar station (Thermo Scientific, USA). Each sample with glial neoplasm was immunohistochemically examined to determine the level (high or low) of its Ki-67 proliferative index using SP6 clone (Thermo Scientific, USA). Immunohistochemical staining was performed using the BOND-MAX immunohistostainer and BOND 5.1 software (Leica Biosystems, York, UK), in line with the standard protocols recommended by the manufacturer. Bond polymer refine detection (Leica Biosystems, UK) was the detection system used in the study. The results of the immunohistochemical study were interpreted using light microscopy with a Leica 2500 (Leica Biosystems, UK) equipped with a ×10 magnification eyepiece and ×10, ×20, and ×40 magnification objectives. Cells were considered positively expressed if they had moderate or significant brown staining of their nuclei (Figure 1). The Ki-67 proliferative index was calculated in 10 fields of view at a magnification of ×400; the fields of view was selected using the hot spot method, with a study of at least 1000 cells within the biopsy sample [10]. The proportion of stained tumor cells to all tumor cells in the studied area was calculated. The resulting numerical values were expressed as average percentages across 10 fields of view [11].

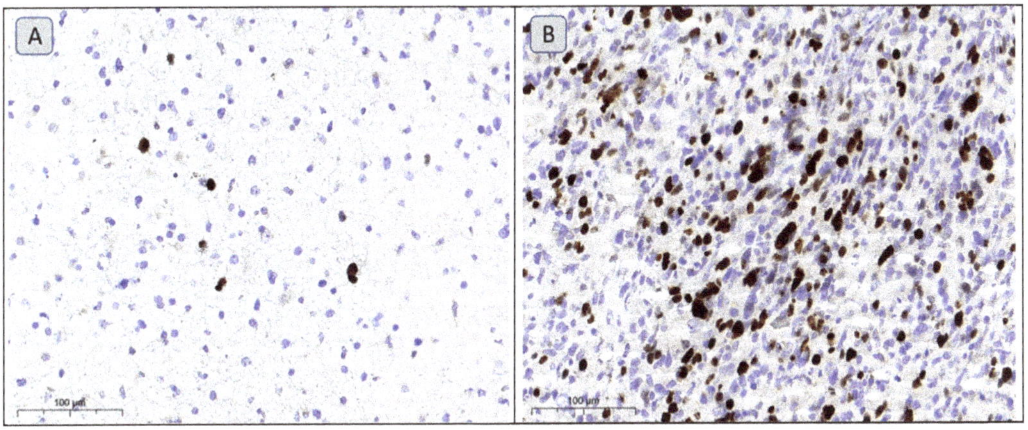

Figure 1. Immunohistochemical study of the Ki-67 marker. Magnification ×200; Bar: 100 µm. The images were used to determine the level of nuclear non-histone protein; the Ki-67 marker of cell proliferation in the material were obtained from gliomas. (**A**) The locally pronounced but overall limited brown staining of nuclei (up to 5% of cells in a few hot spots) shows the low Ki-67 proliferation index in a typical Grade II oligodendroglioma. (**B**) The diffuse but pronounced nuclear brown staining (up to 50% of cells in hot spots) shows a high Ki-67 proliferation index in a Grade IV glioblastoma.

2.3. Analysis of the Biochemical Parameters of Lipid Metabolism

A determination of the concentration of lipid metabolites was carried out colorimetrically using a UV-mini 1240 spectrophotometer (Shimadzu Corporation, Kyoto, Japan). For the analysis of cholesterol concentrations, a set of cholesterol–UTS reagents (Russia (Cat No. B-11061)) was used. The concentration of triacylglycerols (TAGs) was determined using a set of triglyceride–UTS reagents, Russia (Cat. No. 019.002) for low-density lipoprotein (LDL levels), while high-density lipoprotein (HDL) was determined using a set of HDLC3–HDLC4 reagents (Germany (Cat. No. 17005)) on the StatFaks 04 biochemical analyzer (Awareness Technology, Inc., Palm City, Florida, USA).

2.4. Measurement of Levels of Short-Chain Fatty Acids in Blood Samples

Blood from patients was sampled as drops on PerkinElmer 226 blood sample collection filter paper plates, which were then air-dried and stored at −80 °C until analysis.

A Shimadzu LCMS-8050 triple quadrupole liquid chromatograph–mass spectrometer (Shimadzu Corporation, Kyoto, Japan) with a Shimadzu Nexera XR liquid chromatograph system was used for analysis with the LC/MS/MS Method Package for short-chain fatty acids. The package is available online at the following site:

https://www.shimadzu.com/an/products/liquid-chromatograph-mass-spectrometry/lc-ms-system/lcmsms-method-package-for-short-chain-fatty-acids/index.html (accessed on 24 November 2022).

Before analysis, the blood samples were extracted using an internal standard solution (2-ethylbutyric acid in ethanol at 7.92 μmol/L), and 7 calibration solutions of various fatty acid concentrations (from 0.64 nmol/L to 10,000 nmol/L) were prepared and used to construct a calibration curve. Then, the derivatization process was started using a solution of pyridine, ECD (1-ethyl-3-(3-dimethylaminopropyl) carbodiimide hydrochloride), and 3-NPH (3-nitrophenylhydrazine hydrochloride); each resulting mixture was then filtered on a nylon syringe filter with a pore size of 0.22 μm into a vial for the autosampler.

2.5. Statistical Analysis

Statistical processing of the data was carried out using StatPlus, version 6, from AnalystSoft Inc. (www.analystsoft.com/ru/). The results of our biochemical research were presented in the form of medians, percentiles, and quartiles (25%; 75%). The significance of the differences obtained was evaluated using non-parametric criteria (Mann–Whitney U-criterion). Correlation analysis between the immunohistochemical marker Ki-67 and the blood lipid metabolism parameters was carried out through the determination of Spearman's rank coefficient.

3. Results

3.1. Parameters of Lipid Metabolism Corresponding to Different Pathologies

A (significantly) 63.43% higher level of cholesterol was detected in the blood plasma of patients with brain tumors compared to the level found in patients with cerebral atherosclerosis (Mann–Whitney U-test = 4.78×10^{-8}) (Table 1).

The levels of LDLs (low-density lipoproteins that are involved in cholesterol transfer into tissues) were also (significantly) 58.70% higher in atherosclerosis compared to their levels in patients with brain tumors (U-criterion = 0.0003) (Table 1).

When comparing the routine parameters of the blood plasma lipidome of the control group and those of patients with atherosclerosis, a statistically significant difference was seen only for the newly considered coefficient, the neutral lipids to cholesterol ratio; thus, the various ratios of these parameters were calculated. Significant differences were found between the values in atherosclerosis and in the "healthy" control group (Table 1, Figure 2).

The analysis of the levels of metabolically dependent compounds, including SCFAs, appears to be a promising approach for studying the blood lipidome in patients with brain tumors (Table 2). Here, it is important to identify the individual person profiles for these compounds in patients with the various types of tumors.

Table 1. Indicators of lipid metabolism found in blood plasma from patients with brain tumors or atherosclerosis.

Parameters	Healthy Persons		Brain Tumors		Atherosclerosis		U-Criterion of Mann–Whitney Test		
	Median	Quartiles	Median	Quartiles	Median	Quartiles	Difference Between Healthy Persons and Brain Tumors	Difference Between Healthy Persons and Atherosclerosis	Difference Between Brain Tumors and Atherosclerosis
Cholesterol	4.6	(4.5; 4.7)	5.72	(4.88; 6.415)	3.5	(2.9; 4.4)	0.316	0.002 *	4.78×10^{-8} *
Triacylglycerols	0.8	(0.765; 2.01)	1.43	(1.01; 2.02)	1.21	(0.87; 1.59)	0.138	0.714	0.248
Low-density lipoproteins	2.9	(2.8; 2.85)	3.65	(2.67; 4.30)	2.30	(2.1; 2.7)	0.225	0.001 *	0.0003 *
High-density lipoproteins	1.56	(1.43; 1.755)	1.34	(1.22; 1.61)	1.15	(0.945; 1.42)	0.497	0.139	0.135
Triacylglycerols/cholesterol	0.18	(0.16; 0.255)	0.28	(0.17; 0.35)	0.33	(0.25; 0.4)	0.016 *	0.0001 *	0.046 *
Triacylglycerols/Low-density lipoproteins	0.32	(0.27; 0.42)	0.4 (0.31; 0.56)	(0.495; 0.285)	0.03	(0.395; 0.69)	0.089	0.002 *	0.039 *
Cholesterol/High-density lipoproteins	2.965	(2.73; 3.395)	4.34 (2.86; 5.76)	(4.08; 2.735)	2.72	(2.3; 3.33)	0.129	0.158	0.007 *
Low-density lipoproteins/High-density lipoproteins	0.705	(0.56; 1.14)	2.88 (1.79; 3.61)	(4.08; 2.735)	1.52	(1.035; 2.36)	0.003 *	0.0005 *	0.108
Triacylglycerols/High-density lipoproteins	0.55	(0.46; 0.89)	1.54	(1.54; 1.065)	0.82	(0.7; 1.29)	0.011 *	0.019 *	0.769
Cholesterol/Low-density lipoproteins	1.72	(1.61; 1.855)	1.315	(1.315; 0.725)	1.865	(1.49; 2.17)	0.085	0.671	0.213

Legend: *—statistically significant differences.

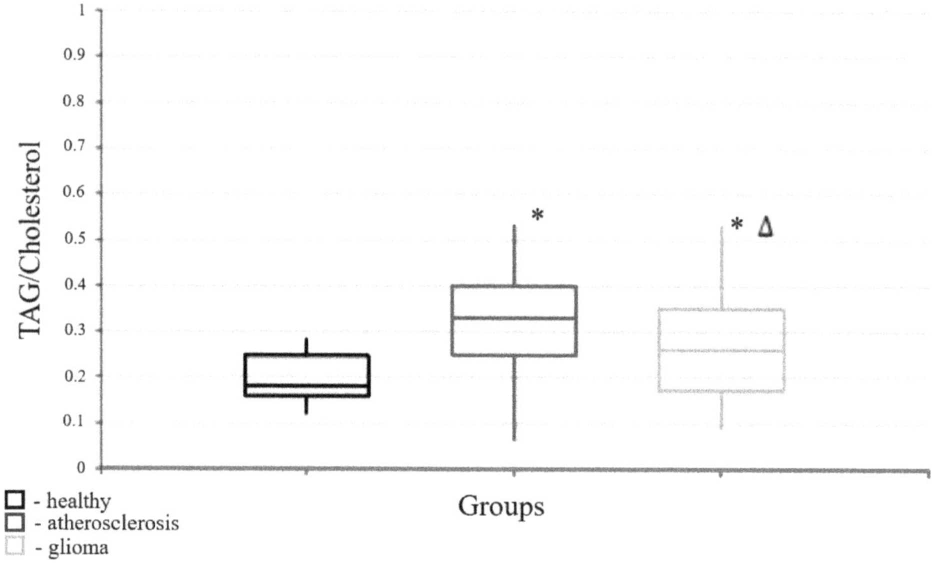

Figure 2. Medians and interquartile ranges (IQRs) of the triacylglycerol/cholesterol ratios in the blood plasma of patients with either atherosclerosis or brain tumors. There are significant differences in the values of the triacylglycerol/cholesterol ratios of blood plasma between the atherosclerosis and brain tumor groups (U-criterion of Mann–Whitney $p = 0.046$) and the brain tumors/healthy persons (U-criterion of Mann–Whitney $p = 0.016$). Legend: *—statistically significant differences from the "healthy" group ($p < 0.05$); Δ—statistically significant differences from the atherosclerosis group ($p < 0.05$).

Table 2. Concentrations of short-chain fatty acids in the blood of patients with brain tumors.

Acid	Healthy Persons (0)		Meningiomas (1)		U-Criterion of Mann–Whitney	Gliomas (2)		U-criterion of Mann–Whitney (Difference from the "Healthy" Control Group)	U-criterion of Mann–Whitney (Difference from the "Healthy" Control Group)
	Median	Quartiles Q1 / Q2 / Q3	Median	Quartiles Q1 / Q2 / Q3		Median	Quartiles Q1 / Q2 / Q3		
Lactic acid	268.1	237.7 268.1 303.3	554.4	298.5 554.4 640.8	0.142	439.1	313.4 439.1 779.1	0.028	1.000
Acetic acid	68.28	64.95 68.28 76.00	30.78	11.08 30.78 55.50	0.014 *	39.43	31.64 39.43 99.93	0.361	0.394
Propionic acid	3.880	2.160 3.880 5.614	3.177	1.489 3.177 3.956	0.624	3.247	1.983 3.247 3.906	0.273	1.000
Isobutyric acid	0.727	0.669 0.727 0.816	0.610	0.349 0.610 0.990	0.806	0.915	0.254 0.915 3.230	1.000	0.670
Butyric acid	2.592	2.289 2.592 2.778	0.515	0.075 0.515 1.303	0.014 *	0.591	0.064 0.591 1.283	0.006 *	0.748

Table 2. *Cont.*

Acid	Healthy Persons (0) Median	Healthy Persons (0) Quartiles Q1 / Q2 / Q3	Meningiomas (1) Median	Meningiomas (1) Quartiles Q1 / Q2 / Q3	U-Criterion of Mann–Whitney	Gliomas (2) Median	Gliomas (2) Quartiles Q1 / Q2 / Q3	U-criterion of Mann–Whitney (Difference from the "Healthy" Control Group)	U-criterion of Mann–Whitney (Difference from the "Healthy" Control Group)
Succinic acid	0.240	0.199	0.477	0.243	0.221	0.377	0.192	0.361	0.670
		0.240		0.477			0.377		
		0.295		0.839			0.646		
Isovaleric acid	0.192	0.116	0.211	0.019	1.000	0.272	0.199	0.272	0.915
		0.192		0.211			0.272		
		0.266		0.581			0.325		
Valeric acid	2.700	2.633	0.670	0.014	0.014	1.169	0.587	0.006 *	0.394
		2.700		0.670			1.169		
		2.753		1.292			1.854		
Maleic acid	49.26	32.43	5.730	3.737	0.014 *	1.906	0.000	0.360	0.521
		49.26		5.730			1.906		
		61.98		16.01			143.3		
Glyoxylic acid	0.390	0.379	0.944	0.738	0.014 *	1.557	0.491	0.067	0.670
		0.390		0.944			1.557		
		0.397		1.745			2.211		
Pyruvic acid	5.350	3.916	2.722	1.965	0.027 *	4.012	3.092	0.584	0.088
		5.350		2.722			4.012		
		6.463		4.001			7.373		
Glycolic acid	1.275	0.808	2.848	1.419	0.086	1.524	0.954	0.361	0.286
		1.275		2.848			1.524		
		1.652		6.029			2.782		

Legend: *—statistically significant differences from the "healthy" group.

The combination of mass spectrometry and high-performance liquid chromatography allowed for the detection of 12 compounds of interest in the blood samples; these compounds were saturated fatty acids with aliphatic tails containing 1–6 carbon atoms configured as either straight or branched chains (acetate, propionate, butyrate, succinate, isovalerate, etc.). An analysis of the study data showed statistically significant differences in the levels of acetic acid, butyric acid, maleic acid, glyoxylic acid, and pyruvic acid between the "healthy" control group and patients with meningiomas (Table 2). In particular, the analysis of the concentrations of SCFAs in the control group and in the patient group with gliomas demonstrated statistically significant differences in the butyric acid and valeric acid indicators (Table 2). By contrast, a comparison of the meningioma and glioma groups revealed no statistically significant differences between the group values of the tested SCFAs (Table 2).

According to the research data, butyric acid is of the greatest interest among these SCFAs, as significant differences from its blood level in the "healthy" group were found in the meningioma and glioma groups. In particular, the level of butyric acid in both neoplasm groups revealed a sharp decrease compared to the "healthy" control group (Figure 3).

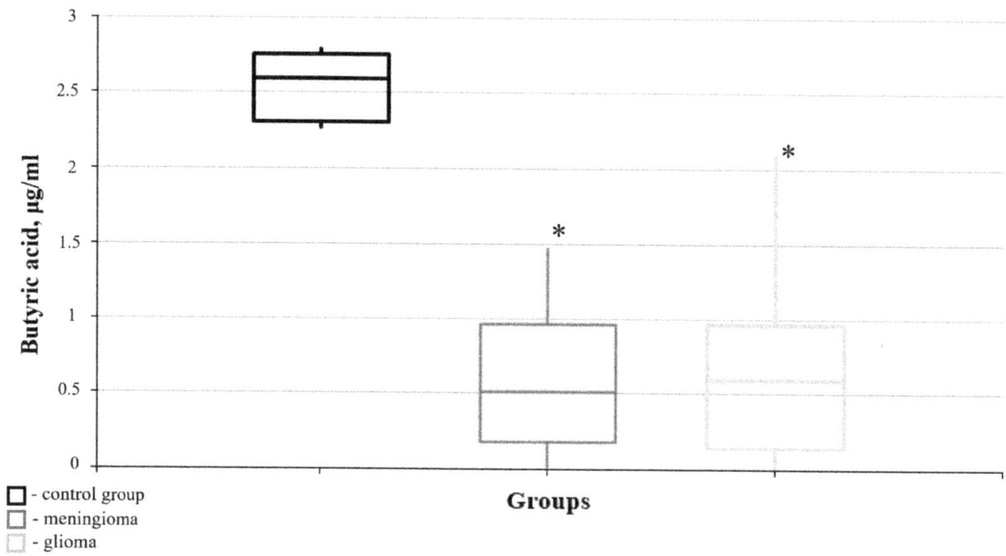

Figure 3. Medians and interquartile ranges (IQRs) of blood butyric acid concentration in the meningioma and glioma groups. There are significant differences in the blood butyric acid content between the meningioma and "healthy" groups (U-criterion of Mann–Whitney $p = 0.016$) and between the glioma and "healthy" groups (U-criterion of Mann–Whitney $p = 0.004$). Legend: *—statistically significant differences from the "healthy" group ($p < 0.05$).

3.2. *Lipid Metabolism Characteristics Corresponding to the Immunohistochemical Profile of Brain Tumors*

To identify the correlations between the Ki-67 marker and the blood lipid spectrum parameters, the Spearman rank correlation coefficient for nonparametric data was calculated (Table 3).

Table 3. Correlations (Spearman's rank coefficient) between the immunohistochemical Ki-67 marker of cell proliferation and lipid metabolism parameters of blood.

Parameter of Lipid Metabolism of Blood Plasma	Statistical Parameters (Spearman's Coefficient)	Ki-67 Marker of Cell Proliferation in Glioblastoma Material
Cholesterol	Rho	0.290
	p	0.191
Triacylglycerols	Rho	−0.169
	p	0.478
Low-density lipoproteins	Rho	0.155
	p	0.567
High-density lipoproteins	Rho	−0.222
	p	0.409
Triacylglycerols/Cholesterol	Rho	−0.524 *
	p	0.039
Triacylglycerols/Low-density lipoproteins	Rho	−0.412
	p	0.127
Cholesterol/High-density lipoproteins	Rho	0.536 *
	p	0.039

Table 3. *Cont.*

Parameter of Lipid Metabolism of Blood Plasma	Statistical Parameters (Spearman's Coefficient)	Ki-67 Marker of Cell Proliferation in Glioblastoma Material
Low-density lipoproteins/High-density lipoproteins	Rho	0.381
	p	0.179
Triacylglycerols/High-density lipoproteins	Rho	0.172
	p	0.494
Cholesterol/Low-density lipoproteins	Rho	−0.014
	p	0.956
Lactic acid	Rho	0.455
	p	0.137
Acetic acid	Rho	−0.158
	p	0.6247
Propionic acid	Rho	−0.322
	p	0.307
Isobutyric acid	Rho	0.112
	p	0.729
Butyric acid	Rho	−0.526 *
	p	0.044
Succinic acid	Rho	0.388
	p	0.212
Isovaleric acid	Rho	−0.123
	p	0.703
Valeric acid	Rho	−0.463
	p	0.130
Maleic acid	Rho	−0.358
	p	0.253
Glyoxylic acid	Rho	0.431
	p	0.162
Pyruvic acid	Rho	0.315
	p	0.318
Glycolic acid	Rho	0.237
	p	0.458
Malic acid	Rho	−0.541
	p	0.069

Legend: *—statistically significant differences from the "healthy" group ($p < 0.05$).

A valid correlation with the Ki-67 level was established for the cholesterol/HDL ratio (Rho = 0.536 and $p = 0.039$) and the neutral lipid/cholesterol ratio (Rho = −0.524 and $p = 0.039$). It is of particular interest that statistically significant correlation coefficients were found for the level of the Ki-67 cell proliferation marker with butyric acid (Rho = −0.526 and $p = 0.049$). Here, the correlation with Ki-67 was seen both in the meningioma group (Rho = −0.516, $p = 0.05$) and in the glioma group (Rho = 0.667, $p = 0.05$).

When the data on the parameters of lipid metabolism were further divided into groups depending on the immunohistochemical profile, taking into account the Ki-67 cell proliferation marker, the figures represented only data with meaningful differences between the groups.

No significant differences in the triacylglycerol/cholesterol ratios of blood plasma were found between the two pathology groups, regardless of whether they had high or low

Ki-67 mitotic index levels (Figure 4). However, a significant increase in this ratio was seen in the group with a low Ki-67 level compared to the "healthy" control group (Figure 4).

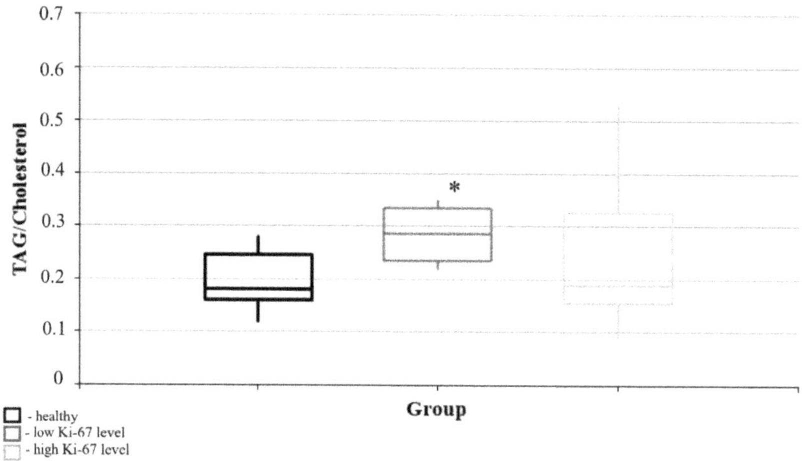

Figure 4. Medians and interquartile ranges (IQRs) of the triacylglycerol/cholesterol ratio in the blood plasma in relation to the magnitude of the Ki-67 mitotic index. There is a significant difference in the triacylglycerol/cholesterol ratio between groups with both low and high Ki-67 mitotic indices (U-criterion of Mann–Whitney $p = 0.05$). Legend: *—statistically significant differences from the "healthy" group ($p < 0.05$).

Reduced butyric acid level in the blood was demonstrated for all levels of the cell proliferation index in the brain tumor tissue when compared to the "healthy" control group. However, no significant differences in the levels of butyric acid in the blood were found between the pathology groups, regardless of whether they had high or low Ki-67 mitotic index levels (Figure 5).

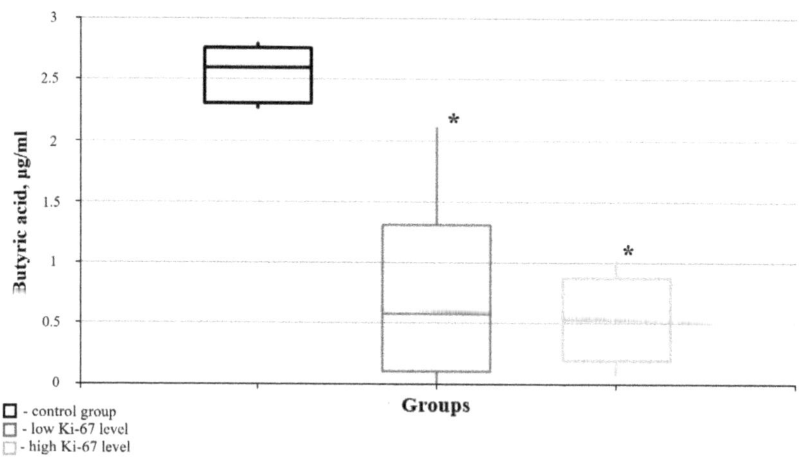

Figure 5. Medians and interquartile ranges (IQRs) of the content of butyric acid in the blood in the case of brain tumors as functions of the value of the Ki-67 mitotic index. There are significant differences between the butyric acid levels in the blood of the "healthy" controls and the low level of marker tumors (Mann–Whitney U-test: $p = 0.008$) and for the high Ki-67 mitotic index groups (Mann–Whitney U-test: $p = 0.016$). Legend: *—statistically significant differences from the "healthy" group ($p < 0.05$).

4. Discussion

Brain tumors are characterized by their increased cholesterol content, as this supports their growth and proliferation. This increase occurs due to both the activation of its synthesis and the absorption of exogenous cholesterol using lipoprotein receptors for LDLs [12]. The Ki-67 cell proliferative activity indicator is a nuclear protein that participates in ribosomal RNA synthesis [13]. Its expression reflects the degree of cell proliferative activity. In glial neoplasms, Ki-67 is used for differential diagnosis of low-grade (I, II) and high-grade (III, IV) gliomas [14], as its values correlate with the degree of malignancy [15].

Glioblastoma cells accumulate lipid droplets to satisfy their needs for rapid growth. These lipid droplets are subcellular organelles that contain numerous types of neutral lipids, including cholesterol and its esters [16]. As a tumor grows, the permeability of the blood–brain tumor barrier increases [17], resulting in an increase in the level of neutral lipids, particularly cholesterol in the blood.

Thus, the revealed correlation between the ratio of overall neutral lipid/cholesterol in blood plasma and the immunohistochemical marker of cell proliferation activity, the Ki-67 nuclear protein, is very probably due to the accumulation of lipid droplets in the malignant cells and due to the increase in the permeability of the blood–brain barrier resulting from the growth and increasing degree of anaplasia of the brain tumor. In addition to cholesterol levels, our study showed interesting changes in saturated fatty acids, the levels of which were also elevated in the group of patients with brain tumors. Moreover, meningiomas and gliomas showed increases in differing parameters, reflecting the different metabolic activity of these tumor groups due to their different states of histogenesis and grades of malignancy. Cross-changes were found only for butyric acid, which may indicate it as a more versatile metabolite in the energy chain during tumorigenesis. This is also consistent with the correlation of the Ki-67 value with both the cholesterol/HDL and neutral lipid/cholesterol ratios, showing the more active lipid metabolism of tumors that have a high proliferative capacity. Somewhat contradictory results concern other parameters. In particular, the absence of differences in the plasma triacylglycerol/cholesterol ratios between groups for both high and low levels of the Ki-67 mitotic index, and the relatively low LDL values in patients with brain tumors indicate that tumor metabolism is very complex and does not always depend directly on proliferative activity.

It is interesting to consider all the above changes from the point of view of interactions with other body systems, especially with the intestines. SCFAs are major microbial metabolites that primarily serve as an energy supply for colonocytes. Recent studies have confirmed that SCFA levels change in the case of many neurological diseases, including neurodegenerative diseases such as Parkinson's disease, Alzheimer's disease, cerebrovascular diseases (stroke, transient ischemic attack), epilepsy, and neuroimmune inflammatory diseases (multiple sclerosis), autism spectrum disorder, and depressive disorders, hence implying that SCFAs may be vitally important for the microbiota–gut–brain axis connections [18].

The gut–brain axis is a newly discovered and under-investigated phenomenon, the understanding of which might be used to improve the efficacy of anti-glioma therapy [19]. Among SCFAs, the most common ones are acetate, propionate, and butyrate [20], together accounting for 95% of all SCFAs [21].

SCFAs penetrate into endothelial cells using MCTs. In human and rodent brain endothelium, MCT1, acting as a proton-dependent cotransporter/exchanger, has a key role in the entry of SCFAs into the brain parenchyma [22]. By forming a special pool of metabolites, malignant neoplasms can probably modulate the production of neurotransmitters using intestinal nerve terminals and their de novo synthesis. Dono A, Patrizz A, McCormack RM et al. (2020) [23] demonstrated that the levels of various SCFAs and neurotransmitters in gliomas change in both experiments in mice and in human patients. Changes in the pool of SCFAs after glioma development seem to be related to CNS damage and to blood–brain barrier destruction, which result in changes in the microbial community through the gut axis interaction, leading to further reduction in bacterial SCFA production [23].

Blood–brain barrier destruction mechanisms may include the loss of tight junction integrity, increased transcytosis, endothelial cell apoptosis, the disruption of the glial limiting membrane, glycocalyx degradation, etc.

Some researchers have suggested that SCFAs function in the brain by means of histone deacetylase (HDAC) inhibition, with butyrate being the most potent class I and class IIa HDAC inhibitor [24]. In the central nervous system, such SCFA inhibition of HDAC histone deacetylase activity increases the acetylation of lysine residues in the nucleosomal histones. These changes in acetylation regulate the genesis and development of carcinomas [25].

Several publications have demonstrated evidence that SCFAs are associated with intracellular cascades that have the effect of helping to repair the impaired blood–brain barrier by increasing the expression of proteins of the intercellular junctional complex [26,27]. According to other studies, SCFA-induced restoration of the integrity of the blood–brain barrier can be mediated by the direct impact of SCFAs on brain endothelial cells, as well as by the suppression of systemic inflammation and the inhibition of activation of the microglia and astrocytes [28,29].

Microglia are very sensitive to SCFAs. Taking into account that three SCFA-specific receptors—GPR43, GPR41, and GPR109A—are expressed in microglial cells and that HDAC involvement in microglial activation has been demonstrated, the SCFA–microglia interaction seems to be even more important for blood–brain barrier preservation than the direct action of SCFAs on endothelial cells [30].

Butyrate also inhibits STAT3 signaling, thus suppressing the expression of bcl-2, bcl-XL, c-myc, cyclin D1, and HIF-1, resulting in decreased cell proliferation and increased apoptosis in both hypoxia and in colon cancer [31]. SCFAs also enhance the methylation of oncogenes, thereby decreasing their expression [32].

Ras signaling activation of STAT3 and the ability of SCFAs strongly to inhibit Ras activity may also reduce STAT3 activation, resulting in decreased expression of the proinflammatory IL-8 and VEGF interleukins and the blocking of angiogenesis [33]. For instance, strong anti-inflammatory effects of SCFAs have been shown particularly for butyrate through the inhibition of the IFNγ signaling pathway and the blocking of JAK2/STAT1 [34,35].

Thus, SCFAs are integral parts of the metabolic cascades of both normal and neoplastic tissues; they form a pool of regulatory molecules involved in neuroendocrine, immune, and neural pathways, requiring more detailed study for prognosis and treatment purposes.

It is important to note several possible limitations of our study. We understand that the lipid spectrum can be influenced by many factors, and we had therefore tried to take these into account as much as possible when selecting patients. However, individual characteristics, such as genetic predisposition, lifestyle, bad habits, and diet variability present difficulties in this respect. Therefore, it is necessary to consider that some values may fluctuate slightly due to the influence of these factors. It is also necessary to realize the potential impact of the heterogeneity of tumors, not only relative to each other, but even within a single tumor. For example, some gliomas can have areas with different levels of proliferation, which can therefore cause difficulties in interpreting the Ki-67 values. We aimed to include the maximum possible number of cases available to our study in order to minimize the influence of random factors, but a larger study will be required for determining patterns with greater certainty. In the future, we plan to expand our study to include a larger number of patients and to incorporate data that could indicate the degree of influence of other relevant factors.

5. Conclusions

The study identified several parameters of the lipid spectrum in blood plasma, the levels of which significantly change as a result of brain tumors and differ from the lipidome changes seen with atherosclerosis. The ratio of neutral lipids to cholesterol can be determined on the basis of routine clinical tests and thus be used for preoperative diagnosis of brain tumors. A decreased level of butyric acid in the blood can be an additional verifying criterion for brain tumor diagnosis, as it is especially significant in the initial stages of anaplasia. However, these parameters cannot serve as a preoperative non-invasive predictor of the level of the Ki-67 mitotic index, as no significant differences between them were found for low-grade and high-grade anaplasia of brain tumors.

Supplementary Materials: The supporting information can be downloaded at https://www.mdpi.com/article/10.3390/cimb47040228/s1.

Author Contributions: Conceptualization, L.O. and O.B.; Methodology, N.S. and I.M.; Software, O.B.; Validation, L.O., N.S., I.M. and K.Y.; Resources, N.S., I.M., K.Y., A.G. and O.B.; Data curation, A.G. and O.B.; Writing—original draft, L.O. and N.S.; Writing—review & editing, I.M., K.Y. and A.G.; Visualization, O.B. and I.A.; Project administration, L.O. All authors have read and agreed to the published version of the manuscript.

Funding: This research received no external funding.

Institutional Review Board Statement: The study was approved by the Ethics Committee of the Federal State Budgetary Educational Institution of Higher Education, the "Privolzhsky Research Medical University" of the Ministry of Health of the Russian Federation (No. 6 of 17.04.2019).

Informed Consent Statement: Informed consent was obtained from all subjects involved in the study.

Data Availability Statement: The original contributions presented in this study are included in the article/Supplementary Materials. Further inquiries can be directed to the corresponding author(s).

Conflicts of Interest: The authors declare no conflict of interest.

References

1. Marien, E.; Meister, M.; Muley, T.; Fieuws, S.; Bordel, S.; Derua, R.; Spraggins, J.; Van de Plas, R.; Dehairs, J.; Wouters, J.; et al. Non-small cell lung cancer is characterized by dramatic changes in phospholipid profiles. *Int. J. Cancer* **2015**, *137*, 1539–1548. [CrossRef] [PubMed]
2. Guo, D.; Bell, E.H.; Chakravarti, A. Lipid metabolism emerges as a promising target for malignant glioma therapy. *CNS Oncol.* **2013**, *2*, 289–299. [CrossRef] [PubMed]
3. Herrington, W.; Lacey, B.; Sherliker, P.; Armitage, J.; Lewington, S. Epidemiology of Atherosclerosis and the Potential to Reduce the Global Burden of Atherothrombotic Disease. *Circ. Res.* **2016**, *118*, 535–546. [CrossRef] [PubMed]
4. Mayer, E.A.; Savidge, T.; Shulman, R.J. Brain–Gut Microbiome Interactions and Functional Bowel Disorders. *Gastroenterology* **2014**, *146*, 1500–1512. [CrossRef]
5. Silva, Y.P.; Bernardi, A.; Frozza, R.L. The Role of Short-Chain Fatty Acids From Gut Microbiota in Gut-Brain Communication. *Front. Endocrinol.* **2020**, *11*, 25. [CrossRef]
6. Louis, D.N.; Perry, A.; Wesseling, P.; Brat, D.J.; Cree, I.A.; Figarella-Branger, D.; Hawkins, C.; Ng, H.K.; Pfister, S.M.; Reifenberger, G.; et al. The 2021 WHO Classification of Tumors of the Central Nervous System: A summary. *Neuro-Oncology* **2021**, *23*, 1231–1251. [CrossRef]
7. Theresia, E.; Malueka, R.G.; Pranacipta, S.; Kameswari, B.; Dananjoyo, K.; Asmedi, A.; Wicaksono, A.S.; Hartanto, R.A.; Dwianingsih, E.K. Association between Ki-67 Labeling index and Histopathological Grading of Glioma in Indonesian Population. *Asian Pac. J. Cancer Prev.* **2020**, *21*, 1063–1068. [CrossRef] [PubMed]
8. Familiari, P.; Relucenti, M.; Lapolla, P.; Palmieri, M.; Antonelli, M.; Cristiano, L.; Barbaranelli, C.; Catalano, M.; D'Angelo, L.; Familiari, G.; et al. Adult IDH Wild-Type Glioblastoma Ultrastructural Investigation Suggests a Possible Correlation between Morphological Biomarkers and Ki-67 Index. *Biomedicines* **2023**, *11*, 1968. [CrossRef]

9. Zhang, Z.; Gu, W.; Hu, M.; Zhang, G.; Yu, F.; Xu, J.; Deng, J.; Xu, L.; Mei, J.; Wang, C.; et al. Based on clinical Ki-67 expression and serum infiltrating lymphocytes related nomogram for predicting the diagnosis of glioma-grading. *Front. Oncol.* **2022**, *12*, 696037. [CrossRef]
10. Leung, S.C.Y.; Nielsen, T.O.; Zabaglo, L.; Arun, I.; Badve, S.S.; Bane, A.L.; Bartlett, J.M.S.; Borgquist, S.; Chang, M.C.; Dodson, A.; et al. Analytical validation of a standardized scoring protocol for Ki67: Phase 3 of an international multicenter collaboration. *NPJ Breast Cancer* **2016**, *2*, 16014. [CrossRef]
11. Nielsen, L.A.G.; Bangso, J.A.; Lindahl, K.H.; Dahlrot, R.H.; Hjelmborg, J.V.B.; Hansen, S.; Kristensen, B.W. Evaluation of the proliferation marker Ki-67 in gliomas: Interobserver variability and digital quantification. *Diagn. Pathol.* **2018**, *3*, 8. [CrossRef]
12. Ríos-Marco, P.; Martín-Fernández, M.; Soria-Bretones, I.; Ríos, A.; Carrasco, M.P.; Marco, C. Alkylphospholipids deregulate cholesterol metabolism and induce cell-cycle arrest and autophagy in U-87 MG glioblastoma cells. *Biochim. Biophys. Acta* **2013**, *1831*, 1322–1334. [CrossRef] [PubMed]
13. Bullwinkel, J.; Baron-Luehr, B.; Lüdemann, A.; Wohlenberg, C.; Gerdes, J.; Scholzen, T. Ki-67 protein is associated with ribosomal RNA transcription in quiescent and proliferating cells. *J. Cell Physiol.* **2006**, *206*, 624–635. [CrossRef]
14. Yoda, R.A.; Marxen, T.; Longo, L.; Ene, C.; Wirsching, H.G.; Keene, C.D.; Holland, E.C.; Cimino, P.J. Mitotic Index Thresholds Do Not Predict Clinical Outcome for IDH-Mutant Astrocytoma. *J. Neuropathol. Exp. Neurol.* **2019**, *78*, 1002–1010. [CrossRef]
15. Dahlrot, R.H.; Bangso, J.A.; Petersen, J.K.; Rosager, A.M.; Sorensen, M.D.; Reifenberger, G.; Hansen, S.; Kristensen, B.W. Prognostic role of Ki-67 in glioblastomas excluding contribution from non-neoplastic cells. *Sci. Rep.* **2021**, *11*, 17918. [CrossRef] [PubMed]
16. Geng, F.; Guo, D. Lipid droplets, potential biomarker and metabolic target in glioblastoma. *Intern. Med. Rev.* **2017**, *3*, 10.18103/imr.v3i5.443. [CrossRef]
17. Sarkaria, J.N.; Hu, L.S.; Parney, I.F.; Pafundi, D.H.; Brinkmann, D.H.; Laack, N.N.; Giannini, C.; Burns, T.C.; Kizilbash, S.H.; Laramy, J.K.; et al. Is the blood-brain barrier really disrupted in all glioblastomas? A critical assessment of existing clinical data. *Neuro-Oncology* **2018**, *20*, 184–191. [CrossRef]
18. Dalile, B.; Van Oudenhove, L.; Vervliet, B.; Verbeke, K. The role of short-chain fatty acids in microbiota-gut-brain communication. *Nat. Rev. Gastroenterol. Hepatol.* **2019**, *16*, 461–478. [CrossRef]
19. Dono, A.; Nickles, J.; Rodriguez-Armendariz, A.G.; McFarland, B.C.; Ajami, N.J.; Ballester, L.Y.; Wargo, J.A.; Esquenazi, Y. Glioma and the gut-brain axis: Opportunities and future perspectives. *Neuro-Oncol. Adv.* **2022**, *4*, vdac054. [CrossRef]
20. Ney, L.M.; Wipplinger, M.; Grossmann, M.; Engert, N.; Wegner, V.D.; Mosig, A.S. Short chain fatty acids: Key regulators of the local and systemic immune response in inflammatory diseases and infections. *Open Biol.* **2023**, *13*, 230014. [CrossRef]
21. O'Riordan, K.J.; Collins, M.K.; Moloney, G.M.; Knox, E.G.; Aburto, M.R.; Fülling, C.; Morley, S.J.; Clarke, G.; Schellekens, H.; Cryan, J.F. Short chain fatty acids: Microbial metabolites for gut-brain axis signalling. *Mol. Cell Endocrinol.* **2022**, *546*, 111572. [CrossRef]
22. Vijay, N.; Morris, M.E. Role of monocarboxylate transporters in drug delivery to the brain. *Curr. Pharm. Des.* **2014**, *20*, 1487–1498. [CrossRef]
23. Dono, A.; Patrizz, A.; McCormack, R.M.; Putluri, N.; Ganesh, B.P.; Kaur, B.; McCullough, L.D.; Ballester, L.Y.; Esquenazi, Y. Glioma induced alterations in fecal short-chain fatty acids and neurotransmitters. *CNS Oncol.* **2020**, *9*, CNS57. [CrossRef]
24. Stilling, R.M.; Van de Wouw, M.; Clarke, G.; Stanton, C.; Dinan, T.G.; Cryan, J.F. The neuropharmacology of butyrate: The bread and butter of the microbiota-gut-brain axis? *Neurochem. Int.* **2016**, *99*, 110–132. [CrossRef]
25. Guo, C.; Huo, Y.J.; Li, Y.; Han, Y.; Zhou, D. Gut-brain axis: Focus on gut metabolites short-chain fatty acids. *World J. Clin. Cases* **2022**, *10*, 1754–1763. [CrossRef] [PubMed]
26. Wu, Q.; Zhang, Y.; Zhang, Y.; Xia, C.; Lai, Q.; Dong, Z.; Kuang, W.; Yang, C.; Su, D.; Li, H.; et al. Potential effects of antibiotic-induced gut microbiome alteration on blood-brain barrier permeability compromise in rhesus monkeys. *Ann. N. Y. Acad. Sci.* **2020**, *1470*, 14–24. [CrossRef] [PubMed]
27. Sun, N.; Hu, H.; Wang, F.; Li, L.; Zhu, W.; Shen, Y.; Xiu, J.; Xu, Q. Antibiotic-induced microbiome depletion in adult mice disrupts blood-brain barrier and facilitates brain infiltration of monocytes after bone-marrow transplantation. *Brain Behav. Immun.* **2021**, *92*, 102–114. [CrossRef] [PubMed]
28. Wen, J.; Ding, Y.; Wang, L.; Xiao, Y. Gut microbiome improves postoperative cognitive function by decreasing permeability of the blood-brain barrier in aged mice. *Brain Res. Bull.* **2020**, *164*, 249–256. [CrossRef]
29. Caetano-Silva, M.E.; Rund, L.; Hutchinson, N.T.; Woods, J.A.; Steelman, A.J.; Johnson, R.W. Inhibition of inflammatory microglia by dietary fiber and short-chain fatty acids. *Sci. Rep.* **2023**, *13*, 2819. [CrossRef]
30. Fock, E.; Parnova, R. Mechanisms of Blood-Brain Barrier Protection by Microbiota-Derived Short-Chain Fatty Acids. *Cells* **2023**, *12*, 657. [CrossRef]
31. Emenaker, N.J.; Calaf, G.M.; Cox, D.; Basson, M.D.; Qureshi, N. Short-chain fatty acids inhibit invasive human colon cancer by modulating uPA, TIMP-1, TIMP-2, mutant p53, Bcl-2, Bax, p21 and PCNA protein expression in an in vitro cell culture model. *J. Nutr.* **2001**, *131* (Suppl. S11), 3041S–3046S. [CrossRef]

32. Chen, J.; Zhao, K.N.; Vitetta, L. Effects of Intestinal Microbial-Elaborated Butyrate on Oncogenic Signaling Pathways. *Nutrients* **2019**, *11*, 1026. [CrossRef] [PubMed]
33. McBrearty, N.; Arzumanyan, A.; Bichenkov, E.; Merali, S.; Merali, C.; Feitelson, M. Short chain fatty acids delay the development of hepatocellular carcinoma in HBx transgenic mice. *Neoplasia* **2021**, *23*, 529–538. [CrossRef] [PubMed]
34. Feitelson, M.A.; Arzumanyan, A.; Medhat, A.; Spector, I. Short-chain fatty acids in cancer pathogenesis. *Cancer Metastasis Rev.* **2023**, *42*, 677–698. [CrossRef]
35. Gao, S.; Chen, C.; Wang, L.; Hong, L.; Wu, J.; Dong, P. Histone deacetylases inhibitor sodium butyrate inhibits JAK2/STAT signaling through upregulation of SOCS1 and SOCS3 mediated by HDAC8 inhibition in myeloproliferative neoplasms. *Exp. Hematol.* **2013**, *41*, 261–270.e4. [CrossRef] [PubMed]

current issues in molecular biology

MDPI

Review

From Molecular to Clinical Implications of Sleep-Related Breathing Disorders on the Treatment and Recovery of Acute Stroke: A Scoping Review

Karol Uscamaita [1,2], Olga Parra Ordaz [2,3], Imán Yazbeck Morell [4], Marta García Pla [5], María-José Sánchez-López [6] and Adrià Arboix [7,*]

1 Neurology Service, Sleep Disorders Unit, Hospital Universitari Sagrat Cor, Grupo Quirónsalud, 08029 Barcelona, Spain; karol.uscamaita@quironsalud.es
2 Medicine Department, University of Barcelona, 08036 Barcelona, Spain; oparra@ub.edu
3 Department of Pneumology, Sleep Disorders Unit, Hospital Universitari Sagrat Cor, Grupo Quirónsalud, 08029 Barcelona, Spain
4 Department of Internal Medicine, Hospital Universitari Sagrat Cor, Grupo Quirónsalud, 08029 Barcelona, Spain; iman.yazbeck@quironsalud.es
5 Emergency Department, Hospital Universitari Sagrat Cor, Grupo Quirónsalud, 08029 Barcelona, Spain; martagarcia.bcn.ics@gencat.cat
6 Medical Library, Hospital Universitari Sagrat Cor, Grupo Quirónsalud, 08029 Barcelona, Spain; biblioteca.hsc@quironsalud.es
7 Cerebrovascular Division, Department of Neurology, Hospital Universitari Sagrat Cor, Grupo Quirónsalud, Universitat de Barcelona, 08029 Barcelona, Spain
* Correspondence: aarboix@quironsalud.es; Tel.: +34-93-494-8940; Fax: +34-93-494-8906

Abstract: (1) Background: The aim of this review is to map research into the molecular mechanisms linking sleep-related breathing disorders (SRBDs) and acute stroke and their clinical and therapeutic implications and to identify existing knowledge gaps to suggest new areas of research. (2) Methods: This review was conducted according to the PRISMA extension for scoping reviews (PRISMA-ScR) and a predetermined protocol shared among all authors. (3) Results: The review of the thirteen studies analyzed provides a focused view of the molecular features about interaction between obstructive sleep apnea (OSA) and acute stroke. Our review identifies and highlights the biomarkers most frequently found to be associated with acute stroke, SRBDs, and their clinical repercussions. (4) Conclusions: The association between the presence of sleep apnea, especially in its severe form, and elevated levels of inflammatory markers in patients with acute stroke is highlighted and new research topics in this area are proposed.

Keywords: stroke; ischemic stroke; CPAP; pathophysiology; sleep apnea

1. Introduction

Sleep-related breathing disorders (SRBD) are characterized by abnormalities of respiration during sleep. These disorders include obstructive sleep apnea (OSA) disorders, central sleep apnea disorders, sleep-related hypoventilation disorders, and sleep related hypoxemia disorder [1]. OSA are clearly the most frequent SRBD in general population, affecting between 23 to 49% of general population [2].

OSA is characterized by repetitive episodes of complete (apnea) or partial (hypopnea) upper airway obstruction occurring during sleep. These events often result in reductions in blood oxyhemoglobin saturation and are usually terminated by brief arousals from sleep. Oxyhemoglobin saturation usually returns to baseline values following resumption of normal breathing but may remain low if the apneic or hypopneic events are very frequent and prolonged, or if there is underlying pulmonary pathology [1].

Worldwide, stroke continues to be leading cause of death in women, the second most common cause of mortality, and the second most common cause of disability in the adult population [3]. Specifically in Spain, according to data extracted from the Annual Report of the National Health System 2020–2021, cerebrovascular disease affects 1.5% of the population, although it affects six out of every hundred people over the age of 65 and ten out of every hundred over the age of 80 [4,5]. While the incidence of stroke and overall stroke-related mortality rate is decreasing, the absolute number of people with stroke, stroke survivors, stroke-related deaths, and the overall burden of stroke-related disability is high and increasing [6–8].

Strokes can be classified into ischemic or hemorrhagic. Ischemic strokes are episodes of neurological dysfunction caused by focal cerebral, spinal, or retinal infarction. Hemorrhagic strokes are focal collections of blood within the brain parenchyma or ventricular system that are not caused by trauma [8]. An acute stroke is a heterogeneous disease, and in clinical studies, it is extremely important to adequately differentiate between the different stroke subtypes (atherothrombotic infarct, cardioembolic stroke, lacunar infarct, infarct of unusual etiology, essential cerebral infarct, intracerebral hemorrhages) among the population with stroke. This recommendation is due to the impact of the stroke subtypes on the distribution of risk factors, stroke severity, and outcome [9].

SRBDs, particularly obstructive sleep apnea (OSA), are frequent in the acute phase of stroke, with approximately 62% of patients experiencing elevated levels of sleep apnea within the first 24 h of stroke onset [10–15]. The presence of SRBDs in acute stroke is highly relevant. Over the last 15 years, we have witnessed a growing interest in the relationship between SRBDs and stroke. Several randomized controlled trials demonstrated that sleep-related breathing disorders are linked to early neurological worsening after stroke [16–22] and an increase in its mortality [23–28].

Some studies have provided pathophysiological information on how SRBDs, especially OSA, worsen the prognosis in acute stroke. It has been demonstrated that during periods of apnea, heart rate decreases; although the specific mechanism is unknown, it is believed to be due to incomplete autonomic injury caused by the intermittent hypoxia. On the other hand, it has also been shown that there is lower cerebral blood flow during episodes of obstructive hypopnea. This has been attributed to the fact that during hypopnea episodes, negative intrathoracic pressure occurs in an attempt to overcome the obstruction of the upper airway. This would increase the heart's afterload and decrease cerebral blood flow. It is also theorized that this may be due to a reflex vasodilation caused by the hypoxia produced by each apnea episode [19,20].

Clinical trials in the general population have showed that treatment of SRBDs with continuous positive airway pressure (CPAP) or other therapies significantly reduces oxy-hemoglobin desaturations and apnea. Moreover, a large number of biomarkers related to sympathetic activation, endothelial dysfunction, hypercoagulability, oxidative stress, inflammation, and metabolic dysregulation have been studied in patients with SRBDs, as these are some of the major mechanisms linking SRBDs to cardiovascular disease. However, to date, limited evidence is available on the efficacy of CPAP in stroke patients due to unclear strategy and timing [29–31].

To date, screening SRBDs or the use of CPAP devices are not indicated in acute stroke [24], as various studies have failed to demonstrate its relevance in improving the functional status of patients or reducing mortality [32–35]. The aim of this review is to systematically map research on the molecular mechanisms linking SRBD and acute stroke, their clinical and therapeutic implications, and to identify existing knowledge gaps to suggest new areas of research.

2. Materials and Methods

This review was conducted according to the PRISMA extension for scoping reviews (PRISMA-ScR) [28,29] and a predetermined protocol shared among all authors. Institutional review board approval was no necessary as this study utilized only published data. Study protocol was register in osf.io with DOI: 10.17605/OSF.IO/375KD.

Eligibility criteria

Studies assessing molecular, clinical, and therapeutic repercussions of SRBDs on acute stroke as primary or secondary outcome variables were selected. Only published articles up to December 2023, written in English, Spanish, French, or German, involving human participants were included. The study design included quantitative, qualitative and mixed-methods studies to consider a broad catchment of the topic. Letters to the editor, editorials, commentaries, published protocols, conference proceedings, studies only published in abstract form, and case reports were excluded. No context limit was applied.

Information sources

The PubMed, Scopus, and Web of Science databases for studies published from inception to 31 December 2023 in English, Spanish, French, or German, were searched. The search query was constructed in PubMed based on key terms "sleep apnea syndrome" and "stroke" combined with key term "biomarkers" and a set of terms that refer to different biomarkers options. Th search was adapted for use in the other databases [Appendix A]. Studies not published in journals, and articles in languages other than those selected were excluded.

Study Selection and Data Extraction

The title/abstract screening was conducted by two authors of the study. Subsequently, two different authors independently reviewed the eligibility of the studies included and extracted data. Full text articles were assessed by two of the authors and disagreement was resolved by consulting a third author. The final selection was shared among all the six coauthors and any discrepancies were resolved through discussion and consensus.

Data items

Information from each included article was summarized using a structured form, including data on article characteristics (study design), settings, detailing the summary of findings and key outcomes.

3. Results

Our scoping review identified 186 articles and, following the removal of 25 duplicated records, we included a total of 162 studies for screening. From these, 60 were found irrelevant, and the remaining 102 were screened based on title and abstract. Of those, a total of 61 studies were excluded because they did not meet the inclusion criteria.

Finally, 41 full-text studies were reviewed. Among these, 24 studies were excluded for not reporting on the topic of the study (n = 24), conference proceedings and letter to the editor (n = 3), and non-English language (n = 1), resulting in 13 studies considered eligible (Figure 1).

The included studies were Ifergane et al., 2016 [30]; Chen et al., 2015 [31], von Känel et al., 2013 [32]; Jensen et al., 2018 [33]; Yeh et al., 2017 [34]; Lévy et al., 2009 [35]; Gregori-Pla et al., 2018 [36]; Tsai et al., 2010 [37]; Fei et al., 2021 [38] Mashaqi et al., 2021 [39]; Kunz et al., [40]; Azarbarzin et al., 2012 [41]; and Medeiros et al., 2012 [42].

Of the 13 studies reviewed, 7 (53.8%) were prospective observational studies, mostly cross-sectional, 4 (30.7%) were reviews of other studies, and 2 (15.3%) were prospective clinical trials. Almost all studies investigate the molecular relationships of ischemic stroke with sleep-related breathing disorders (SRBDs), with only one focusing on hemorrhagic stroke.

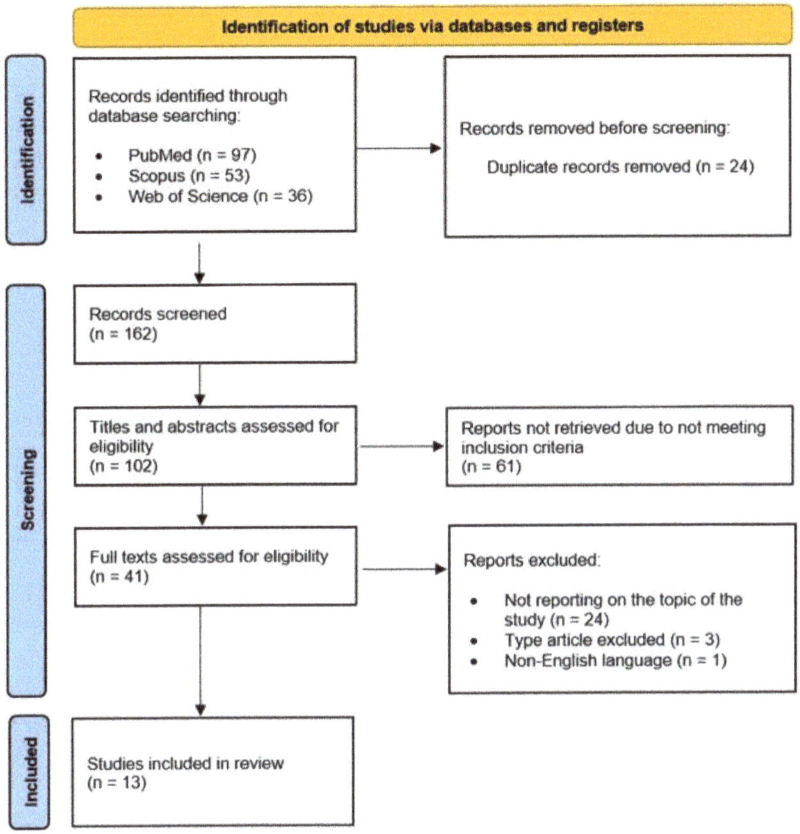

Figure 1. Preferred reporting items for systematic reviews and meta-analyses extension for scoping reviews (PRISMA-ScR) diagram for included studies [28,29].

In 7 of the 13 reviewed studies [30,32,35,38–40,42], an elevation of pro-inflammatory factors was demonstrated, among which interleukin 6 (IL-6) was found elevated most frequently; there is also significant evidence regarding the elevation of matrix metalloproteinase-9. Other pro-inflammatory factors that have been found elevated include C-reactive protein (CRP), caspase-3, Bax, cytokines IL-1β, TNF-α, NF-κB, and soluble tumor necrosis factor receptors -1 and -2. One study highlights that elevated levels of soluble intercellular adhesion molecule-1 (sICAM-1) are related to sleep fragmentation after stroke rehabilitation, serving as an inflammatory marker that links functional improvement from stroke to sleep quality [34].

One of the studies focused on investigating the variation of prothrombotic factors in patients with SRBD treated with CPAP and without CPAP; this study found no differences in these biomarkers [32].

Two studies focused on measuring cerebral blood flow in patients with sleep apnea using different methodologies; both concluded that patients with obstructive sleep apnea (OSA) have lower cerebral blood flow during apneas and exhibit altered vasoreactivity that correlates with the severity of OSA [33,36].

Details of each study are included in Table 1.

Table 1. Summary table of included studies.

References	Study Type	No. Participants or Reported Case	Factors Assessed (Methodology)	Summary of Outcomes	Conclusions
Lévy et al., 2009 [35]	Literature review study	96 articles	- Review of overall studies on mechanisms of atherosclerosis - Data applicable to humans, animal models, and the therapeutic perspectives	Key factors for promoting atherosclerosis were: - Intermittent hypoxia (IH) - Blood pressure alterations - Hemodynamic strains on the vascular wall - Impairment in vascular reactivity - Lipid metabolism dysregulation - Activation of proinflammatory transcription factors at the vascular wall level	There are animal and clinical data supporting the role of IH in triggering aterosclerosis related to stroke.
Fei et al., 2021 [38]	Animal model study	159 mice	- Animal study on neuronal damage induced by OSA - Mechanisms by which intracerebral hemorrhage (ICH)-induced early brain injury regulates neural apoptosis	OSA induction produced: - Decreased survival rate, neurological score and neuron survival - Upregulation of protein expression levels of Caspase-3, Bax, cytokines IL-1β, IL-6, TNF-α and NF-κB - OSA-mediated induction of apoptosis and neuroinflammation-aggravated neuronal death following ICH	- OSA induced apoptosis and neuroinflammation aggravating neuronal death following ICH. - The molecular mechanism was partly dependent on the activating transcription factor/CHOP pathway.
Kunz et al., 2011 [40]	Population prospective study	76 patients	Comparison of serum inflammatory biomarkers in patients with acute stroke and OSA vs. stroke patients without OSA, including: - Soluble tumor necrosis factor receptor-1 and -2 (sTNF-R1 and sTNF-R2) - Tumor necrosis factor-β (TNF-β) - Soluble intercellular cell adhesion molecule-1 (sICAM-1) - Soluble vascular cell adhesion molecule-1 (sVCAM-1)	In patients with apnea-hypopnea index (AHI) $\geq 10/h$: - sTNF-R1 and sTNF-R2 levels were significantly higher. - TNF-β, sICAM-1 and sVCAM-1. showed no significant differences.	- sTNF-R1 and sTNF-R2 may be part of the pathophysiological pathway linking OSA to stroke.

Table 1. *Cont.*

References	Study Type	No. Participants or Reported Case	Factors Assessed (Methodology)	Summary of Outcomes	Conclusions
Medeiros et al., 2012 [42]	Cross-sectional prospective study	31 participants	Comparison of serum inflammatory biomarkers in patients with acute stroke (first week) OSA vs. stroke patients without OSA vs. controls, including: - Interleukin-6(IL-6) - Interleukin-1β (IL-1β) - TNF-α	- IL-6 increased in stroke patients with SA. - IL-6 is correlated with oxyhemoglobin desaturation and with desaturation index	Inflammation is an important unifying mechanism between OSA and stroke.
Ifergane et al., 2016 [30]	Cross-sectional prospective study	43 patients	Comparison of serum inflammatory biomarkers in patients with acute stroke and OSA vs. stroke patients without OSA, including: - IL-6 - TNF - Plasminogen activator inhibitor-1 (PAI-1)	- Levels of all 3 biomarkers were higher among patients with AHI $\geq$ 15	OSA was associated with significantly increased levels of inflammatory biomarkers
Yeh et al., 2017 [34]	Cross-sectional prospective study	20 patients	Comparison of inflammatory and oxidative stress biomarkers, sleep awakenings (messured by acelerometer) and rehabilitation outcomes in subacute stroke patients, including: - sICAM-1 - Glutathione peroxidase (GPx) - Malondialdehyde (MDA)	Positive correlations were observed between baseline level of sICAM-1 and number of awakenings at post-treatment ($\rho = 0.51$, $p < 0.05$).	- Patients with higher baseline levels of inflammation had higher numbers of awakenings at rehabilitation post-treatment.
Tsai et al., 2010 [37]	Literature review study	70 articles	Review of cognitive and neurobehavioral sequelae in patients with stroke and OSA	- AHI is the most common measure of OSA and is variably related to stroke and cognitive impairment. - There is significant overlap in the neuropsychological profiles of OSA and stroke patients.	- Sleepiness is the most common neurobehavioral sequela in patients with OSA and stroke. - CPAP treatment for as little as 2 weeks improves daytime sleepiness in most controlled trials

Table 1. *Cont.*

References	Study Type	No. Participants or Reported Case	Factors Assessed (Methodology)	Summary of Outcomes	Conclusions
Chen et al., 2015 [31]	Cross-sectional prospective study	72 patients	Comparison of serum and urine biomarkers of oxidative stress and inflammation in patients with stable ischemic stroke and OSA vs. without OSA, including: - Serum C-reactive protein (CRP) - Serum IL-6 - Serum total antioxidant capacity (TAC) - Urinary 8-hydroxy-2-deoxyguanosine	- Participants with AHI > 30 had a significantly lower level of CRP. - TAC levels were significantly and negatively correlated with mean SaO2 levels.	- Adaptive antioxidative response to hypoxia emerges, and the role of OSA with respect to inflammatory reaction is attenuated in ischemic stroke patients with OSA.
von Känel et al., 2013 [32]	Cross-sectional prospective study	51 participants	Comparison of prothrombotic markers in OSA (AHI $\geq$ 10) (before and after CPAP treatment) and controls, including: - Plasminogen activator inhibitor (PAI)-1 antigen - D-dimer - Von Willebrand factor (VWF) antigen - soluble tissue factor (sTF) antigen	- OSA patients showed lower mean and amplitude of D-dimer. - OSA patients showed higher mean of plasminogen activator inhibitor-1.	There may be altered day/night rhythm of some prothrombotic markers in OSA. - CPAP treatment did not affect day/night rhythm of prothrombotic marker.
Jensen et al., 2018 [33]	Cross-sectional prospective study	28 patients	Comparison of: - Cerebral blood flow (CBF) - Cerebral metabolic rate of oxygen (CMRO2) - Lactate serum levels - In OSA patients (before and after CPAP treatment) vs. controls.	- In hypoxia, CBF significantly increased and blood oxygen concentration decreasing in control group. - CBF was unchanged in OSA patients.	- OSA patients had a reduced CBF in response to hypoxia. - CPAP treatment normalized hypoxia response.
Gregori-Pla et al., 2018 [36]	Cross-sectional prospective study	78 patients	CBF measured at supine to 30° to supine position on bed, comparing - Non-treated OSA group - CPAP-treated OSA group - Control subjects	- Moderate and severe OSA did not recover CBF when patients were tilted back to the supine position. - CBF response was normalized after CPAP treatment.	- Moderate and severe patients with OSA have altered cerebral vasoreactivity related to OSA severity.

Table 1. *Cont.*

References	Study Type	No. Participants or Reported Case	Factors Assessed (Methodology)	Summary of Outcomes	Conclusions
Azarbarzin et al., 2012 [41]	Cross-sectional prospective study	113 patients	- Pulse rate response to respiratory events (ΔHR) was measured in Coronary Arterial Disease (CAD) patients with OSA (AIH $\geq$ 15) comparing groups treated with CPAP or not. Primary outcomes were: - Repeat revascularization - Myocardial infarction - Stroke - Cardiovascular mortality	- CPAP reduced risk of outcomes while increasing progressively pulse rate response to respiratory events.	- There is a protective effect (including against stroke) of CPAP in patients with CAD and OSA modified by the ΔHR. - Patients with higher ΔHR exhibit greater cardiovascular benefit from CPAP.
Mashaqi et al., 2021 [39]	Literature review study	50 articles	- Review of MEDLINE database (PubMed) for publications on Matrix metalloproteinase-9 (MMP-9), OSA, and cardiovascular disease	MMP-9 levels are elevated in patients with: - Stroke - OSA - Systemic hypertension - Myocardial infarction - Postmyocardial infarction heart failure	- MMP-9 levels are positively correlated with stroke severity at 12, 24, and 48 h from the onset of stroke - MMP-9 levels positive correlated with the size of the brain infarcted area

4. Discussion

The relationship between sleep-related breathing disorders (SRBDs) and stroke has gained significant attention in recent decades, with research beginning in the 1990s [12,43]. Over the past 30 years, numerous studies have explored the connections between various types of stroke (both ischemic and hemorrhagic) in acute, chronic, and preventive phases. Our review focused specifically on molecular alterations in patients in the acute phase of stroke (both ischemic and hemorrhagic) with SRBD and its clinical repercussions, limiting our analysis to 13 studies. The heterogeneity of these studies precludes a systematic review that could combine results and highlight biases in each study.

Regarding acute hemorrhagic stroke, the interplay between obstructive SRBD and intracerebral hemorrhage (ICH) presents a complex pathophysiological scenario with significant clinical implications. Recent studies have elucidated the molecular mechanisms underlying this relationship, focusing on apoptotic mediators and inflammatory cytokines [44–50]. However, a critical limitation in current animal models is the inability to distinguish between inflammatory markers produced by ICH and those resulting from OSA.

Our review shows that SRBD induction decreases survival rates, neurological scores, and neuronal survival while upregulating protein expression levels of caspase-3, Bax, and pro-inflammatory cytokines such as IL-1β, IL-6, and TNF-α [38]. These findings suggest that OSA-mediated induction of apoptosis and neuroinflammation exacerbates neuronal death following ICH. However, the inflammatory response observed in these models may be confounded by the combined effects of ICH and SRBD, due to the overlap of inflammatory biomarkers produced by both conditions, making it challenging to delineate the specific contribution of each to the overall inflammatory state. This limitation underscores the need for more refined experimental approaches to elucidate the distinct roles of ICH and SRBD in neuroinflammation and subsequent brain injury.

Regarding acute ischemic stroke (AIS), our review also highlights a complex interplay with SRBDs, revealing a shared inflammatory profile. While interleukin-6 (IL-6) is elevated in both conditions [51–54], several biomarkers appear to be specifically associated with the comorbidity of SRBDs and acute ischemic stroke:

1. Soluble tumor necrosis factor receptors (sTNF-R1 and sTNF-R2)
2. Plasminogen activator inhibitor-1 (PAI-1) [30,40,42]

These SRBD-specific inflammatory markers in AIS patients suggest a unique pathophysiological pathway linking the two conditions. The identification of these biomarkers opens up new possibilities for targeted immunomodulatory therapies, which could potentially mitigate the detrimental effects of SRBDs on acute ischemic stroke treatment outcomes and offer novel approaches to stroke management in patients with comorbid SRBDs.

Serum total antioxidant capacity (TAC) plays a crucial role in acute ischemic stroke by reflecting the overall ability of the body to counteract oxidative stress [31]. Our review found that SRBDs and AIS present a paradoxical scenario in terms of oxidative stress and antioxidant capacity. In patients with severe SRBDs (AHI > 30), elevated TAC levels have been observed, suggesting an adaptive response to chronic intermittent hypoxia. This increase in TAC appears to attenuate the inflammatory reaction typically associated with SRBD in ischemic stroke patients. Conversely, AIS patients generally exhibit lower TAC levels, which correlate with poorer functional outcomes and increased neurological deficits.

This apparent contradiction may explain the exacerbated outcomes observed in patients with comorbid AIS and SRBD. We propose that the cumulative oxidative stress burden from both conditions may overwhelm the compensatory increase in TAC associated with SRBD alone. The insufficient antioxidant capacity to counteract the combined oxidative insult could be a key factor in the poor prognosis of these patients. Therefore, aggressive management of SRBD in the acute stroke setting could be crucial for improving outcomes.

In the context of acute ischemic stroke, MMP-9 is implicated in blood-brain barrier (BBB) disruption and subsequent brain injury. Elevated MMP-9 levels are associated with increased infarct size and worse clinical outcomes due to its role in neuroinflammation and tissue destruction. MMP-9 facilitates the breakdown of the BBB, allowing for the infiltration of inflammatory cells and further exacerbating ischemic damage [52]. In our review, MMP-9 has been implicated in AIS associated with SRBD. Higher MMP-9 levels in SRBD patients are linked to AIS [39], likely due to the hypoxia and oxidative stress characteristic of SRBD. This is a strong link mechanism in both pathologies that deserve deeper research regarding its role in acute stroke and its treatment.

Regarding procoagulant biomarkers, SRBD patients showed lower mean and amplitude of D-dimer and higher mean of PAI-1. There may be altered day/night rhythm of some prothrombotic markers in SRBD. Continuous positive airway pressure (CPAP) treatment did not affect the day/night rhythm of prothrombotic markers [32]. While CPAP therapy has shown efficacy in managing SRBD symptoms, our review suggests that it may not fully address the prothrombotic state associated with this condition. Elevated levels of PAI-1 and D-dimer in SRBD patients indicate a persistent prothrombotic environment, which may contribute to increased stroke risk and poorer outcomes. While CPAP therapy effectively manages respiratory symptoms and improves cardiovascular outcomes in SRBD patients, its impact on these prothrombotic markers remains unclear. This suggests that even with optimal CPAP adherence, SRBD patients may remain at elevated risk for thrombotic events. Future research should focus on developing targeted therapies to address this residual prothrombotic state, potentially combining CPAP with anticoagulant or antifibrinolytic treatments to comprehensively manage stroke risk in SRBD patients.

Regarding cerebral blood flow, alterations and impaired cerebrovascular reactivity in SRBD patients [36], may contribute to increased stroke risk, although some studies show conflicting results. These shared molecular pathways and biomarkers not only enhance our understanding of the SRBD–stroke relationship but also offer potential targets for future diagnostic and therapeutic interventions.

5. Strengths and Limitations

The methodology of this review exhibits several key strengths that enhance its credibility and rigor. By following the PRISMA-ScR guidelines, the review ensures a systematic and transparent approach, which is further reinforced by a predetermined protocol shared among all authors. The inclusion of diverse study designs—quantitative, qualitative, and mixed-methods—allows for a comprehensive exploration of the impact of sleep apnea syndrome (SBD) on acute stroke. A meticulous screening process conducted by multiple authors minimizes bias and enhances the quality of included studies. Additionally, utilizing reputable databases like PubMed, Scopus, and Web of Science ensures access to high-quality data. Collectively, these strengths provide a solid foundation for understanding the complex relationship between SBD and acute stroke outcomes.

The study also has limitations, including potential selection bias from excluding letters to the editor and abstracts, which may omit valuable insights. Language restrictions to English, Spanish, French, and German could introduce bias by excluding relevant studies in other languages. The focus on the molecular, clinical, and therapeutic aspects of SRBDs and stroke may overlook other important factors. Additionally, the inclusion of diverse study methodologies could lead to variability in data, complicating comparisons. Publication bias is a concern, as relying solely on published studies may exclude negative results. The lack of a systematic quality assessment for included studies affects reliability, and searching only three databases may have missed relevant research. Finally, the review is limited to studies published until December 2023, leaving out any subsequent findings.

6. Conclusions

The relationship between SRBDs and cardiovascular conditions is significant, encompassing stroke (ischemic and hemorrhagic), hypertension, myocardial infarction, post-myocardial infarction heart failure, and atherosclerosis.

Markers including matrix metalloproteinase-9 (MMP-9) and tissue inhibitor of metalloproteinase-1 are prevalent in individuals with obstructive sleep apnea (OSA), hypertension, myocardial infarction, and heart failure.

The existence of severe obstructive sleep apnea in stroke patients may affect the inflammatory response and the severity of the stroke.

The impact of SRBDs on cerebral and cardiovascular responses includes potential alterations in brain response to positional changes and cerebral vasoreactivity, which may influence the regulation of cerebral blood flow.

Identifying inflammatory and oxidative stress markers in patients with OSA may be essential for prognostic prediction and therapeutic guidance.

The influence of SRBDs on inflammatory and cardiovascular response during acute stroke may alter the inflammatory response in acute stroke patients, thereby affecting the severity and prognosis of the cerebrovascular incident.

The paucity of studies investigating the beneficial effects of CPAP in the acute phase of stroke, particularly in relation to its impact on the pathophysiological mechanisms identified in various articles, represents a significant gap in current research. This lack of evidence underscores the need for further investigation into how CPAP intervention might modulate the acute inflammatory and prothrombotic states associated with comorbid OSA and stroke, potentially improving outcomes in this high-risk patient population.

The prompt recognition of sleep apnea in acute stroke patients could be important for a possible therapy and for forecasting potential complications and long-term outcomes.

7. Future Research Topics

Future studies should focus on longitudinal designs to establish causality and explore the potential of novel therapeutic approaches targeting the identified molecular pathways.

It is important to highlight that studies with a stronger methodology, such as randomized clinical trials, are needed. Based on the data obtained, we know that both inflammation and altered cerebral blood flow are the main candidates to justify the worse functional and vital prognosis of patients with acute stroke and SRBDs. Therefore, an area of interest would be to investigate anti-inflammatory treatments, particularly those aimed at inhibiting interleukin-6 and matrix metalloproteinase-9.

Regarding the role of CPAP, we can hypothesize that its use should eliminate the negative pressure effect in the thorax and thus normalize cerebral blood flow. However, this theory has not been confirmed by previous studies, which also suggest that there may be a pre-existing dysautonomic alteration due to chronic intermittent hypoxia that would not be quickly corrected with CPAP use [19,20]. Therefore, the timing of introducing CPAP therapy is crucial and needs to be studied. It is likely necessary to identify patients at risk of stroke who should be treated with CPAP to prevent developing a stroke, and if it occurs, to evaluate whether if outcomes are better. It is also essential to study the possibility of initiating SRBD therapy from the very first moment of the patient's sleep after the onset of stroke, as the first hours after stroke onset are critical and when treatments have the most significant impact on future outcomes.

It is also important to conduct studies classifying strokes according to their different types, first dividing them into hemorrhagic and ischemic strokes, and within the latter, investigating whether there are differences between lacunar and non-lacunar ischemic stroke patients. This is due to the fact that the pathophysiology, prognosis, and clinical features of lacunar ischemic strokes differ from all other ischemic stroke subtypes [55]. It is also necessary to study the use of other devices different from CPAP, as we know that within sleep-related breathing disorders (SRBD), we also have central apnea disorders, which are not treated with CPAP. In this case, adaptive servo-ventilation could play an important role.

Funding: This research received no external funding.

Institutional Review Board Statement: Institutional review board approval was not necessary as this study utilized only published data.

Data Availability Statement: The data that support the findings of this study are available on request from the corresponding author. The data are not publicly available due to privacy or ethical restrictions.

Conflicts of Interest: The authors declare no conflicts of interest.

Appendix A

This table contents the complete search strategy ordered by data bases.

Databases	Key Terms	Results
PubMed	"Sleep apnea syndromes" AND "acute stroke" AND "biomarkers-2" AND (1000/01/01:2023/12/31[Date—Publication] AND ("english"[Language] OR "french"[Language] OR "german"[Language] OR "spanish"[Language])) "sleep apnoea syndromes"[All Fields] OR "sleep apnea syndromes"[MeSH Terms] OR ("sleep"[All Fields] AND "apnea"[All Fields] AND "syndromes"[All Fields]) OR "sleep apnea syndromes"[All Fields] AND "stroke"[MeSH Terms] OR "stroke"[All Fields] OR ("acute"[All Fields] AND "stroke"[All Fields]) OR "acute stroke"[All Fields] AND OR "biomarkers"[MeSH Terms] OR "biomarkers"[All Fields] OR "biomarker"[All Fields] OR "lymphotoxin-alpha"[MeSH Terms] OR "lymphotoxin-alpha"[All Fields] OR ("lymphotoxin"[All Fields] AND "alpha"[All Fields]) OR "lymphotoxin alpha"[All Fields] OR "receptors, interleukin-4"[MeSH Terms] OR ("receptors"[All Fields] AND "interleukin-4"[All Fields]) OR "interleukin-4 receptors"[All Fields] OR "interleukin 4 receptors"[All Fields] OR "tumor necrosis factor"[All Fields] OR "tumor necrosis factor-alpha"[MeSH Terms] OR ("tumor"[All Fields] AND "necrosis"[All Fields] AND "factor-alpha"[All Fields]) OR "tumor necrosis factor-alpha"[All Fields] OR ("tumor"[All Fields] AND "necrosis"[All Fields] AND "factor"[All Fields]) OR "tumor necrosis factor"[All Fields] OR "receptors, tumor necrosis factor, type ii"[MeSH Terms] OR ("receptors"[All Fields] AND "tumor"[All Fields] AND "necrosis"[All Fields] AND "factor"[All Fields] AND "type"[All Fields] AND "ii"[All Fields]) OR "type ii tumor necrosis factor receptors"[All Fields] OR "receptors, tumor necrosis factor, type ii"[All Fields] OR "vascular cell adhesion molecule-1"[MeSH Terms] OR ("vascular"[All Fields] AND "cell"[All Fields] AND "adhesion"[All Fields] AND "molecule-1"[All Fields]) OR "vascular cell adhesion molecule-1"[All Fields] OR "vascular cell adhesion molecule 1"[All Fields] OR "intercellular adhesion molecule-1"[MeSH Terms] OR ("intercellular"[All Fields] AND "adhesion"[All Fields] AND "molecule-1"[All Fields]) OR "intercellular adhesion molecule-1"[All Fields] OR "intercellular adhesion molecule 1"[All Fields] OR "fibrinogen"[MeSH Terms] OR "fibrinogen"[All Fields] OR "fibrinogens"[All Fields] OR "fibrinogen's"[All Fields] OR "fibrinogene"[All Fields] OR "genome"[MeSH Terms] OR "genome"[All Fields] OR "genomes"[All Fields] OR "genome's"[All Fields] OR "genomically"[All Fields] OR "genomics"[MeSH Terms] OR "genomics"[All Fields] OR "genomic"[All Fields] OR "proteom"[All Fields] OR "proteome"[MeSH Terms] OR "proteome"[All Fields] OR "proteomes"[All Fields] OR "proteomical"[All Fields] OR "proteomically"[All Fields] OR "proteomics"[MeSH Terms] OR "proteomics"[All Fields] OR "proteomic"[All Fields]]	References = 97

Databases	Key Terms	Results
Scopus	(TITLE-ABS-KEY (biological AND marker)) AND (TITLE-ABS-KEY (sleep AND disordered AND breathing) OR TITLE-ABS-KEY (sleep AND obstructive AND apnea)) AND TITLE-ABS-KEY (cardiovascular AND disease) AND TITLE-ABS-KEY (acute)) AND PUBYEAR < 2024 AND (LIMIT-TO (LANGUAGE, "English") OR LIMIT-TO (LANGUAGE, "French") OR LIMIT-TO (LANGUAGE, "German") OR LIMIT-TO (LANGUAGE, "Spanish"))	References = 53
Web of Science	(((TS = (biomarker OR biological marker)) AND TS = (sleep apn?ea syndrome* OR sleep breathing disorder*)) AND TS = (cardiovascular disease* OR stroke) AND TS = (acute) AND (LA == ("ENGLISH" OR "GERMAN" OR "FRENCH" OR "SPANISH")) NOT PY = ("2024"))	References = 36

References

1. American Academy of Sleep Medicine. *International Classification of Sleep Disorders Version*; American Assocation of Sleep Medicine: Darien, IL, USA, 2014; Volume 3.
2. Heinzer, R.; Vat, S.; Marques-Vidal, P.; Marti-Soler, H.; Andries, D.; Tobback, N.; Mooser, V.; Preisig, M.; Malhotra, A.; Waeber, G.; et al. Prevalence of sleep-disordered breathing in the general population: The HypnoLaus study. *Lancet Respir. Med.* **2015**, *3*, 310–318. [CrossRef] [PubMed]
3. GBD 2017 Disease and Injury Incidence and Prevalence Collaborators. Global, regional, and national incidence, prevalence, and years lived with disability for 354 diseases and injuries for 195 countries and territories, 1990–2017: A systematic analysis for the Global Burden of Disease Study 2017. *Lancet* **2018**, *392*, 1789–1858. [CrossRef] [PubMed]
4. Calcerrada Díaz-Santos, N.; Estirado Gómez, A.; Lozano Sánchez, J.A. *Informe Anual del Sistema Nacional de Salud 2020–2021*; Ministerio de Sanidad: Madrid, Spain, 2022.
5. Norrving, B.; Barrick, J.; Davalos, A.; Dichgans, M.; Cordonnier, C.; Guekht, A.; Caso, V. Action plan for stroke in Europe 2018–2030. *Eur. Stroke J.* **2018**, *3*, 309–336. [CrossRef] [PubMed]
6. Vangen-Lønne, A.M.; Wilsgaard, T.; Johnsen, S.H. Declining Incidence of Ischemic Stroke: What Is the Impact of Changing Risk Factors? The Tromsø Study 1995 to 2012. *Stroke* **2017**, *48*, 544–550. [CrossRef]
7. Madsen, T.E.; Khoury, J.C.; Leppert, M. Temporal Trends in Stroke Incidence Over Time by Sex and Age in the GCNKSS. *Stroke* **2020**, *51*, 1070–1076. [CrossRef]
8. Sacco, R.L.; Kasner, S.E.; Broderick, J.P.; Caplan, L.R.; Connors, J.J.; Culebras, A.; Elkind, M.S.V.; George, M.G.; Hamdan, A.D.; Higashida, R.T.; et al. An Updated Definition of Stroke for the 21st Century: A Statement for Healthcare Professionals from the American Heart Association/American Stroke Association. *Stroke* **2013**, *44*, 2064–2089. [CrossRef] [PubMed]
9. Arboix, A.; Alvarez-Sabín, J.; Ictus, S.L. Clasificación y criterios diagnósticos. Comité de Redacción ad hoc del Grupo de Estudio de Enfermedades Cerebrovasculares de la SEN [Stroke. Classification and diagnostic criteria. Ad hoc Editorial Committee of the Task Force on Cerebrovascular Diseases of SEN]. *Neurologia* **1998**, *3*, 3–10.
10. Iranzo, A.; Santamaría, J.; Berenguer, J.; Sánchez, M.; Chamorro, A. Prevalence and clinical importance of sleep apnea in the first night after cerebral infarction. *Neurology* **2002**, *58*, 911–916. [CrossRef]
11. Mohsenin, V.; Valor, R. Sleep apnea in patients with hemispheric stroke. *Arch. Phys. Med. Rehabil.* **1995**, *76*, 71–76. [CrossRef]
12. Dyken, M.E.; Somers, V.K.; Yamada, T.; Ren, Z.Y.; Zimmerman, M.B. Investigating the relationship between stroke and obstructive sleep apnea. *Stroke* **1996**, *27*, 401–407. [CrossRef]
13. Bassetti, C.; Aldrich, M.S. Sleep apnea in acute cerebrovascular diseases: Final report on 128 patients. *Sleep* **1999**, *22*, 217–223. [CrossRef] [PubMed]
14. Senaratna, C.V.; Perret, J.L.; Lodge, C.J.; Lowe, A.J.; Campbell, B.E.; Matheson, M.C.; Hamilton, G.S.; Dharmage, S.C. Prevalence of obstructive sleep apnea in the general population: A systematic review. *Sleep Med. Rev.* **2017**, *34*, 70–81. [CrossRef] [PubMed]
15. Turkington, P.M.; Allgar, V.; Bamford, J.; Wanklyn, P.; Elliott, M.W. Effect of upper airway obstruction in acute stroke on functional outcome at 6 months. *Thorax* **2004**, *59*, 367–371. [CrossRef]
16. Sahlin, C.; Sandberg, O.; Gustafson, Y.; Bucht, G.; Carlberg, B.; Stenlund, H.; Franklin, K.A. Obstructive sleep apnea is a risk factor for death in patients with stroke: A 10-year follow-up. *Arch. Intern. Med.* **2008**, *168*, 297–301. [CrossRef]
17. Nogueira, R.C.; Beishon, L.; Bor-Seng-Shu, E.; Panerai, R.B.; Robinson, T.G. Cerebral Autoregulation in Ischemic Stroke: From Pathophysiology to Clinical Concepts. *Brain Sci.* **2021**, *11*, 511. [CrossRef] [PubMed]
18. Guilleminault, C.; Connolly, S.J.; Winkle, R.A. Cardiac arrhythmia and conduction disturbances during sleep in 400 patients with sleep apnea syndrome. *Am. J. Cardiol.* **1983**, *52*, 490–494. [CrossRef]
19. Guilleminault, C.; Connolly, S.; Winkle, R.; Melvin, K.; Tilkian, A. Cyclical variation of the heart rate in sleep apnoea syndrome. Mechanisms, and usefulness of 24 h electrocardiography as a screening technique. *Lancet* **1984**, *323*, 126–131. [CrossRef] [PubMed]

20. Netzer, N.; Werner, P.; Jochums, I.; Lehmann, M.; Strohl, K.P. Blood flow of the middle cerebral artery with sleep-disordered breathing: Correlation with obstructive hypopneas. *Stroke* **1998**, *29*, 87–93. [CrossRef] [PubMed]
21. Kohler, M.; Stradling, J. Mechanisms of vascular damage in obstructive sleep apnea. *Nat. Rev. Cardiol.* **2010**, *7*, 677–685. [CrossRef] [PubMed]
22. Lal, C.; Hardiman, G.; Kumbhare, S.; Strange, C. Proteomic biomarkers of cognitive impairment in obstructive sleep apnea syndrome. *Sleep Breath.* **2019**, *23*, 251–257. [CrossRef]
23. Baril, A.A.; Carrier, J.; Lafrenière, A.; Warby, S.; Poirier, J.; Osorio, R.S. Biomarkers of dementia in obstructive sleep apnea. *Sleep Med. Rev.* **2018**, *42*, 139–148. [CrossRef] [PubMed]
24. Powers, W.J.; Rabinstein, A.A.; Ackerson, T.; Adeoye, O.M.; Bambakidis, N.C.; Becker, K.; Biller, J.; Brown, M.; Demaerschalk, B.M.; Hoh, B.; et al. Guidelines for the Early Management of Patients with Acute Ischemic Stroke: 2019 Update to the 2018 Guidelines for the Early Management of Acute Ischemic Stroke: A Guideline for Healthcare Professionals from the American Heart Association/American Stroke Association. *Stroke* **2019**, *50*, e344–e418. [CrossRef] [PubMed]
25. Parra, O.; Sánchez-Armengol, Á.; Capote, F.; Bonnin, M.; Arboix, A.; Campos-Rodríguez, F.; Pérez-Ronchel, J.; Durán-Cantolla, J.; Martínez-Null, C.; Peña, M.; et al. Efficacy of continuous positive airway pressure treatment on 5-year survival in patients with ischaemic stroke and obstructive sleep apnea: A randomized controlled trial. *J. Sleep Res.* **2015**, *24*, 47–53. [CrossRef] [PubMed]
26. McEvoy, R.D.; Antic, N.A.; Heeley, E.; Luo, Y.; Ou, Q.; Zhang, X.; Mediano, O.; Chen, R.; Drager, L.F.; Liu, Z.; et al. CPAP for Prevention of Cardiovascular Events in Obstructive Sleep Apnea. *N. Engl. J. Med.* **2016**, *375*, 919–931. [CrossRef]
27. Bravata, D.M.; Sico, J.; Vaz Fragoso, C.A.; Miech, E.J.; Matthias, M.S.; Lampert, R.; Williams, L.S.; Concato, J.; Ivan, C.S.; Fleck, J.D.; et al. Diagnosing and Treating Sleep Apnea in Patients With Acute Cerebrovascular Disease. *J. Am. Heart Assoc.* **2018**, *7*, e008841. [CrossRef] [PubMed]
28. Scoping—[Internet]. PRISMA statement. Available online: https://www.prisma-statement.org/scoping (accessed on 19 November 2024).
29. Tricco, A.C.; Lillie, E.; Zarin, W.; O'Brien, K.K.; Colquhoun, H.; Levac, D. PRISMA Extension for Scoping Reviews (PRISMA-ScR): Checklist and Explanation. *Ann. Intern. Med.* **2018**, *169*, 467–473. [CrossRef] [PubMed]
30. Ifergane, G.; Ovanyan, A.; Toledano, R.; Goldbart, A.; Abu-Salame, I.; Tal, A.; Stavsky, M.; Novack, V. Obstructive Sleep Apnea in Acute Stroke: A Role for Systemic Inflammation. *Stroke* **2016**, *47*, 1207–1212. [CrossRef] [PubMed]
31. Chen, C.-Y.; Chen, C.-L.; Yu, C.-C.; Chen, T.-T.; Tseng, S.-T.; Ho, C.-H. Association of inflammation and oxidative stress with obstructive sleep apnea in ischemic stroke patients. *Sleep Med.* **2015**, *16*, 113–118. [CrossRef] [PubMed]
32. Von Känel, R.; Natarajan, L.; Ancoli-Israel, S.; Mills, P.J.; Wolfson, T.; Gamst, A.C.; Loredo, J.S.; Dimsdale, J.E. Effect of continuous positive airway pressure on day/night rhythm of prothrombotic markers in obstructive sleep apnea. *Sleep Med.* **2013**, *14*, 58–65. [CrossRef] [PubMed]
33. Jensen, M.L.F.; Vestergaard, M.B.; Tønnesen, P.; Larsson, H.B.W.; Jennum, P.J. Cerebral blood flow, oxygen metabolism, and lactate during hypoxia in patients with obstructive sleep apnea. *Sleep* **2018**, *41*, zsy001. [CrossRef]
34. Yeh, T.; Hsieh, Y.; Wu, C.; Wang, J.; Lin, K.; Chen, C. A Preliminary Investigation of the Association of Sleep with Inflammation and Oxidative Stress Biomarkers and Functional Outcomes After Stroke Rehabilitation. *Sci. Rep.* **2017**, *7*, 8634. [CrossRef] [PubMed]
35. Lévy, P.; Pépin, J.-L.; Arnaud, C.; Baguet, J.-P.; Dematteis, M.; Mach, F. Obstructive Sleep Apnea and Atherosclerosis. *Prog. Cardiovasc. Dis.* **2009**, *51*, 400–410. [CrossRef] [PubMed]
36. Gregori-Pla, C.; Cotta, G.; Blanco, I.; Zirak, P.; Giovannella, M.; Mola, A.; Fortuna, A.; Durduran, T.; Mayos, M. Cerebral vasoreactivity in response to a head-of-bed position change is altered in patients with moderate and severe obstructive sleep apnea. *PLoS ONE* **2018**, *13*, e0194204. [CrossRef]
37. Tsai, J.C.G. Neurological and neurobehavioral sequelae of obstructive sleep apnea. *NeuroRehabilitation* **2010**, *26*, 85–94. [CrossRef] [PubMed]
38. Fei, W.; Jiao, W.; Feng, X.; Chen, X.; Wang, Y. Intermittent hypoxia mimicking obstructive sleep apnea aggravates early brain injury following ICH via neuroinflammation and apoptosis. *Mol. Med. Rep.* **2021**, *24*, 824. [CrossRef] [PubMed]
39. Mashaqi, S.; Mansour, H.M.; Alameddin, H.; Combs, D.; Patel, S.; Estep, L.; Parthasarathy, S. Matrix metalloproteinase-9 as a messenger in the cross talk between obstructive sleep apnea and comorbid systemic hypertension, cardiac remodeling, and ischemic stroke: A literature review. *J. Clin. Sleep Med.* **2021**, *17*, 567–591. [CrossRef]
40. Kunz, A.B.; Kraus, J.; Young, P.; Reuss, R.; Wipfler, P.; Oschmann, P.; Blaes, F.; Dziewas, R. Biomarkers of Inflammation and Endothelial Dysfunction in Stroke with and without Sleep Apnea. *Cerebrovasc. Dis.* **2012**, *33*, 453–460. [CrossRef]
41. Azarbarzin, A.; Zinchuk, A.; Peker, Y.; Sands, S.A. Reply to: Heart Rate Response in Obstructive Sleep Apnea: A Clue to Reveal Cardiovascular Benefit from Continuous Positive Airway Pressure? *Am. J. Respir. Crit. Care Med.* **2022**, *206*, 1181–1182. [CrossRef]
42. Medeiros, C.A.M.; De Bruin, V.M.S.; Andrade, G.M.; Coutinho, W.M.; De Castro-Silva, C.; De Bruin, P.F.C. Obstructive sleep apnea and biomarkers of inflammation in ischemic stroke: Interleukin-6 in sleep apnea and acute stroke. *Acta Neurol. Scand.* **2012**, *126*, 17–22. [CrossRef]

43. Palomäki, H.; Partinen, M.; Erkinjuntti, T.; Kaste, M. Snoring, sleep apnea syndrome, and stroke. *Neurology* **1992**, *42* (Suppl. 6), 75–81; discussion 82.
44. Xu, H.; Cao, J.; Xu, J. GATA-4 Regulates Neuronal Apoptosis After Intracerebral Hemorrhage via the NF-κB/ Bax/ Caspase-3 Pathway Both in Vivo and in Vitro. *Exp. Neurol.* **2019**, *315*, 21–31. [CrossRef] [PubMed]
45. Shen, X.; Ma, L.; Dong, W. Autophagy Regulates Intracerebral Hemorrhage Induced Neural Damage via Apoptosis and NF-κB Pathway. *Neurochem. Int.* **2016**, *96*, 100–112. [CrossRef]
46. Zhu, H.; Wang, Z.; Yu, J. Role and Mechanisms of Cytokines in the Secondary Brain Injury After Intracerebral Hemorrhage. *Prog. Neurobiol.* **2019**, *178*, 101610. [CrossRef] [PubMed]
47. Wang, Y.X.; Yan, A.; Ma, Z.H. Nuclear Factor-κB and Apoptosis in Patients With Intracerebral Hemorrhage. *J. Clin. Neurosci. Off. J. Neurosurg. Soc. Australas.* **2011**, *18*, 1392–1395. [CrossRef] [PubMed]
48. Leasure, A.C.; Kuohn, L.R.; Vanent, K.N. Association of Serum IL-6 (Interleukin 6) With Functional Outcome After Intracerebral Hemorrhage. *Stroke* **2021**, *52*, 1733–1740. [CrossRef] [PubMed]
49. Tapia-Pérez, J.H.; Karagianis, D.; Zilke, R.; Koufuglou, V.; Bondar, I.; Schneider, T. Assessment of Systemic Cellular Inflammatory Response After Spontaneous Intracerebral Hemorrhage. *Clin. Neurol. Neurosurg.* **2016**, *150*, 72–79. [CrossRef]
50. Zhang, Z.; Song, Y.; Li, F.; Xu, Z.; Huang, Q. Inhibiting Nuclear Factor-κB at Different Stages After Intracerebral Hemorrhage Can Influence the Hemorrhage Induced Brain Injury in Experimental Models In Vivo. *Brain Res. Bull.* **2020**, *155*, 159–165. [CrossRef]
51. Kowalski, R.G.; Ledreux, A.; Violette, J.E. Rapid Activation of Neuroinflammation in Stroke: Plasma and Extracellular Vesicles Obtained on a Mobile Stroke Unit. *Stroke* **2023**, *54*, e52–e57. [CrossRef]
52. Lasek-Bal, A.; Jedrzejowska-Szypulka, H.; Student, S. The Importance of Selected Markers of Inflammation and Blood-Brain Barrier Damage for Short-Term Ischemic Stroke Prognosis. *J. Physiol. Pharmacol. Off. J. Pol. Physiol. Soc.* **2019**, *70*, 209–217. [CrossRef]
53. Pawluk, H.; Kołodziejska, R.; Grześk, G. Selected Mediators of Inflammation in Patients With Acute Ischemic Stroke. *Int. J. Mol. Sci.* **2022**, *23*, 10614. [CrossRef]
54. Zeng, L.; Wang, Y.; Liu, J. Pro-Inflammatory Cytokine Network in Peripheral Inflammation Response to Cerebral Ischemia. *Neurosci. Lett.* **2013**, *548*, 4–9. [CrossRef]
55. Arboix, A.; Massons, J.; García-Eroles, L.; Targa, C.; Comes, E.; Parra, O. Clinical predictors of lacunar syndrome not due to lacunar infarction. *BMC Neurol.* **2010**, *10*, 31. [CrossRef] [PubMed]

current issues in
molecular biology

MDPI

Review

Cerebral Small Vessel Disease: Therapeutic Approaches Targeting Neuroinflammation, Oxidative Stress, and Endothelial Dysfunction

Habibe Yılmaz [1,2,*] and Ulvi Bayraktutan [1,*]

1 Academic Unit of Mental Health and Clinical Neurosciences, University of Nottingham, Nottingham NG7 2UH, UK
2 Department of Pharmaceutical Biotechnology, Faculty of Pharmacy, Trakya University, Edirne 22030, Türkiye
* Correspondence: habibe.yilmaz@nottingham.ac.uk (H.Y.); ulvi.bayraktutann@nottingham.ac.uk (U.B.)

Abstract: Cerebral small vessel disease (cSVD) is a common cause of stroke and dementia. Ageing, hypertension, hyperglycaemia, and smoking make up the biggest risk factors for cSVD. They individually or collectively increase the levels of reactive oxygen species, pro-inflammatory cytokines and matrix metalloproteinases, decrease the bioavailability of nitric oxide, and, in the process, compromise the structural integrity and function of the vascular endothelium, blood–brain barrier, and brain parenchyma. These then appear as white matter hyperintensities, enlarged perivascular spaces, cerebral microbleeds, and atrophy in cerebral imaging. As there is currently no curative therapy for cSVD, prevention or delay of cSVD remains of particular importance to preserve quality of life for as long as possible. Bearing that in mind, this review explores whether drugs used for other neurovascular conditions may prevent neuroinflammation and oxidative damage and effectively maintain endothelial function and blood–brain barrier integrity. It also examines whether potential benefits may be extended to cSVD. The list of drugs includes anti-anginal drugs, acetylcholine esterase inhibitors, β-hydroxy β-methylglutaryl-CoA reductase inhibitors, lithium drugs, phosphodiesterase inhibitors, oral antihyperglycaemic drugs, and tetracycline antibiotics. This review discusses the mechanisms of action of these agents and critically evaluates preclinical, translational, and clinical research pertaining to cSVD.

Keywords: brain; small vessel disease; anti-anginal drugs; acetylcholine esterase inhibitors; HMG-CoA reductase inhibitors; lithium drugs; phosphodiesterase inhibitors; oral antihyperglycaemic drugs; tetracycline antibiotics

1. Introduction

Using the available data from 2021, the World Health Organisation (WHO) has recently released the list of the top 10 causes of death in 2024. While ischaemic heart disease continued to be the leading cause of death, stroke came third in the list after COVID-19. Alzheimer's disease (AD) and other dementias also ranked amongst the leading causes of death [1]. The statistics from the WHO indicate that each year, globally, approximately 15 million people suffer a stroke. While one-third of these patients recover from it, another one-third are left permanently disabled and the remaining one-third sadly die from stroke [2]. According to data from the Stroke Association, stroke is also one of the leading causes of death in the UK [3]. The future projections concerning the stroke incidence and prevalence in the UK predict about 60% and 120% increases by 2035, respectively [4].

Stroke is characterised by pathological alterations resulting from disrupted cerebral blood flow and divided into two main types: haemorrhagic or ischaemic [5]. Cerebrovascular haemorrhages occur when an artery within or on the surface of the brain is ruptured. One this happens, the blood is extravasated into the brain parenchyma, ventricles, or the subarachnoid space. This may be precipitated by elevated blood pressure, the use of anticoagulants or antithrombotic agents, or head trauma [6]. Ischaemic strokes are, on the other hand, characterised by an obstruction in the vessels leading to or in the brain due to thrombus or embolism. Ischaemic strokes account for about 85% of all stroke cases in Western populations [7]. Large artery atherosclerosis, cardiac embolism, and cerebral small vessel disease (cSVD) represent the most common causes of ischaemic strokes. cSVD arises from a variety of pathological alterations that impair the cerebrovascular network, including the cerebral arteries, capillaries, arterioles, and venules [8]. It accounts for ~25% of all ischaemic stroke cases and about 45% of all dementia cases. The incidence of this condition escalates with advancing age, irrespective of differences in race or sex [9]. Hypertension, smoking, diabetes mellitus (DM), obstructive sleep apnoea, chronic kidney disease, and branch atheromatous disease are amongst the main risk factors involved in the pathogenesis of cSVD [10].

There is currently no cure for cSVD and available therapeutic strategies focus on addressing associated risk factors and promoting lifestyle modifications. Hence, exploration of the potential pharmaceutical interventions that could modify the progression of cSVD has been at the forefront of recent research. In this regard, scavengers of reactive oxygen species (ROS), drugs that enhance endothelial nitric oxide synthase (eNOS) activity and restore normal endothelial function, matrix metalloproteinase inhibitors, and agents that strengthen the integrity of tight junctions (TJs) and the blood–brain barrier (BBB) have attracted some attention [11–13]. Similarly, cytokine inhibitors and microglial modulators that aim to regulate neuroinflammation, calcium channel blockers that modulate cerebral blood flow, and cyclic guanosine monophosphate (cGMP) enhancers have also been under investigation [14–18].

All relevant studies from all years, including both reviews and original articles, were identified in the PubMed database by a literature search using the key MeSH terms "cerebral small vessel disease", "ischaemic stroke", "tetracycline antibiotics", "anti anginal drugs", "acetylcholine esterase inhibitors", "HMG-CoA reductase inhibitors", "lithium drugs", "phosphodiesterase inhibitors", "in vitro blood-brain barrier models", and "oral antihyperglycaemic drugs". A Nottingham University search, ClinicalTrials.Gov, and Google Scholar were also used to collect pertinent studies.

2. Cerebral Small Vessel Disease

Cerebral SVD is a cerebrovascular syndrome characterised by the progressive deterioration of small penetrating arteries, arterioles, capillaries, and venules. It encompasses a broad spectrum of conditions including endothelial dysfunction, BBB disruption, neuroinflammation, oxidative stress, and impaired cerebral autoregulation. These lead to white matter hyperintensities (WMHs), lacunar infarctions, microhaemorrhages, and cerebral atrophy and, as a consequence, trigger stroke, vascular dementia, and accompanying cognitive impairments [19,20].

cSVD markers can be categorised as imaging, non-genetic biomarkers from fluids, and genetic markers. Magnetic resonance imaging (MRI) is considered the gold standard for imaging-based diagnosis. WMHs can be detected via T2-weighted MRI. Quantitative and volumetric analyses of these lesions provide valuable insights into the progression and severity of the underlying cSVD [21,22]. MRI commonly shows small, cavitated lesions known as lacunes, resulting from damage to deep penetrating arteries. The presence of lacunes is associated with a high risk of stroke and dementia [23]. Cerebral microhaemorrhages (CMHs) may be detected using susceptibility-weighted imaging. CMHs indicate microvascular damage and haemorrhage-prone vessels, mostly in amyloid-associated angiopathies and hypertensive cases [24]. Enlarged perivascular spaces (EPSs), which signify microvascular impairment, impaired glymphatic clearance, and neuroinflammation, may also be present [25].

Non-genetic biomarkers of fluids include neurofilament light chain (NfL), amyloid-β and tau proteins, and inflammatory endothelial dysfunction markers. Increased levels of NfL in the cerebrospinal fluid (CSF) and blood are indicative of axonal damage and neurodegeneration [26]. Elevated concentrations of amyloid-β and tau proteins in the brain correlate with Alzheimer's disease and indicate vascular impairment [27]. Increased plasma levels of interleukin-6 (IL-6), C-reactive protein (CRP), and tumour necrosis factor-α (TNF-α) pinpoint systemic inflammation and contribute to endothelial dysfunction in cSVD [28–30]. Similarly, circulating vascular endothelial growth factor (VEGF) and soluble intercellular adhesion molecule-1 (SICAM-1) also indicate the presence of endothelial dysfunction [31].

Cerebral autosomal dominant arteriopathy with subcortical infarcts and leukoencephalopathy (CADASIL) is the most common genetic risk factor for cSVD. Genetic mutations in the neurogenic locus notch homolog protein 3 (NOTCH3) gene, which produces a transmembrane protein found in vascular smooth muscle cells, lead to the formation of misfolded NOTCH3 protein and granular osmiophilic material. This in turn results in impaired vascular function and alterations to white matter [32]. Cerebral autosomal recessive arteriopathy with subcortical infarcts and leukoencephalopathy (CARASIL) is associated with the HtrA serine peptidase 1 (HTRA1) gene mutation, which encodes a serine protease. Transforming growth factor-β (TGF-β) accumulation is observed in individuals with HTRA1 gene mutations, leading to vascular fibrosis and progressive ischaemic injury [33]. Retinal vasculopathy with cerebral leukodystrophy and systemic manifestations (RVCL-S) is caused by mutations in the three prime repair exonuclease 1 (TREX1) gene, which encodes a DNA exonuclease. Consequently, there is an accumulation of cytosolic deoxyribonucleic acid (DNA) and subsequent activation of the innate immune system, which leads to vascular inflammation and endothelial dysfunction [34]. Collagen α-1(IV) chain (COL4A1) and collagen α-2(IV) chain (COL4A2) genes encode type IV collagen. Genetic alterations in this region disrupt the integrity of the basal membrane. The vessels become susceptible to haemorrhage and ischaemia, leading to recurrent strokes, white matter lesions, and CMBs [35]. Fabry disease is a genetically inherited disorder linked to the X chromosome and arises from pathogenic variants in the galactosidase α (GLA) gene. The accumulation of globotriaosylceramide within endothelial cells can lead to the occlusion of small vessels, alterations in white matter, and the occurrence of stroke [36]. Familial idiopathic basal ganglia calcification, known as Fahr's disease, is caused by mutations in the platelet-derived growth factor receptor-β (PDGFRB) gene. As a result of this mutation, pericyte function is impaired and the BBB integrity is disrupted [37].

Impaired function of the endothelium, including disturbances in the small penetrating arteries, can result in reduced cerebral perfusion and ischaemia, which are associated with white matter abnormalities [19,38,39]. Endothelial dysfunction emerges from and leads to reduced bioavailability of NO, which impairs vasodilatation and promotes vascular inflammation. The increased vascular permeability and disruption of the BBB further contribute to inflammation and oedema and correlate with CMHs [11,38,39]. Vascular smooth muscle dysfunction, on the other hand, is linked to the severity of cSVD [39]. The reduced bioavailability of NO, derived in part from the downregulation of eNOS, elicits a marked elevation in superoxide anion levels and other ROS [40]. It is of note that the diminished expression of claudin-5 and occludin proteins within the TJs, which regulate the selective permeability of substances, also disrupts the function and integrity of the BBB and thus facilitates the infiltration of inflammatory mediators into the brain parenchyma. Degradation of the extracellular matrix and myelin-supporting proteins by enhanced activities of matrix metalloproteinases (MMPs), in particular MMP-2 and MMP-9, further exacerbate BBB and brain tissue disruption following the onset of endothelial dysfunction. Heat shock protein 90α (HSP90α), synthesised by dysfunctional endothelial cells (ECs), also prevents oligodendrocyte maturation, causing myelin loss and white matter changes. Taken together, all these factors contribute to BBB disruption, inflammatory cell infiltration, and the observation of the so-called WMHs on MRI [41–43]. Figure 1 summarises the information discussed in this section.

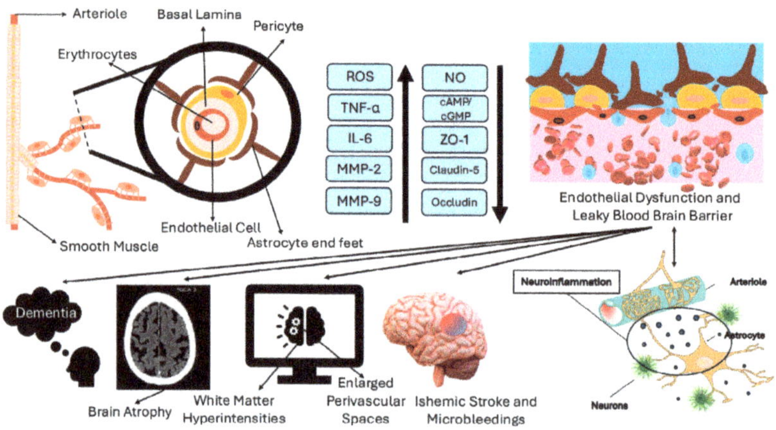

Figure 1. The blood–brain barrier consists of endothelial cells, pericytes, astrocytic end feet, and the basal lamina. Increased availability or activity of ROS, pro-inflammatory cytokines such as TNF-α and IL-6, and MMPs along with decreases in NO, cAMP/cGMP, and tight junction proteins ZO-1, claudin-5, and occludin levels cause endothelial dysfunction. This in turn increases the permeability of the BBB, allows the passage of inflammatory compounds to brain parenchyma, and exacerbates BBB damage. As a result, ischemic stroke, cerebral microhaemorrhages, WMH, and EPSs are observed. Unless medically intervened with, brain atrophy and dementia will emerge. ROS: Reactive oxygen species, TNF-α: Tumour necrosis factor α, IL-6: Interleukin-6, MMP: Matrix metalloproteinase, NO: Nitric oxide, cAMP: Cyclic adenosine monophosphate, cGMP: Cyclic guanosine monophosphate, ZO-1: Zonula occludens-1.

3. BBB Integrity in Cerebral Small Vessel Disease and In Vitro Models of the BBB

The BBB is one of the three barriers in the central nervous system (CNS). ECs, pericytes, and astrocytes make up the main cellular components of the BBB. The ECs in the CNS cover the entire inner surface of all cerebral blood vessels and, in the process, form a continuous and complete basement membrane without pores. As well as being a physical barrier, the BBB is also a biochemical barrier [44]. The continuous tight junctions, formed by proteins like claudins and occludin, supported by accessory proteins like zonula occludens-1 (ZO-1), between capillary ECs, the basement membrane, and the extracellular matrix; perivascular neurons; pericytes; and astrocytic foot processes maintain the normal BBB function and limit the entry of potentially harmful chemicals into the CNS [45,46]. While ECs form a highly selective permeable barrier, astrocytes contribute to BBB integrity by regulating ion homeostasis, vascular function, and neuroimmune responses. Pericytes communicate with ECs to regulate TJ formation, vascular stability, and neuroinflammation. As pericytes are embedded in the EC lining, any pathology affecting their numbers and function can also affect the integrity of the BBB [46,47].

As alluded to above, BBB hyperpermeability constitutes an important part of brain damage. It is also regarded as an early pathology of cSVD in that decreased expression and/or mislocalisation of key TJ proteins occludin and claudin-5 due to oxidative stress, inflammation, and vascular damage play a key role [48]. Breakdown of the BBB facilitates the infiltration of plasma proteins, immune cells, and pro-inflammatory cytokines to the brain parenchyma, which activate microglia and astrocytes and promote the generation of more cytokines and ROS. In addition to the BBB, neurovascular coupling (NVC) is also disrupted during cSVD. This culminates in chronic hypoperfusion, which exacerbates oxidative stress, damages ECs, and inflicts damage on the white matter [49]. Oxidative stress arises when the balance between ROS production and the body's antioxidant defence is disrupted. Elevated ROS levels target polyunsaturated fatty acids in endothelial cell membranes, triggering lipid peroxidation and the release of reactive aldehydes such as malondialdehyde and 4-hydroxynonenal, which, in turn, bind to DNA and cause strand breaks. Through oxidizing cysteine residues and disrupting disulphide bonds in tight junction proteins like ZO-1, occludin, and claudin-5, ROS destabilize them. Additionally, ROS can activate MMPs and further compromise the structural integrity of these proteins. As a consequence, the breakdown of tight junctions creates gaps between neighbouring endothelial cells and promotes BBB permeability. As the barrier weakens, harmful proteins such as fibrinogen, thrombin, and albumin, along with immune cells, infiltrate the brain. The cytokines released from these immune cells, including TNF-α and IL-6, activate the NF-κB and MAPK pathways, amplifying pro-inflammatory cytokine production. This escalation in neuroinflammation accelerates tissue damage and exacerbates disease progression [50,51].

In cSVD, the extent of changes in BBB disruption correlates positively with the magnitude of the changes in WMH and CMH on MRI. It is noteworthy here that the level of BBB disruption varies from one region to another in different cSVD types [52,53]. Taken together, these findings indicate that therapeutic strategies aimed at preserving the integrity and function of the BBB are more efficacious in reducing the burden of disease compared to a symptomatic approach.

The use of in vitro models of the BBB in cerebrovascular research is highly informative. The models have evolved greatly in recent years and now include both dynamic and static models. The static two-dimensional (2D) models represent the most accessible, cost-effective, and widely used models. Triple cell culture models, enabling intercellular interactions amongst ECs, astrocytes, and pericytes on transwell inserts, remain the most commonly used models of the human BBB in laboratory conditions [54–56]. Through incorporation of neurons into the system, these models have since become a quadruple cell system [57]. In recent years, static three-dimensional (3D) models using horizontal microfluidics support, organoid models, and models utilising bioprinted scaffolds to mimic the lumen structure have also become available. For dynamic models, single- or multi-channel vertical microfluidic systems have been used to generate both 2D and 3D models [58,59]. To simulate shear stress as a dynamic effect, the cone and plate model, which can be adapted to transwell systems, has emerged [60]. More sophisticated dynamic in vitro BBB (DIV-BBB) systems using hollow-fibre cartridges, a flow path, and a reservoir have also been developed [61]. In recent years, viable perfusion-based systems have been explored as an alternative to the use of hollow fibres for the preparation of the DIV-BBB [62]. Figure 2 illustrates the commonly used in vitro models of the human BBB.

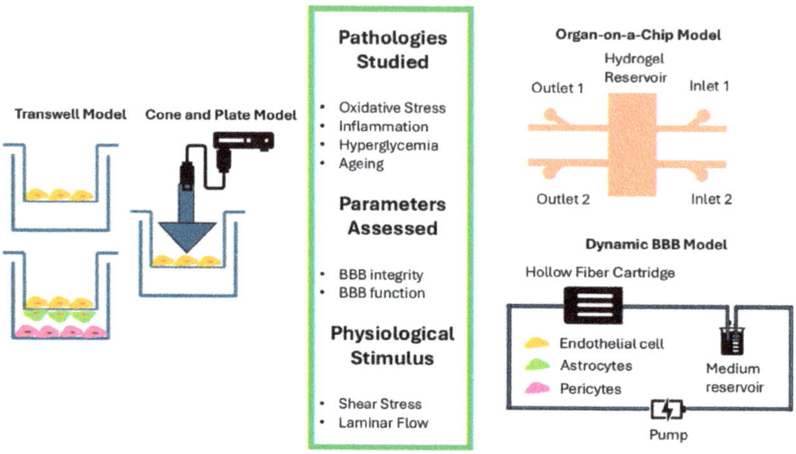

Figure 2. In vitro BBB models used to investigate the impact of pathologies associated with cSVD on the blood–brain barrier. All models illustrated in the figure can effectively assess the impact of different pathologies on barrier integrity and function. The cone and plate, BBB-on-a-chip, and dynamic models allow the assessment of these pathologies in the presence of key physiological stimuli, notably shear stress and blood flow.

Static 2D BBB models remain among the most effective tools for investigating the molecular mechanisms underlying cSVD. They can specifically assess the impact of various cSVD-related pathologies such as hyperglycaemia, oxidative stress, inflammation, and ageing on barrier integrity and function in the absence or presence of various so-called therapeutics. Recent studies using triple cell culture models have attributed hyperglycaemia-mediated impairments in BBB integrity to tight junction protein dissolution, stress fibre formation, and increased release of endothelin-1, IL-8, and basic fibroblast growth factor [45]. Using the same model, another study has coupled the TNF-α-induced BBB dysfunction to the upregulation of endothelial Rho kinase expression [29]. Previous studies have reported that TNF-α also increases MMP-2 activity, actin stress fibre formation, caspase activity, NADPH oxidase activity, and O_2^- production in both HBMEC and HA [63]. Oxidative stress emerging from ischemia-reperfusion injury, mimicked experimentally by oxygen–glucose deprivation alone or followed by reperfusion (OGD $\pm$ R), has also appeared to compromise cerebral barrier integrity through increases in MMP-2/9 and NADPH oxidase activity, TNF-α and O_2^- secretion, apoptosis, and actin stress fibre formation [64]. These findings were supported by the findings of another study subjecting 2D or 3D BBB models to different concentrations of H_2O_2 to investigate the effect of acute and chronic oxidative stress [65]. Finally, BBB models set up with stress-induced or replicative senescent endothelial cells have also corroborated the barrier-disruptive effects of ageing on BBB characteristics, where alterations in the expression and subcellular localisation of tight junction proteins ZO-1 and claudin-5 and MMP-2 activity appeared to be instrumental [66–68]. Organ-on-a-chip models enable the simulation of shear stress and fluctuating blood flow, key factors linked to hypertension—a major risk factor for cSVD—while also allowing the examination of molecular-level changes observed in cSVD pathology [69,70].

The literature concerning the barrier-restorative effects of the outlined specific pharmacological agents within the context of cSVD or associated risk factors is extremely scant. However, there are studies highlighting the molecular mechanisms involved in their barrier-protective effects. For instance, oral antidiabetic metformin (10 mM) has been shown to attenuate BBB disruption in rodents after middle cerebral artery occlusion [71], which may be explained by its ability to activate AMP-activated protein kinase in endothelial cells in a dose-dependent manner [72]. Similarly, lithium carbonate (1.25–10 mM) has also been shown to alleviate BBB breakdown after cerebral ischaemic injury in mice by upregulating Wnt/β-catenin signalling [73] and MAPK/ERK1/2 activity [74]. The effects of atorvastatin also appear to be dose-dependent, where higher doses (0.2 to 25 μM) evoke endothelial dysfunction via the disruption of mitochondrial respiration in endothelial cells [75] and BBB failure at a dose range of 45 to 90 μM [76]. Interestingly, the doses applied in these studies were higher than the pharmacological dose ranges for metformin (20–30 μM), lithium (0.4–1.2 mM), and atorvastatin (4–15 nM) [77–79]. However, the effective cerebrovascular protective dose for minocycline HCl (400 nM) following OGD-R injury [80] was considerably lower than the recommended serum concentration of 6.1 to 17.7 mM [81]. It is noteworthy here that in cell culture conditions and animal experiments, the doses of most pharmaceutical agents tend to be higher than their corresponding plasma or serum concentrations in humans. This is in part because only minute amounts of the drug remain in the circulation when a drug is highly tissue-bound. In other words, a low plasma concentration indicates a high volume of distribution and vice versa [82]. Secondly, in most cases, in vitro concentrations of drugs have to be markedly higher than the corresponding in vivo plasma concentration in order to produce similar biological effects in the target cells [83]. Finally, the rate of drug elimination is faster in small laboratory animals than humans, a well-documented difference in pharmacokinetics [84].

The BBB models developed over time have aimed to better reflect human physiology or pathophysiology. However, there is a standardisation problem for dynamic and sophisticated models in drug research. As sophisticated models demand considerable expertise and time to establish, they are not widely used in laboratory studies. However, selecting an appropriate model is crucial for the research question to be addressed. For instance, triple culture systems are widely used for screening drugs and investigating the molecular mechanisms involved in neurovascular pathologies [56]. The use of human primary cells in these models increases the translatability of the results to clinical settings [85]. Although in vitro models are not sufficient to exactly mimic in vivo conditions, it is anticipated that for studies like cSVD, where the pathology and molecular mechanisms are not yet fully understood, the use of cell culture models of the human BBB will be highly beneficial.

4. Current Pharmacological Interventions for cSVD

Considering the pathologies and molecular pathways involved in cSVD, some medicines are likely to be effective in its management. The mechanisms of action for each drug group and their relationship with the well-documented causes of cSVD are summarized below to provide the reader with an overview.

4.1. Acetylcholine Esterase Inhibitors

Acetylcholine (ACh), an essential excitatory neurotransmitter synthesised by cholinergic neurons, is closely linked to learning, memory, and various behaviours. ACh deficiency is therefore closely associated with the pathogenesis of Alzheimer's disease, a form of dementia [86]. ACh is found in nerves, muscles, and the central and peripheral nervous systems. In the peripheral nervous system, it operates at the connections between motor nerves and skeletal muscle. In the CNS, it has been identified primarily between neurons and in several important long-axon cholinergic pathways. Within the cholinergic system, acetylcholinesterase hydrolyses ACh to choline. The released choline is then reuptaken into the presynaptic terminal via the high-affinity choline uptake process. It is subsequently converted back to ACh by the choline acetyltransferase enzyme using choline and acetyl coenzyme A. In the presence of acetylcholinesterase (AChE) inhibitors, the degradation of ACh is prevented and its accumulation in the synaptic cleft is ensured, thus keeping postsynaptic receptors activated and prolonging cholinergic receptor signalling [87].

In cerebral ischaemia, cholinergic system dysfunction and post-ischaemic inflammation have crucial effects on vascular dementia. It has been shown that in cases of arterial occlusion, there is a loss of cholinergic nerves, a decrease in choline acetyltransferase and AChE activities, and a reduction in ACh levels in the neocortex, hippocampus, and CSF. Similar findings have been reported in post-mortem patient studies. As a result of the inflammatory processes following cerebral ischaemia, immune cells secrete ROS and various toxic components such as pro-inflammatory cytokines, which damage neurons and disrupt the BBB integrity. BBB disruption can aggravate the inflammatory state by causing a secondary ischaemic attack. However, ACh secreted by functional cholinergic neurons can bind to the α-7 nicotinic acetylcholine receptors of nearby immune cells and suppress cytokine release in a reaction termed the "cholinergic anti-inflammatory pathway" [88]. In brief, the activation of the vagus nerve by pro-inflammatory cytokines leads to the release of ACh from efferent vagal nerve terminals. ACh then interacts with α7nAChR on macrophages, glial cells, and other non-neuronal cells, blocking the nuclear translocation of NF-κB, which induces the expression of many pro-inflammatory genes, including those encoding cytokines and chemokines. Additionally, this interaction triggers the phosphorylation of JAK2 and STAT3, promoting the movement of STAT3 into the nucleus, which then lowers the expression of various cytokines, including TNF-α, macrophage inflammatory protein 2, and IL-6 [89,90]. By inhibiting AChE, the breakdown of ACh can be prevented, potentially increasing its levels and facilitating its interaction with α7nAChR. By binding to α7nAChR, ACh may reduce the production of both local and peripheral pro-inflammatory cytokines, thereby decreasing local inflammation as well as the leakage of pro-inflammatory cytokines from the blood in conditions where the BBB integrity is compromised. Consequently, inflammation and the associated tissue damage are reduced.

NVC, defined as the tight relationship between neurons and haemodynamic responses, is crucial for brain function and is utilised in functional magnetic resonance imaging (fMRI), positron emission tomography (PET), and single-photon emission computed tomography (SPECT) brain imaging techniques, where neural activity is examined alongside vascular responses [91]. ACh has been shown to regulate cerebral blood flow in NVC. ACh exerts this effect through its vasodilatory properties, either by activating eNOS or indirectly by stimulating interneurons containing neuronal nitric oxide synthase (nNOS), which secrete NO [92]. This pathway appears to be one of the important factors in the preservation of BBB integrity in vascular dementia.

AChE inhibitors are divided into two categories: reversible and irreversible. Irreversible AChE inhibitors encompass compounds such as diazinon, diazoxon, and malathion diethylmaleate, which are organophosphorus compounds. As these irreversible inhibitors are generally associated with toxic effects, it is unlikely that they possess any therapeutic properties [93]. In contrast, reversible AChE inhibitors, such as donepezil, rivastigmine, and galantamine, improve ACh-mediated vasodilation by modulating eNOS activity and, as a consequence, enhance cerebral perfusion [94]. By increasing ACh levels, they may contribute to the activation of the cholinergic-anti-inflammatory pathway and attenuate or prevent neuroinflammation, a hallmark of cSVD. This effect is largely achieved by decreasing the release of pro-inflammatory cytokines such as TNF-α and IL-1β [95]. Moreover, ACh helps to preserve BBB integrity by reducing oxidative stress and modulating tight junctional complex formation [96].

4.2. *β-Hydroxy β-Methylglutaryl-CoA (HMG-CoA) Reductase Inhibitors*

HMG-CoA reductase inhibitors, a group of anti-hyperlipidaemic drugs also known as statins, are medications used to control the deleterious effects of lipids and overall cardiovascular disease risk. They act by competitively binding to the β-Hydroxy β-methylglutaryl-CoA reductase enzyme, which is involved in the rate-limiting step of the cholesterol synthesis pathway, thereby preventing the conversion of HMG-CoA to mevalonate. They reduce the intracellular cholesterol level and increase the expression of low-density lipoprotein (LDL) receptors on the cell surface, thus reducing plasma LDL levels [97,98]. Decreased plasma cholesterol levels activate sterol regulatory element binding protein-2 (SREBP-2), leading to upregulation of LDL receptor gene expression, which further diminishes LDL-C levels in the circulation [99]. In addition, inhibition of mevalonate synthesis reduces the synthesis of isoprenoid derivatives that are involved in post-translational modification of proteins involved in cell signalling. This accounts for the well-known pleiotropic effects of statins, including both anti-inflammatory and anti-proliferative effects [100]. Indeed, statins can reduce systemic inflammation and cytokine production by inhibiting the nuclear factor-kappa B (NF-κB) pathway. The pleiotropic benefits of statins also include a reduction in oxidative stress, improvement of endothelial function, and prevention of thrombus formation and platelet aggregation through modulation of isoprenoid intermediates [101,102]. Increased NO production by eNOS and decreased production of ROS by NADPH oxidase, a pro-oxidant enzyme, play a key role in statin-mediated improvements observed in endothelial function [103–105]. Due to their suppressive effects on ROS and ensuing oxidative damage, statins help maintain TJs between ECs [106,107].

4.3. *Lithium Drugs*

Lithium salts, especially lithium carbonate and lithium citrate, have been used as mood stabilisers in bipolar disorder for many years. They show these effects by regulating neural signalling and neuroplasticity through complex and multiple pathways [108]. Lithium inhibits inositol monophosphatase, the key enzyme in the phosphatidyl inositol pathway. As a result of this inhibition, it reduces the conversion of free inositol to inositol-1-phosphate, thus reducing the substrate required for phosphatidylinositol-4,5-bisphosphate synthesis. This diminishes inositol triphosphate and diacylglycerol phosphate activity, modulating calcium signalling and protein kinase C activity, which are known to elevate during early phases of ischaemic stroke due to BBB permeability [109,110]. Lithium also directly inhibits glycogen synthase kinase-3 (GSK-3), which is involved in many cellular processes such as the circadian rhythm, neurogenesis, and apoptosis. GSK-3 phosphorylation often interferes with the activity or stability of its targets, including glycogen synthase and β-catenin. Inhibiting GSK-3 enhances BBB function by stabilizing proteins that extend the half-life of tight junction proteins occludin and claudin-5. Furthermore, GSK-3 inhibition reduces the release of pro-inflammatory cytokines and adhesion molecules like ICAM-1 and VCAM-1 from endothelial cells. This also leads to a decrease in GSK-3-mediated immune-endothelial activity and leukocyte trans-endothelial migration. As indicated above, lithium enhances the activity of several transcription factors such as β-catenin and thus promotes cell survival and neuroplasticity [111,112]. Lithium also up-regulates brain-derived neurotrophic factor, which promotes hippocampal neurogenesis, neural survival, and synaptic plasticity [113,114]. By increasing oxidative phosphorylation and adenosine triphosphate production in mitochondria, lithium regulates cellular energy homoeostasis, as required for high neural activity [115,116]. In addition, it reduces ROS production by inducing superoxide dismutase activity [117,118] and improves vascular relaxation and CBF by enhancing the bioavailability of NO through a mechanism involving an increase in β-catenin levels [119]. Through mediating a decrease in occludin and claudin-5 turnover, via the inhibition of GSK-3, lithium has also been coupled with BBB integrity and overall vascular homeostasis [120].

4.4. Phosphodiesterase Inhibitors

Phosphodiesterase (PDE) inhibitors are used in the treatment of many diseases such as cardiovascular diseases, erectile dysfunction, pulmonary hypertension, and inflammatory disorders. PDE inhibitors selectively target each isoform of the phosphodiesterase enzyme family, which includes 11 isoforms [121]. Drugs that target the PDE3 isoform in cardiovascular tissue, such as cilostazol and milrinone, increase cyclic adenosine monophosphate (cAMP) levels, whereby they evoke protein kinase A (PKA) activity and augment vasodilatory responses [122]. The PDE4 isoform is primarily localised within immune and inflammatory cells. As anticipated, PDE4 inhibitors trigger anti-inflammatory responses by elevating cAMP levels and diminishing the production of major pro-inflammatory cytokines, including TNF-α and IL-6. PDE4 inhibitors prevent chronic inflammation in NVU and are commonly preferred therapeutics for chronic obstructive pulmonary disease and asthma [123]. The PDE5 isoform is highly expressed in vascular smooth muscle and promotes cGMP hydrolysis. It provides relaxation and vasodilation of smooth muscle by stimulating the NO-cGMP signalling pathway. It is widely used in the treatment of erectile dysfunction [124]. By increasing cGMP levels, PDE5 inhibitors also stimulate microvascular perfusion and decrease the risk of ischaemic damage [125,126]. While PDE1 and PDE2 inhibitors are believed to have beneficial effects on neural signalling and cardiac function by modulating the cAMP and cGMP pathways due to their dual-specific functions, PDE7 and PDE8 inhibitors are considered for the treatment of autoimmune diseases and neuroinflammatory conditions because they selectively regulate cAMP levels in immune cells [127,128]. PDE inhibitors that are effective on PDE3, PDE4, and PDE5 also support neurogenesis and synaptic plasticity by modulating the cAMP-responsive element-binding protein (CREB) pathway [129].

Data pertaining to the effects of PDE inhibitors on WMHs remain inconclusive. Indeed, while treatment with ibudilast, a PDE4 inhibitor, has been reported to reduce lesions in rats subjected to chronic hypoperfusion by about 50% and 70% at doses of 30 mg/kg and 60 mg/kg, respectively [130], treatment with another PDE4 inhibitor, rolipram, has appeared to exacerbate periventricular WMHs in rat pups [131]. Despite not affecting WMH progression in stroke- and dementia-free subjects, the PDE3 inhibitor cilostazol has demonstrated an acceptable safety profile [132]. In a small-sized clinical study, tadalafil, a PDE5 inhibitor, has been shown to improve perfusion in the cerebral microvasculature in patients with cerebral small vessel disease stroke. The increased oxygenation positions tadalafil as an attractive new therapeutic target in cSVD [133]. However, well-designed future studies are needed to evaluate the effects of long-term PDE inhibitor treatment on cerebrovascular reactivity, endothelial function, and general neurovascular changes, including WMHs in cSVD and stroke.

4.5. Oral Antihyperglycaemic Drugs

Diabetes mellitus (DM), characterised by chronic hyperglycaemia, causes oxidative stress, vascular inflammation, and endothelial dysfunction and constitutes one of the main risk factors for cSVD [134]. Oral forms of sodium–glucose cotransporter-2 (SGLT2) inhibitors and glucagon-like peptide-1 receptor agonists (GLP-1 RAs) represent the commonly prescribed medications to control this condition [135,136]. High levels of blood glucose impair endothelial NO production, reduce vascular reactivity, increase permeability, and potentiate the production of several pro-inflammatory cytokines such as TNF-α and IL-6 [45]. Protein glycation and the formation of advanced glycation end products

(AGEs) play a critical role in the development of various hyperglycaemia-induced medical complications like retinopathy, cardiomyopathy, and neuropathy. Increased production of ROS, mediated in a large part by the activation of PKC and NADPH oxidase, are also known to contribute to neurovascular degeneration [137–140]. Impaired insulin signalling and associated insulin resistance in diabetics further exacerbate cerebral and neuronal damage [134].

Metformin, a commonly prescribed biguanide, has an effect by activating the AMP-activated protein kinase. Activation of this enzyme leads to decreased expression of gluconeogenic enzymes like phosphoenolpyruvate carboxykinase and glucose-6-phosphatase, preventing hepatic glucose output, and augments insulin sensitivity in peripheral tissues [141]. Therefore, it aids in restoring endothelial function by lowering the levels of ROS and inflammatory substances. These concurrent reductions in oxidative stress and inflammation enhance cerebral perfusion and abate WMHs [142]. Similar to GLP-1 RAs, SGLT2 inhibitors also exert their beneficial effects on neurovasculature by reducing oxidative stress and inflammation. Reduced neuroinflammation in turn leads to diminished WMH levels and BBB permeability [135,143]. Another group of oral antidiabetic agents, namely dipeptidyl peptidase-4 (DPP-4) inhibitors, also elevate insulin sensitivity by increasing endogenous GLP-1 levels. As a result, they reduce inflammation and improve endothelial function [144].

4.6. Tetracycline Antibiotics

Beyond their antimicrobial effects, tetracycline antibiotics show anti-inflammatory properties and can modulate MMP activity in the host organism. By chelating the zinc and calcium ions, essential minerals for MMP function, tetracyclines inhibit MMP activity [145]. Due to this inhibitory effect on MMPs and their modulatory effects on mitochondrial function, tetracyclins exhibit anti-inflammatory properties and help successfully reduce chronic neuroinflammation by suppressing pro-inflammatory mediators such as TNF-α and IL-6 [146]. By inhibiting MMPs, tetracyclines prevent the degradation of extracellular matrix components and TJ proteins and thus help preserve BBB integrity [147,148]. Tetracyclines tend to accumulate in host mitochondria owing to their similarity to bacterial ribosomes and mitochondrial ribosomes. They can affect mitochondrial respiration and ATP production by inhibiting mitochondrial protein synthesis. As a result, they reduce mitochondrial ROS production, prevent oxidative damage, and attenuate microglial activation [149,150]. Additionally, they contribute to the preservation of white matter integrity by regulating apoptotic pathways [151].

4.7. Anti-Anginal Drugs

cSVD and coronary artery disease share many aetiological similarities, including endothelial dysfunction, impaired autoregulation, and ischaemia. Hence, pharmacological interventions used to treat angina are expected to produce positive responses in case of cSVD [152,153]. Pharmacological agents utilised for the treatment of angina pectoris encompass a variety of drug classes, including nitrates, calcium channel blockers, and β-adrenergic antagonists, as well as specific medications like ranolazine, nicorandil, and ivabradine.

4.7.1. Nitrates

NO donors, such as nitrates, enhance vasodilation and reduce vascular resistance. In smooth muscle, NO activates soluble guanylyl cyclase, stimulating the production of cGMP. Increased levels of cGMP activate cGMP-dependent protein kinase (PKG). cGMP and PKG reduce the Ca^{2+} sensitivity of smooth muscle. Ca^{2+} causes a decrease in the activity of calmodulin-activated myosin light-chain kinase and muscle relaxation [154,155]. Thus, they regulate CBF and mitigate ischaemic damage [156]. Additionally, nitrates may have a protective effect by reducing the excessive Ca^{2+} influx seen in nerve damage [157].

4.7.2. Calcium Channel Blockers

Calcium channel blockers (CCBs) reduce the influx of Ca^{2+} into vascular smooth muscle, decrease vascular tone, and, consequently, increase blood flow [158]. By preventing the entry of extracellular Ca^{2+} into cells through voltage-gated Ca^{2+} channels, CCBs markedly decrease peripheral vascular resistance and blood pressure. While CCBs like nifedipine and felodipine exert vasodilator effects by inhibiting high-voltage-activated L-type channels, those such as amiodipine and clinidipine block high-voltage-activated N-type channels and reduce the release of norepinephrine, a vasoconstrictor, from sympathetic nerve endings. CCBs such as efonidipine and azeinidipine, on the other hand, inhibit low-voltage-activated channels. It is therefore thought that CCBs that inhibit L- and N-type channels may have organ-protective effects [159]. By blocking the entry of basal Ca^{2+} into cerebral arteries, they can increase CBF and protect the neurovascular system against various cerebral degenerative diseases, such as subarachnoid haemorrhage, known to be associated with excessive availability of intracellular Ca^{2+} [158].

4.7.3. β-Adrenergic Receptor Antagonists

The β-adrenergic receptor antagonists (β-blockers) are a family of agents that are used to treat hypertension and cardiac arrhythmias. The β-blockers compete with β-adrenergic agonists, e.g., epinephrine and norepinephrine, to bind beta-receptor sites. They are often divided into "selective" and "non-selective" agents based on their ability to block β-1 receptors that are found in cardiac muscle and β-2 receptors found in bronchial and smooth muscles. While β-1 selective blockers are used to treat heart disease, the nonselective β-blockers are predominantly used to control recurrent haemorrhages observed in patients with cirrhosis and portal hypertension [160,161].

Blockade of β1-adrenergic receptors aids vascular relaxation by preventing hyperpolarization of intermediate-conductance Ca^{2+}-activated potassium channels (IKCa) in vascular ECs [162]. Another study investigating the effects of selective and non-selective α-1, β-1, α-2, and β-2 blockers on regional CBF responses to hyperbaric hyperoxia concluded that only propranolol, a non-selective β-1 and β-2 blocker, regulated CBF [163].

4.7.4. Others

Ranolazine, an agent used to treat chronic angina, has been shown to mitigate myocardial infarction by suppressing late sodium ion channel activity, which leads reduced intracellular and mitochondrial Ca^{2+}. As a result, it prevents cell damage by inhibiting the suppression of the ROS-scavenging feature of mitochondria. This property suggests that it may enhance mitochondrial function and attenuate neuroinflammation in cerebral ischaemia [164]. Ivabradine, a hyperpolarization activated cyclic nucleotide-gated channel blocker, specifically suppresses the electrical current in the sinoatrial node to slow the heart rate, whereby it enhances diastolic perfusion and, as a consequence, improves cerebral perfusion [165]. Nicorandil, a vasodilator used to treat angina, functions as both a nitrate and a potassium channel activator. Its nitrate characteristic facilitates vasodilation by increasing NO levels, while its potassium channel opening action enhances smooth muscle relaxation through cellular hyperpolarization [166].

5. cSVD and Clinical Trials

The findings of the clinical trials conducted with the drugs mentioned in this review and the cSVD keywords are summarised in Table 1. Clinical studies mostly investigate the efficacy of drugs. However, a few of them also investigate the basic pathologies of the disease. The number of studies on cSVD appears to have risen significantly since 2020.

Table 1. Pharmacological interventions with agents targeting cSVD or associated cognitive impairments.

Clinical Trial ID	Study Focus	Findings	Ref.
ChiCTR-IOR-17013557/ChiCTR-EOC-017013598	Statin therapy in adults $\geq$ 75 years with cSVD	Statin therapy alleviated the progression of WMH, lacunes, and EPS without increasing the risk of CMHs, suggesting it as an efficient and safe intervention for cSVD in older adults.	[167]
NCT00103948	Efficacy, safety, and tolerability of donepezil HCl in patients with CADASIL who have cognitive impairment	While no change was observed in the vascular AD assessment scale cognitive subscale, an improvement was observed in the Trail Making Test Part B.	[168]
NCT02444637	Effectiveness of Rivastigmine in subjects with AD and cerebrovascular disease	No results posted	[1]
NCT00163150	Evaluation of the efficacy of atorvastatin treatment for 3 months (80 mg/day) on cerebral vasoreactivity in lacunar patients	There was no positive effect of 3-month treatment with atorvastatin on severe cerebral microvasculature endothelial dysfunction in patients with lacunar stroke.	[169]
NCT03855332	Examines the effects of sildenafil, a PDE5 inhibitor, on cerebrovascular pulsatility and reactivity, and compares its efficacy to that of cilostazol, a PDE3 inhibitor	Lowering cerebral pulsatility and improving cerebrovascular reactivity with sildenafil treatment may indicate potential benefits in preventing small vessel disease progression.	[170]
NCT06175663	Studies abnormalities in brain structure and vascular function in young hypertensive patients. Also determines the effects of blood pressure changes, during a period of withdrawal and restart of blood pressure medications, on brain structure and vascular function	No results posted	[1]
NCT05169021	Evaluates the impact of intensive antihypertensive therapy versus standard antihypertensive therapy in reducing the risk of combined cardiovascular and cerebrovascular events in cSVD patients with hypertension and elevated homocysteine levels, using two common antihypertensive drugs: amlodipine 5 mg tablets or amlodipine–folic acid 5.8 mg tablets	No results posted	[1]
NCT03451591	Investigates the feasibility, safety, and effects of a 1-year treatment using ISMN and cilostazol on blood vessel health, daily functioning, and cognitive abilities in patients who have had a lacunar stroke	The LACI-2 trial was successful, and ISMN and cilostazol were safe and well-tolerated. These medications may help prevent additional strokes, disability, and cognitive issues after lacunar stroke. They can also avoid other negative outcomes in cerebral small vessel disease.	[171]

Table 1. *Cont.*

Clinical Trial ID	Study Focus	Findings	Ref.
NCT06715007	Evaluates the effect of different antiplatelet agents (e.g., cilostazol) on cSVD and the retina in patients with cSVD (recent small subcortical infarcts or WMH, respectively)	No results posted	[1]
NCT04753970	Tests whether impaired cerebral and retinal vasoreactivity may serve as biomarkers for SVD progression, and evaluates the safety and efficacy of cilostazol (antiplatelet agent with vasodilatory and anti-inflammatory properties) for the treatment of SVD	No results posted	[1]
NCT02481323	Investigates if short-term ISMN and cilostazol, alone or in combination, improves magnetic resonance imaging-measured cerebrovascular function in patients with lacunar ischemic stroke	It is demonstrated that measuring cerebrovascular function with magnetic resonance imaging is feasible in clinical trials and that ISMN and cilostazol may improve cerebrovascular function.	[172]
NCT01932203	Assesses if there is difference in the efficacy of cilostazol and aspirin in impacting the progression of white matter changes in cSVD	No results posted	[1]
NCT00741286	Investigates if cilostazol decrease the vascular resistance in patients with acute lacunar infarction and decreases the PIs in patients with acute lacunar infarction	Cilostazol reduced transcranial Doppler PIs more than placebo at 90 days in acute lacunar infarction. This may be due to its additional effects, like vasodilation, beyond just its antiplatelet action.	[173]
NCT06649240	Investigates the optimal LDL-cholesterol target for patients with small vessel occlusion stroke	No results posted	[1]
NCT06031610	Determines if carotid artery stenosis revascularization, compared to standard medication treatment alone, can effectively reduce the progression of cSVD burden and improve the severity of retinal issues and cognitive impairment	No results posted	[1]
NCT05356104	Investigates the safety and efficacy of the GLP-1R agonist in patients with moderate-to-severe cSVD	No results posted	[1]
NCT05710367	Tests dapagliflizone in Fabry patients for kidney and heart function improvement	No results posted	[1]

[1] Relevant clinical trial information can be found at www.clinicaltrials.gov (accessed on 15 February 2025) via the clinical trial ID. cSVD: cerebral small vessel disease, CMHs: Cerebral microhaemorrhages, WMH: White matter hyperintensity, AD: Alzheimer's disease, ISMN: Isosorbide mononitrate, EPS: Enlarged perivascular space, PDE: Phosphodiesterase, PI: Pulsatility index, LDL: Low-density lipoprotein, LACI-2: Lacunar Intervention Trial-2, GLP-1R: Glucagon-like peptide-1 receptor.

6. Discussion and Conclusions

Cardiovascular diseases remain the leading cause of death globally, with dementia now following closely. Due to an increase in life expectancy, dementia has actually become the primary cause of death in some countries. Many risk factors such as hypertension, hyperglycaemia, smoking, and sleep apnoea contribute to the rise of both cardiovascular diseases and dementia, which appear to share similar changes in vascular structure and function. Lacunar stroke is one of the subtypes of ischemic stroke. It is characterised by the appearance of lacunar infarcts, which may represent the first recognisable sign of cSVD, affecting the small perforating arteries, capillaries, and venules of the brain. Owing to the spatial resolution issues in neurovascular imaging, infrequency of pathological studies, and lack of appropriate experimental model availability, the aetiology of a perforating artery occlusion remains unclear [174]. Even so, the pathological changes observed in cSVD are known to include oxidative stress, inflammation, ischaemia due to nitric oxide deficiency, loss of tight junction proteins in ECs caused by MMP activation, and, ultimately, endothelial dysfunction, an early change in cSVD. Endothelial dysfunction primarily disrupts the BBB permeability and accelerates the leakage of toxic components from the blood to the brain parenchyma. This triggers neuroinflammation and neurodegeneration and, as a consequence, promotes neurological damage and dementia. Scrutiny of the therapeutic options that can preserve BBB integrity and normal endothelial function is of crucial importance for patients diagnosed with cSVD.

Advancements in drug repositioning studies and emerging technologies have led to a better understanding of the mechanisms underlying both diseases and the available drugs. This review discusses medications currently used to treat dementia and neurological disorders, as well as those preferred for managing vascular diseases and hyperglycaemia, which is closely associated with cSVD. As summarised in Table 2, these compounds have been found to promote vasodilation by increasing NO levels through various pathways, and to decrease levels of ROS and pro-inflammatory cytokines. Tetracycline antibiotics and lithium drugs have also been observed to preserve BBB integrity by inhibiting MMPs. Investigating the direct relationship between these drugs and cSVD is necessary. Determining their effects on endothelial function and the NVU presents a valuable first step in this regard. In vitro BBB models are among the most valuable tools for investigating diseases like cSVD, whose underlying molecular mechanisms remain incompletely elucidated, as well as for examining the effects of pharmacological interventions. Transwell-based triple or quadruple co-culture systems derived from human primary cells represent the optimal models in this context, as they are time-efficient, cost-effective, and enable detailed investigation of the molecular pathways involved. The data obtained from these studies will provide a foundation for a healthier and more ethically grounded transition to translational and clinical research. In conclusion, there is a need to improve our understanding of the molecular pathologies underlying cSVD and explore potential pharmacological interventions through well-designed preclinical studies. The data generated will help inform the design of future clinical studies. For an update on lacunar stroke pathophysiology and a clinical perspective on managing lacunar strokes, readers are advised to read recent reviews discussing the relevant imaging and translational studies [174].

Table 2. The effects of pharmacological agents included in the review on cerebral small vessel disease (cSVD).

Drug Category	Major Effects
Acetylcholine Esterase Inhibitors	• Enhance ACh levels in synaptic cleft. • Improve cholinergic signalling. • Reduce neuroinflammation (via cholinergic anti-inflammatory pathway). • Improve BBB integrity and reduce oxidative stress. • Examples: Donepezil, Rivastigmine, Galantamine.
HMG-CoA Reductase Inhibitors (Statins)	• Lower LDL cholesterol levels. • Anti-inflammatory and anti-proliferative effects. • Improve endothelial function and NO bioavailability. • Reduce oxidative stress and enhance BBB integrity. • Examples: Atorvastatin, Simvastatin.

Table 2. *Cont.*

Drug Category	Major Effects
Lithium Drugs	• Regulate neural signalling via GSK-3 inhibition. • Promote neuroplasticity and neurogenesis. • Reduce ROS production and oxidative stress. • Enhance mitochondrial function and BBB integrity. • Examples: Lithium Carbonate, Lithium Citrate.
Phosphodiesterase Inhibitors	• Increase cAMP/cGMP levels. • Promote NO-mediated vasodilation. • Reduce inflammation by lowering TNF-α and IL-6. • Preserve BBB integrity. • Examples: PDE3 (Cilostazol), PDE4 (Roflumilast), PDE5 (Sildenafil).
Oral Antihyperglycaemic Drugs	• Reduce oxidative stress and vascular inflammation. • Improve endothelial function and insulin sensitivity. • Examples: Metformin (AMPK activator), SGLT2 inhibitors, GLP-1 receptor agonists, DPP-4 inhibitors.
Tetracycline Antibiotics	• Anti-microbial. • Anti-inflammatory via MMP inhibition. • Reduce ROS and preserve BBB integrity. • Prevent microglial activation and oxidative damage. • Examples: Doxycycline, Minocycline.
Anti-Anginal Drugs	• Enhance vasodilation via NO and cGMP pathways. • Improve cerebral perfusion and reduce ischaemic damage. • Examples: Nitrates (Nitroglycerin), Calcium Channel Blockers, β-blockers, Ranolazine, Ivabradine, Nicorandil.

ACh: Acetylcholine, LDL: Low-density lipoprotein, NO: Nitric oxide, BBB: Blood–brain barrier, GSK-3: Glycogen synthase kinase-3, ROS: Reactive oxygen species, cAMP: Cyclic adenosine monophosphate, cGMP: Cyclic guanosine monophosphate, PDE: Phosphodiesterase, AMPK: AMP-activated protein kinase, SGLT2: Sodium–glucose cotransporter-2, GLP-1: Glucagon-like peptide-1, DPP-4: Dipeptidyl peptidase-4.

Author Contributions: Visualization H.Y. and U.B.; writing—original draft H.Y.; supervision, conceptualization, and writing—review and editing, U.B. All authors have read and agreed to the published version of the manuscript.

Funding: This research received no external funding.

Acknowledgments: Habibe Yılmaz would like to thank the Scientific and Technological Research Council of Türkiye (TÜBİTAK) for supporting her postdoctoral research activities at the University of Nottingham Academic Unit of Mental Health and Clinical Neurosciences within the framework of the 2219 programme.

Conflicts of Interest: The authors declare no conflicts of interest.

References

1. Available online: https://www.who.int/news-room/fact-sheets/detail/the-top-10-causes-of-death (accessed on 29 October 2024).
2. Available online: https://www.emro.who.int/health-topics/stroke-cerebrovascular-accident/index.html (accessed on 29 October 2024).
3. Available online: https://www.stroke.org.uk/stroke/statistics (accessed on 29 October 2024).
4. King, D.; Wittenberg, R.; Pate, A.; Quayyum, Z.; Berdunov, V.; Knapp, M. The future incidence, prevalence and costs of stroke in the UK. *Age Ageing* **2020**, *49*, 277–282. [CrossRef] [PubMed]
5. Carroll, K.A.L.; Chataway, J. Understanding stroke: Pathophysiology, presentation, and investigation. *BMJ* **2006**, *333*, 0609319. [CrossRef]
6. Maida, C.D.; Norrito, R.L.; Rizzica, S.; Mazzola, M.; Scarantino, E.R.; Tuttolomondo, A. Molecular Pathogenesis of Ischemic and Hemorrhagic Strokes: Background and Therapeutic Approaches. *Int. J. Mol. Sci.* **2024**, *25*, 6297. [CrossRef] [PubMed]
7. Kadir, R.R.A.; Alwjwaj, M.; Bayraktutan, U. MicroRNA: An Emerging Predictive, Diagnostic, Prognostic and Therapeutic Strategy in Ischaemic Stroke. *Cell Mol. Neurobiol.* **2022**, *42*, 1301–1319. [CrossRef]

8. Markus, H.S.; de Leeuw, F.E. Cerebral small vessel disease: Recent advances and future directions. *Int. J. Stroke* **2023**, *18*, 4–14. [CrossRef]
9. Ya, J.; Whitby, A.; Bayraktutan, U. Metabolites and Metabolic Functional Changes-Potential Markers for Endothelial Cell Senescence. *Biomolecules* **2024**, *14*, 1476. [CrossRef]
10. Cannistraro, R.J.; Badi, M.; Eidelman, B.H.; Dickson, D.W.; Middlebrooks, E.H.; Meschia, J.F. CNS small vessel disease: A clinical review. *Neurology* **2019**, *92*, 1146–1156. [CrossRef]
11. Higashi, Y. Roles of Oxidative Stress and Inflammation in Vascular Endothelial Dysfunction-Related Disease. *Antioxidants* **2022**, *11*, 1958. [CrossRef]
12. Bath, P.M.; Wardlaw, J.M. Pharmacological treatment and prevention of cerebral small vessel disease: A review of potential interventions. *Int. J. Stroke* **2015**, *10*, 469–478. [CrossRef]
13. Al-Thani, M.; Goodwin-Trotman, M.; Bell, S.; Patel, K.; Fleming, L.K.; Vilain, C.; Abramowicz, M.; Allan, S.M.; Wang, T.; Cader, M.Z.; et al. A novel human iPSC model of COL4A1/A2 small vessel disease unveils a key pathogenic role of matrix metalloproteinases. *Stem Cell Rep.* **2023**, *18*, 2386–2399. [CrossRef]
14. Zhang, W.; Xiao, D.; Mao, Q.; Xia, H. Role of neuroinflammation in neurodegeneration development. *Signal Transduct. Target. Ther.* **2023**, *8*, 267. [CrossRef] [PubMed]
15. Ridker, P.M.; Lüscher, T.F. Anti-inflammatory therapies for cardiovascular disease. *Eur. Heart J.* **2014**, *35*, 1782–1791. [CrossRef] [PubMed]
16. Goldstein, E.D.; Wolcott, Z.; Garg, G.; Navarro, K.; Delic, A.; Yaghi, S.; Sederholm, B.; Prabhakaran, S.; Wong, K.H.; McLean, K.; et al. Effect of Antihypertensives by Class on Cerebral Small Vessel Disease: A Post Hoc Analysis of SPRINT-MIND. *Stroke* **2022**, *53*, 2435–2440. [CrossRef]
17. Inzitari, D.; Poggesi, A. Calcium channel blockers and stroke. *Aging Clin. Exp. Res.* **2005**, *17* (Suppl. S4), 16–30. [PubMed]
18. Nelissen, E.; Schepers, M.; Ponsaerts, L.; Foulquier, S.; Bronckaers, A.; Vanmierlo, T.; Sandner, P.; Prickaerts, J. Soluble guanylyl cyclase: A novel target for the treatment of vascular cognitive impairment? *Pharmacol. Res.* **2023**, *97*, 106970. [CrossRef]
19. Singh, A.; Bonnell, G.; De Prey, J.; Buchwald, N.; Eskander, K.; Kincaid, K.J.; Wilson, C.A. Small-vessel disease in the brain. *Am. Heart J. Plus* **2023**, *27*, 100277. [CrossRef] [PubMed]
20. Shi, Y.; Wardlaw, J.M. Update on cerebral small vessel disease: A dynamic whole-brain disease. *Stroke Vasc. Neurol.* **2016**, *1*, 83–92. [CrossRef]
21. Liu, X.; Sun, P.; Yang, J.; Fan, Y. Biomarkers involved in the pathogenesis of cerebral small-vessel disease. *Front. Neurol.* **2022**, *13*, 969185. [CrossRef]
22. Debette, S.; Markus, H.S. The clinical importance of white matter hyperintensities on brain magnetic resonance imaging: Systematic review and meta-analysis. *BMJ* **2010**, *341*, c3666. [CrossRef]
23. Pantoni, L. Cerebral small vessel disease: From pathogenesis and clinical characteristics to therapeutic challenges. *Lancet Neurol.* **2010**, *9*, 689–701. [CrossRef]
24. Charidimou, A.; Imaizumi, T.; Moulin, S.; Biffi, A.; Samarasekera, N.; Yakushiji, Y.; Peeters, A.; Vandermeeren, Y.; Laloux, P.; Baron, J.C.; et al. Brain hemorrhage recurrence, small vessel disease type, and cerebral microbleeds: A meta-analysis. *Neurology* **2017**, *89*, 820–829. [CrossRef] [PubMed]
25. Wardlaw, J.M.; Benveniste, H.; Nedergaard, M.; Zlokovic, B.V.; Mestre, H.; Lee, H.; Doubal, F.N.; Brown, R.; Ramirez, J.; MacIntosh, B.J.; et al. Perivascular spaces in the brain: Anatomy, physiology and pathology. *Nat. Rev. Neurol.* **2020**, *16*, 137–153. [CrossRef] [PubMed]
26. Khalil, M.; Teunissen, C.E.; Otto, M.; Piehl, F.; Sormani, M.P.; Gattringer, T.; Barro, C.; Kappos, L.; Comabella, M.; Fazekas, F.; et al. Neurofilaments as biomarkers in neurological disorders. *Nat. Rev. Neurol.* **2018**, *14*, 577–589. [CrossRef]
27. Snyder, H.M.; Corriveau, R.A.; Craft, S.; Faber, J.E.; Greenberg, S.M.; Knopman, D.; Lamb, B.T.; Montine, T.J.; Nedergaard, M.; Schaffer, C.B.; et al. Vascular contributions to cognitive impairment and dementia including Alzheimer's disease. *Alzheimers Dement.* **2015**, *11*, 710–717. [CrossRef]
28. Wan, S.; Dandu, C.; Han, G.; Guo, Y.; Ding, Y.; Song, H.; Meng, R. Plasma inflammatory biomarkers in cerebral small vessel disease: A review. *CNS Neurosci. Ther.* **2023**, *29*, 498–515. [CrossRef] [PubMed]
29. Gao, X.; Bayraktutan, U. TNF-α evokes blood-brain barrier dysfunction through activation of Rho-kinase and neurokinin 1 receptor. *Immunobiology* **2023**, *228*, 152706. [CrossRef]
30. Kadir, R.R.A.; Alwjwaj, M.; Rakkar, K.; Othman, O.A.; Sprigg, N.; Bath, P.M.; Bayraktutan, U. Outgrowth Endothelial Cell Conditioned Medium Negates TNF-α-Evoked Cerebral Barrier Damage: A Reverse Translational Research to Explore Mechanisms. *Stem Cell Rev. Rep.* **2023**, *19*, 503–515. [CrossRef]
31. Poggesi, A.; Pasi, M.; Pescini, F.; Pantoni, L.; Inzitari, D. Circulating biologic markers of endothelial dysfunction in cerebral small vessel disease: A review. *J. Cereb. Blood Flow. Metab.* **2016**, *36*, 72–94. [CrossRef]
32. Chabriat, H.; Joutel, A.; Dichgans, M.; Tournier-Lasserve, E.; Bousser, M.G. Cadasil. *Lancet Neurol.* **2009**, *8*, 643–653. [CrossRef]

33. Liu, J.Y.; Zhu, Y.C.; Zhou, L.X.; Wei, Y.P.; Mao, C.H.; Cui, L.Y.; Peng, B.; Yao, M. HTRA1-related autosomal dominant cerebral small vessel disease. *Chin. Med. J.* **2020**, *134*, 178–184. [CrossRef]
34. Wilms, A.E.; de Boer, I.; Terwindt, G.M. Retinal Vasculopathy with Cerebral Leukoencephalopathy and Systemic manifestations (RVCL-S): An update on basic science and clinical perspectives. *Cereb. Circ. Cogn. Behav.* **2022**, *3*, 100046. [CrossRef]
35. Guey, S.; Hervé, D. Main features of COL4A1-COL4A2 related cerebral microangiopathies. *Cereb. Circ. Cogn. Behav.* **2022**, *3*, 100140. [CrossRef]
36. Tapia, D.; Floriolli, D.; Han, E.; Lee, G.; Paganini-Hill, A.; Wang, S.; Zandihaghighi, S.; Kimonis, V.; Fisher, M. Prevalence of cerebral small vessel disease in a Fabry disease cohort. *Mol. Genet. Metab. Rep.* **2021**, *29*, 100815. [CrossRef] [PubMed]
37. Yektay Farahmand, M.; Wasselius, J.; Englund, E.; Braverman, I.; Puschmann, A.; Ilinca, A. Small vessel disease in primary familial brain calcification with novel truncating PDGFB variants. *Neurol. Neurochir. Pol.* **2024**, *58*, 94–105. [CrossRef] [PubMed]
38. Evans, L.E.; Taylor, J.L.; Smith, C.J.; Pritchard, H.A.T.; Greenstein, A.S.; Allan, S.M. Cardiovascular comorbidities, inflammation, and cerebral small vessel disease. *Cardiovasc. Res.* **2021**, *117*, 2575–2588. [CrossRef]
39. Nezu, T.; Hosomi, N.; Aoki, S.; Kubo, S.; Araki, M.; Mukai, T.; Takahashi, T.; Maruyama, H.; Higashi, Y.; Matsumoto, M. Endothelial dysfunction is associated with the severity of cerebral small vessel disease. *Hypertens. Res.* **2015**, *38*, 291–297. [CrossRef] [PubMed]
40. Srivastava, K.; Bath, P.M.W.; Bayraktutan, U. Current Therapeutic Strategies to Mitigate the eNOS Dysfunction in Ischaemic Stroke. *Cell Mol. Neurobiol.* **2012**, *32*, 319–336. [CrossRef]
41. Quick, S.; Moss, J.; Rajani, R.M.; Williams, A. A Vessel for Change: Endothelial Dysfunction in Cerebral Small Vessel Disease. *Trends Neurosci.* **2021**, *44*, 289–305. [CrossRef]
42. Bai, T.; Yu, S.; Feng, J. Advances in the Role of Endothelial Cells in Cerebral Small Vessel Disease. *Front. Neurol.* **2022**, *13*, 861714. [CrossRef]
43. Abdullah, Z.; Bayraktutan, U. Suppression of PKC-α attenuates TNF-α-evoked cerebral barrier breakdown via regulations of MMP-2 and plasminogen-plasmin system. *Biochim. Biophys. Acta* **2016**, *1862*, 1354–1366. [CrossRef]
44. Mason, W.P. Blood-brain barrier-associated efflux transporters: A significant but underappreciated obstacle to drug development in glioblastoma. *Neuro Oncol.* **2015**, *17*, 1181–1182. [CrossRef] [PubMed]
45. Hashmat, A.; Ya, J.; Kadir, R.; Alwjwaj, M.; Bayraktutan, U. Hyperglycaemia perturbs blood-brain barrier integrity through its effects on endothelial cell characteristics and function. *Tissue Barriers* **2024**, *7*, 2350821. [CrossRef]
46. Wu, D.; Chen, Q.; Chen, X.; Han, F.; Chen, Z.; Wang, Y. The blood-brain barrier: Structure, regulation, and drug delivery. *Signal Transduct. Target. Ther.* **2023**, *8*, 217. [CrossRef] [PubMed]
47. Kadry, H.; Noorani, B.; Cucullo, L. A blood-brain barrier overview on structure, function, impairment, and biomarkers of integrity. *Fluids Barriers CNS* **2020**, *17*, 69. [CrossRef] [PubMed]
48. Wardlaw, J.M.; Makin, S.J.; Valdés Hernández, M.C.; Armitage, P.A.; Heye, A.K.; Chappell, F.M.; Muñoz-Maniega, S.; Sakka, E.; Shuler, K.; Dennis, M.S.; et al. Blood-brain barrier failure as a core mechanism in cerebral small vessel disease and dementia: Evidence from a cohort study. *Alzheimers Dement.* **2017**, *13*, 634–643. [CrossRef]
49. Jia, R.; Solé-Guardia, G.; Kiliaan, A.J. Blood-brain barrier pathology in cerebral small vessel disease. *Neural Regen. Res.* **2024**, *19*, 1233–1240. [CrossRef]
50. Kim, S.; Jung, U.J.; Kim, S.R. Role of Oxidative Stress in Blood–Brain Barrier Disruption and Neurodegenerative Diseases. *Antioxid* **2024**, *13*, 1462. [CrossRef] [PubMed]
51. Song, K.; Li, Y.; Zhang, H.; An, N.; Wei, Y.; Wang, L.; Tian, C.; Yuan, M.; Sun, Y.; Xing, Y.; et al. Oxidative Stress-Mediated Blood-Brain Barrier (BBB) Disruption in Neurological Diseases. *Oxid. Med. Cell Longev.* **2020**, *2020*, 4356386. [CrossRef]
52. Li, Y.; Li, M.; Zuo, L.; Shi, Q.; Qin, W.; Yang, L.; Jiang, T.; Hu, W. Compromised Blood-Brain Barrier Integrity Is Associated With Total Magnetic Resonance Imaging Burden of Cerebral Small Vessel Disease. *Front. Neurol.* **2018**, *9*, 221. [CrossRef]
53. Ying, Y.; Li, Y.; Yao, T.; Shao, X.; Tang, W.; Montagne, A.; Chabriat, H.; Wang, D.J.J.; Wang, C.; Yang, Q.; et al. Heterogeneous blood-brain barrier dysfunction in cerebral small vessel diseases. *Alzheimers Dement.* **2024**, *20*, 4527–4539. [CrossRef]
54. Helms, H.C.; Abbott, N.J.; Burek, M.; Cecchelli, R.; Couraud, P.O.; Deli, M.A.; Förster, C.; Galla, H.J.; Romero, I.A.; Shusta, E.V.; et al. In vitro models of the blood-brain barrier: An overview of commonly used brain endothelial cell culture models and guidelines for their use. *J. Cereb. Blood Flow. Metab.* **2016**, *36*, 862–890. [CrossRef]
55. Allen, C.L.; Bayraktutan, U. Antioxidants attenuate hyperglycaemia-mediated brain endothelial cell dysfunction and blood-brain barrier hyperpermeability. *Diabetes Obes. Metab.* **2009**, *11*, 480–490. [CrossRef] [PubMed]
56. Kadir, R.R.A.; Alwjwaj, M.; Bayraktutan, U. Establishment of an In Vitro Model of Human Blood-Brain Barrier to Study the Impact of Ischemic Injury. *Methods Mol. Biol.* **2022**, *2492*, 143–155. [CrossRef]
57. Stone, N.L.; England, T.J.; O'Sullivan, S.E. A Novel Transwell Blood Brain Barrier Model Using Primary Human Cells. *Front. Cell Neurosci.* **2019**, *13*, 230. [CrossRef] [PubMed]
58. Tran, M.; Heo, C.; Lee, L.P.; Cho, H. Human mini-blood-brain barrier models for biomedical neuroscience research: A review. *Biomater. Res.* **2022**, *26*, 82. [CrossRef] [PubMed]

59. Soliman, Y.; Al-Khodor, J.; Yildirim Köken, G.; Mustafaoglu, N. A guide for blood-brain barrier models. *FEBS Lett.* **2024**, *599*, 599–644. [CrossRef]
60. Bussolari, S.R.; Dewey, C.F.; Gimbrone, M.A., Jr. Apparatus for subjecting living cells to fluid shear stress. *Rev. Sci. Instrum.* **1982**, *53*, 1851–1854. [CrossRef]
61. Santaguida, S.; Janigro, D.; Hossain, M.; Oby, E.; Rapp, E.; Cucullo, L. Side by side comparison between dynamic versus static models of blood-brain barrier in vitro: A permeability study. *Brain Res.* **2006**, *1109*, 1–13. [CrossRef]
62. lbakary, B.; Badhan, R.K.S. A dynamic perfusion based blood-brain barrier model for cytotoxicity testing and drug permeation. *Sci. Rep.* **2020**, *10*, 3788. [CrossRef]
63. Abdullah, Z.; Rakkar, K.; Bath, P.M.W.; Bayraktutan, U. Inhibition of TNF-α protects in vitro brain barrier from ischaemic damage. *Mol. Cell Neurosci.* **2015**, *69*, 65–79. [CrossRef]
64. Alwjwaj, M.; Kadir, R.R.A.; Bayraktutan, U. Outgrowth endothelial progenitor cells restore cerebral barrier function following ischaemic damage: The impact of NOX2 inhibition. *Eur. J. Neurosci.* **2022**, *55*, 1658–1670. [CrossRef] [PubMed]
65. Chung, T.D.; Linville, R.M.; Guo, Z.; Ye, R.; Jha, R.; Grifno, G.N.; Searson, P.C. Effects of acute and chronic oxidative stress on the blood–brain barrier in 2D and 3D in vitro models. *Fluids Barriers CNS* **2022**, *19*, 33. [CrossRef]
66. Ya, J.; Bayraktutan, U. Senolytics and Senomorphics Targeting p38MAPK/NF-κB Pathway Protect Endothelial Cells from Oxidative Stress-Mediated Premature Senescence. *Cells* **2024**, *13*, 1292. [CrossRef] [PubMed]
67. Ya, J.; Kadir, R.R.A.; Bayraktutan, U. Delay of endothelial cell senescence protects cerebral barrier against age-related dysfunction: Role of senolytics and senomorphics. *Tissue Barriers* **2023**, *11*, 2103353. [CrossRef]
68. Kadir, R.R.A.; Alwjwaj, M.; Ahmad Othman, O.; Rakkar, K.; Sprigg, N.; Bath, P.M.; Bayraktutan, U. Inhibition of oxidative stress delays senescence and augments functional capacity of endothelial progenitor cells. *Brain Res.* **2022**, *1787*, 147925. [CrossRef]
69. Adriani, G.; Ma, D.; Pavesi, A.; Kamm, R.D.; Goh, E.L.K. A 3D neurovascular microfluidic model consisting of neurons, astrocytes and cerebral endothelial cells as a blood–brain barrier. *Lab A Chip* **2017**, *17*, 448–459. [CrossRef] [PubMed]
70. Cucullo, L.; Couraud, P.O.; Weksler, B.; Romero, I.A.; Hossain, M.; Rapp, E.; Janigro, D. Immortalized human brain endothelial cells and flow-based vascular modeling: A marriage of convenience for rational neurovascular studies. *J. Cereb. Blood Flow. Metab.* **2008**, *28*, 312–328. [CrossRef]
71. Liu, Y.; Tang, G.; Li, Y.; Wang, Y.; Chen, X.; Gu, X.; Zhang, Z.; Wang, Y.; Yang, G.Y. Metformin attenuates blood-brain barrier disruption in mice following middle cerebral artery occlusion. *J. Neuroinflamm.* **2014**, *11*, 177. [CrossRef]
72. Takata, F.; Dohgu, S.; Matsumoto, J.; Machida, T.; Kaneshima, S.; Matsuo, M.; Sakaguchi, S.; Takeshige, Y.; Yamauchi, A.; Kataoka, Y. Metformin induces up-regulation of blood-brain barrier functions by activating AMP-activated protein kinase in rat brain microvascular endothelial cells. *Biochem. Biophys. Res. Commun.* **2013**, *433*, 586–590. [CrossRef]
73. Ji, Y.B.; Gao, Q.; Tan, X.X.; Huang, X.W.; Ma, Y.Z.; Fang, C.; Wang, S.N.; Qiu, L.H.; Cheng, Y.X.; Guo, F.Y.; et al. Lithium alleviates blood-brain barrier breakdown after cerebral ischemia and reperfusion by upregulating endothelial Wnt/β-catenin signaling in mice. *Neuropharmacology* **2021**, *186*, 108474. [CrossRef]
74. Haupt, M.; Zechmeister, B.; Bosche, B.; Lieschke, S.; Zheng, X.; Zhang, L.; Venkataramani, V.; Jin, F.; Hein, K.; Weber, M.S.; et al. Lithium enhances post-stroke blood-brain barrier integrity, activates the MAPK/ERK1/2 pathway and alters immune cell migration in mice. *Neuropharmacology* **2020**, *181*, 108357. [CrossRef] [PubMed]
75. Tan, Q.; Yu, D.; Song, L. Atorvastatin disrupts primary human brain microvascular endothelial cell functions via prenylation-dependent mitochondrial inhibition and oxidative stress. *Fundam. Clin. Pharmacol.* **2021**, *35*, 341–350. [CrossRef]
76. Lübtow, M.M.; Oerter, S.; Quader, S.; Jeanclos, E.; Cubukova, A.; Krafft, M.; Haider, M.S.; Schulte, C.; Meier, L.; Rist, M.; et al. In Vitro Blood-Brain Barrier Permeability and Cytotoxicity of an Atorvastatin-Loaded Nanoformulation Against Glioblastoma in 2D and 3D Models. *Mol. Pharm.* **2020**, *17*, 1835–1847. [CrossRef]
77. Du, Y.; Zhu, Y.J.; Zhou, Y.X.; Ding, J.; Liu, J.Y. Metformin in therapeutic applications in human diseases: Its mechanism of action and clinical study. *Mol. Biomed.* **2022**, *3*, 41. [CrossRef]
78. Fountoulakis, K.N.; Tohen, M.; Zarate, C.A., Jr. Lithium treatment of Bipolar disorder in adults: A systematic review of randomized trials and meta-analyses. *Eur. Neuropsychopharmacol.* **2022**, *54*, 100–115. [CrossRef]
79. Björkhem-Bergman, L.; Lindh, J.D.; Bergman, P. What is a relevant statin concentration in cell experiments claiming pleiotropic effects? *Br. J. Clin. Pharmacol.* **2011**, *72*, 164–165. [CrossRef] [PubMed]
80. Yenari, M.A.; Xu, L.; Tang, X.N.; Qiao, Y.; Giffard, R.G. Microglia potentiate damage to blood-brain barrier constituents: Improvement by minocycline in vivo and in vitro. *Stroke* **2006**, *37*, 1087–1093. [CrossRef]
81. Asadi, A.; Abdi, M.; Kouhsari, E.; Panahi, P.; Sholeh, M.; Sadeghifard, N.; Amiriani, T.; Ahmadi, A.; Maleki, A.; Gholami, M. Minocycline, focus on mechanisms of resistance, antibacterial activity, and clinical effectiveness: Back to the future. *J. Glob. Antimicrob. Resist.* **2020**, *22*, 161–174. [CrossRef]
82. Smith, D.A.; Beaumont, K.; Maurer, T.S.; Di, L. Volume of Distribution in Drug Design. *J. Med. Chem.* **2015**, *58*, 5691–5698. [CrossRef]
83. Albrecht, W. Which concentrations are optimal for in vitro testing? *Excli J.* **2020**, *19*, 1172–1173. [CrossRef]

84. Zhao, M.; Lepak, A.J.; Andes, D.R. Animal models in the pharmacokinetic/pharmacodynamic evaluation of antimicrobial agents. *Bioorg. Med. Chem.* **2016**, *24*, 6390–6400. [CrossRef] [PubMed]
85. Bagchi, S.; Chhibber, T.; Lahooti, B.; Verma, A.; Borse, V.; Jayant, R.D. In-vitro blood-brain barrier models for drug screening and permeation studies: An overview. *Drug Des. Dev. Ther.* **2019**, *13*, 3591–3605. [CrossRef]
86. Chen, Z.R.; Huang, J.B.; Yang, S.L.; Hong, F.F. Role of Cholinergic Signaling in Alzheimer's Disease. *Molecules* **2022**, *27*, 1816. [CrossRef] [PubMed]
87. Pope, C.; Karanth, S.; Liu, J. Pharmacology and toxicology of cholinesterase inhibitors: Uses and misuses of a common mechanism of action. *Environ. Toxicol. Pharmacol.* **2005**, *19*, 433–446. [CrossRef]
88. Wang, J.; Zhang, H.Y.; Tang, X.C. Cholinergic deficiency involved in vascular dementia: Possible mechanism and strategy of treatment. *Acta Pharmacol. Sin.* **2009**, *30*, 879–888. [CrossRef]
89. Gallowitsch-Puerta, M.; Pavlov, V.A. Neuro-immune interactions via the cholinergic anti-inflammatory pathway. *Life Sci.* **2007**, *80*, 2325–2329. [CrossRef]
90. Han, B.; Li, X.; Hao, J. The cholinergic anti-inflammatory pathway: An innovative treatment strategy for neurological diseases. *Neurosci. Biobehav. Rev.* **2017**, *77*, 358–368. [CrossRef] [PubMed]
91. Lecrux, C.; Sandoe, C.H.; Neupane, S.; Kropf, P.; Toussay, X.; Tong, X.K.; Lacalle-Aurioles, M.; Shmuel, A.; Hamel, E. Impact of Altered Cholinergic Tones on the Neurovascular Coupling Response to Whisker Stimulation. *J. Neurosci.* **2017**, *37*, 1518–1531. [CrossRef]
92. Nizari, S.; Wells, J.A.; Carare, R.O.; Romero, I.A.; Hawkes, C.A. Loss of cholinergic innervation differentially affects eNOS-mediated blood flow, drainage of Aβ and cerebral amyloid angiopathy in the cortex and hippocampus of adult mice. *Acta Neuropathol. Commun.* **2021**, *9*, 12. [CrossRef]
93. Colović, M.B.; Krstić, D.Z.; Lazarević-Pašti, T.D.; Bondžić, A.M.; Vasić, V.M. Acetylcholinesterase inhibitors: Pharmacology and toxicology. *Curr. Neuropharmacol.* **2013**, *11*, 315–335. [CrossRef]
94. Toda, N.; Okamura, T. Cerebral blood flow regulation by nitric oxide in Alzheimer's disease. *J. Alzheimers Dis.* **2012**, *32*, 569–578. [CrossRef] [PubMed]
95. Richardson, C.; Gard, P.R.; Klugman, A.; Isaac, M.; Tabet, N. Blood pro-inflammatory cytokines in Alzheimer's disease in relation to the use of acetylcholinesterase inhibitors. *Int. J. Geriatr. Psychiatry* **2013**, *28*, 1312–1317. [CrossRef]
96. Lin, X.; Chen, Y.; Zhang, P.; Chen, G.; Zhou, Y.; Yu, X. The potential mechanism of postoperative cognitive dysfunction in older people. *Exp. Gerontol.* **2020**, *130*, 110791. [CrossRef]
97. Endo, A. The discovery and development of HMG-CoA reductase inhibitors. *J. Lipid Res.* **1992**, *33*, 1569–1582. [CrossRef] [PubMed]
98. Goldstein, J.L.; Brown, M.S. A century of cholesterol and coronaries: From plaques to genes to statins. *Cell* **2015**, *161*, 161–172. [CrossRef]
99. Field, F.J.; Born, E.; Murthy, S.; Mathur, S.N. Regulation of sterol regulatory element-binding proteins by cholesterol flux in CaCo-2 cells. *J. Lipid Res.* **2001**, *42*, 1687–1698. [CrossRef] [PubMed]
100. Fang, S.Y.; Roan, J.N.; Luo, C.Y.; Tsai, Y.C.; Lam, C.F. Pleiotropic vascular protective effects of statins in perioperative medicine. *Acta Anaesthesiol. Taiwanica* **2013**, *51*, 120–126. [CrossRef]
101. Nakata, S.; Tsutsui, M.; Shimokawa, H.; Yamashita, T.; Tanimoto, A.; Tasaki, H.; Ozumi, K.; Sabanai, K.; Morishita, T.; Suda, O.; et al. Statin treatment upregulates vascular neuronal nitric oxide synthase through Akt/NF-kappaB pathway. *Arterioscler. Thromb. Vasc. Biol.* **2007**, *27*, 92–98. [CrossRef]
102. Blanco-Colio, L.M.; Tuñón, J.; Martín-Ventura, J.L.; Egido, J. Anti-inflammatory and immunomodulatory effects of statins. *Kidney Int.* **2003**, *63*, 12–23. [CrossRef]
103. Chen, W.H.; Chen, C.H.; Hsu, M.C.; Chang, R.W.; Wang, C.H.; Lee, T.S. Advances in the molecular mechanisms of statins in regulating endothelial nitric oxide bioavailability: Interlocking biology between eNOS activity and L-arginine metabolism. *Biomed. Pharmacother.* **2024**, *171*, 116192. [CrossRef]
104. Demirci, B.; McKeown, P.P.; Bayraktutan, U. The bimodal regulation of vascular function by superoxide anion: Role of endothelium. *BMB Rep.* **2008**, *41*, 223–229. [CrossRef] [PubMed]
105. Mahadevan, V.S.; Campbell, M.; McKeown, P.P.; Bayraktutan, U. Internal mammary artery smooth muscle cells resist migration and possess high antioxidant capacity. *Cardiovasc. Res.* **2006**, *72*, 60–68. [CrossRef] [PubMed]
106. Liu, J.C.; Lei, S.Y.; Zhang, D.H.; He, Q.Y.; Sun, Y.Y.; Zhu, H.J.; Qu, Y.; Zhou, S.Y.; Yang, Y.; Li, C.; et al. The pleiotropic effects of statins: A comprehensive exploration of neurovascular unit modulation and blood-brain barrier protection. *Mol. Med.* **2024**, *30*, 256. [CrossRef]
107. Bouitbir, J.; Charles, A.L.; Echaniz-Laguna, A.; Kindo, M.; Daussin, F.; Auwerx, J.; Piquard, F.; Geny, B.; Zoll, J. Opposite effects of statins on mitochondria of cardiac and skeletal muscles: A 'mitohormesis' mechanism involving reactive oxygen species and PGC-1. *Eur. Heart J.* **2012**, *33*, 1397–1407. [CrossRef]

108. Akkouh, I.A.; Skrede, S.; Holmgren, A.; Ersland, K.M.; Hansson, L.; Bahrami, S.; Andreassen, O.A.; Steen, V.M.; Djurovic, S.; Hughes, T. Exploring lithium's transcriptional mechanisms of action in bipolar disorder: A multi-step study. *Neuropsychopharmacology* **2020**, *45*, 947–955. [CrossRef] [PubMed]
109. Malhi, G.S.; Tanious, M.; Das, P.; Coulston, C.M.; Berk, M. Potential mechanisms of action of lithium in bipolar disorder. Current understanding. *CNS Drugs* **2013**, *27*, 135–153. [CrossRef]
110. Rakkar, K.; Bayraktutan, U. Increases in intracellular calcium perturb blood-brain barrier via protein kinase C-alpha and apoptosis. *Biochim. Biophys. Acta* **2016**, *1862*, 56–71. [CrossRef]
111. Snitow, M.E.; Bhansali, R.S.; Klein, P.S. Lithium and Therapeutic Targeting of GSK-3. *Cells* **2021**, *10*, 255. [CrossRef]
112. Reichenbach, N.L.; Gofman, L.; Persidsky, Y. Inhibition of glycogen synthase kinase 3β promotes tight junction stability in brain endothelial cells by half-life extension of occludin and claudin-5. *PLoS ONE* **2013**, *8*, e55972. [CrossRef]
113. Hashimoto, R.; Takei, N.; Shimazu, K.; Christ, L.; Lu, B.; Chuang, D.M. Lithium induces brain-derived neurotrophic factor and activates TrkB in rodent cortical neurons: An essential step for neuroprotection against glutamate excitotoxicity. *Neuropharmacology* **2002**, *43*, 1173–1179. [CrossRef]
114. Fiorentini, A.; Rosi, M.C.; Grossi, C.; Luccarini, I.; Casamenti, F. Lithium improves hippocampal neurogenesis, neuropathology and cognitive functions in APP mutant mice. *PLoS ONE* **2010**, *5*, e14382. [CrossRef] [PubMed]
115. Maurer, I.C.; Schippel, P.; Volz, H.P. Lithium-induced enhancement of mitochondrial oxidative phosphorylation in human brain tissue. *Bipolar Disord.* **2009**, *11*, 515–522. [CrossRef] [PubMed]
116. Osete, J.R.; Akkouh, I.A.; de Assis, D.R.; Szabo, A.; Frei, E.; Hughes, T.; Smeland, O.B.; Steen, N.E.; Andreassen, O.A.; Djurovic, S. Lithium increases mitochondrial respiration in iPSC-derived neural precursor cells from lithium responders. *Mol. Psychiatry* **2021**, *26*, 6789–6805. [CrossRef] [PubMed]
117. Shukla, G.S. Mechanism of lithium action: In vivo and in vitro effects of alkali metals on brain superoxide dismutase. *Pharmacol. Biochem. Behav.* **1987**, *26*, 235–240. [CrossRef]
118. Kerr, F.; Bjedov, I.; Sofola-Adesakin, O. Molecular Mechanisms of Lithium Action: Switching the Light on Multiple Targets for Dementia Using Animal Models. *Front. Mol. Neurosci.* **2018**, *11*, 297. [CrossRef]
119. de Sousa, R.T.; Zanetti, M.V.; Busatto, G.F.; Mouro, M.G.; Zarate, C.A.; Gattaz, W.F., Jr.; Higa, E.M.; Machado-Vieira, R. Lithium increases nitric oxide levels in subjects with bipolar disorder during depressive episodes. *J. Psychiatr. Res.* **2014**, *55*, 96–100. [CrossRef]
120. Greene, C.; Hanley, N.; Campbell, M. Claudin-5: Gatekeeper of neurological function. *Fluids Barriers CNS* **2019**, *16*, 3. [CrossRef]
121. Boswell-Smith, V.; Spina, D.; Page, C.P. Phosphodiesterase inhibitors. *Br. J. Pharmacol.* **2006**, *147* (Suppl. S1), S252–S257. [CrossRef]
122. Schudt, C.; Hatzelmann, A.; Beume, R.; Tenor, H. Phosphodiesterase Inhibitors: History of Pharmacology. In *Phosphodiesterases as Drug Targets*; Francis, S., Conti, M., Houslay, M., Eds.; Handbook of Experimental Pharmacology; Springer: Berlin/Heidelberg, Germany, 2011; Volume 204, pp. 1–46. [CrossRef]
123. Spina, D. PDE4 inhibitors: Current status. *Br. J. Pharmacol.* **2008**, *155*, 308–315. [CrossRef]
124. Samidurai, A.; Xi, L.; Das, A.; Kukreja, R.C. Beyond Erectile Dysfunction: cGMP-Specific Phosphodiesterase 5 Inhibitors for Other Clinical Disorders. *Annu. Rev. Pharmacol. Toxicol.* **2023**, *63*, 585–615. [CrossRef]
125. Ölmestig, J.N.E.; Marlet, I.R.; Hainsworth, A.H.; Kruuse, C. Phosphodiesterase 5 inhibition as a therapeutic target for ischemic stroke: A systematic review of preclinical studies. *Cell. Signal.* **2017**, *38*, 39–48. [CrossRef] [PubMed]
126. Nelson, M.D.; Mehta, P.K.; Wei, J.; Sharif, B.; Thomson, L.E.J.; Berman, D.; Li, D.; Merz, C.N.B. Phosphodiesterase type 5 inhibition may reduce diastolic function in women with ischemia but no obstructive coronary artery disease. *J. Med. Case Rep.* **2017**, *11*, 144. [CrossRef] [PubMed]
127. Zuo, H.; Cattani-Cavalieri, I.; Musheshe, N.; Nikolaev, V.O.; Schmidt, M. Phosphodiesterases as therapeutic targets for respiratory diseases. *Pharmacol. Ther.* **2019**, *197*, 225–242. [CrossRef] [PubMed]
128. Delhaye, S.; Bardoni, B. Role of phosphodiesterases in the pathophysiology of neurodevelopmental disorders. *Mol. Psychiatry* **2021**, *26*, 4570–4582. [CrossRef]
129. Peixoto, C.A.; Nunes, A.K.; Garcia-Osta, A. Phosphodiesterase-5 Inhibitors: Action on the Signaling Pathways of Neuroinflammation, Neurodegeneration, and Cognition. *Mediat. Inflamm.* **2015**, *2015*, 940207. [CrossRef]
130. Wakita, H.; Tomimoto, H.; Akiguchi, I.; Lin, J.X.; Ihara, M.; Ohtani, R.; Shibata, M. Ibudilast, a phosphodiesterase inhibitor, protects against white matter damage under chronic cerebral hypoperfusion in the rat. *Brain Res.* **2003**, *992*, 53–59. [CrossRef] [PubMed]
131. Chang, Y.C.; Huang, C.C.; Hung, P.L.; Huang, H.M. Rolipram, a phosphodiesterase type IV inhibitor, exacerbates periventricular white matter lesions in rat pups. *Pediatr. Res.* **2008**, *64*, 234–239. [CrossRef]
132. Ip, B.Y.M.; Lam, B.Y.K.; Hui, V.M.H.; Au, L.W.C.; Liu, M.W.T.; Shi, L.; Lee, V.W.Y.; Chu, W.C.W.; Leung, T.W.; Ko, H.; et al. Efficacy and safety of cilostazol in decreasing progression of cerebral white matter hyperintensities-A randomized controlled trial. *Alzheimers Dement.* **2022**, *8*, e12369. [CrossRef]

133. Ölmestig, J.; Marlet, I.R.; Hansen, R.H.; Rehman, S.; Krawcyk, R.S.; Rostrup, E.; Lambertsen, K.L.; Kruuse, C. Tadalafil may improve cerebral perfusion in small-vessel occlusion stroke—A pilot study. *Brain Commun.* **2020**, *2*, fcaa020. [CrossRef]
134. van Sloten, T.T.; Sedaghat, S.; Carnethon, M.R.; Launer, L.J.; Stehouwer, C.D.A. Cerebral microvascular complications of type 2 diabetes: Stroke, cognitive dysfunction, and depression. *Lancet Diabetes Endocrinol.* **2020**, *8*, 325–336. [CrossRef]
135. Lv, Y.; Cheng, X.; Dong, Q. SGLT1 and SGLT2 inhibition, circulating metabolites, and cerebral small vessel disease: A mediation Mendelian Randomization study. *Cardiovasc. Diabetol.* **2024**, *23*, 157. [CrossRef] [PubMed]
136. Yang, X.; Qiang, Q.; Li, N.; Feng, P.; Wei, W.; Hölscher, C. Neuroprotective Mechanisms of Glucagon-Like Peptide-1-Based Therapies in Ischemic Stroke: An Update Based on Preclinical Research. *Front. Neurol.* **2022**, *13*, 844697. [CrossRef]
137. Dobi, A.; Rosanaly, S.; Devin, A.; Baret, P.; Meilhac, O.; Harry, G.J.; d'Hellencourt, C.L.; Rondeau, P. Advanced glycation end-products disrupt brain microvascular endothelial cell barrier: The role of mitochondria and oxidative stress. *Microvasc. Res.* **2021**, *133*, 104098. [CrossRef] [PubMed]
138. Kadir, R.R.A.; Alwjwaj, M.; Bayraktutan, U. Protein kinase C-β distinctly regulates blood-brain barrier-forming capacity of Brain Microvascular endothelial cells and outgrowth endothelial cells. *Metab. Brain Dis.* **2022**, *37*, 1815–1827. [CrossRef]
139. Shao, B.; Bayraktutan, U. Hyperglycaemia promotes cerebral barrier dysfunction through activation of protein kinase C-β. *Diabetes Obes. Metab.* **2013**, *15*, 993–999. [CrossRef] [PubMed]
140. Shao, B.; Bayraktutan, U. Hyperglycaemia promotes human brain microvascular endothelial cell apoptosis via induction of protein kinase C-ßI and prooxidant enzyme NADPH oxidase. *Redox Biol.* **2014**, *2*, 694–701. [CrossRef]
141. Rena, G.; Hardie, D.G.; Pearson, E.R. The mechanisms of action of metformin. *Diabetologia* **2017**, *60*, 1577–1585. [CrossRef]
142. Salvatore, T.; Pafundi, P.C.; Galiero, R.; Rinaldi, L.; Caturano, A.; Vetrano, E.; Aprea, C.; Albanese, G.; Di Martino, A.; Ricozzi, C.; et al. Can Metformin Exert as an Active Drug on Endothelial Dysfunction in Diabetic Subjects? *Biomedicines* **2020**, *9*, 3. [CrossRef]
143. Pawlos, A.; Broncel, M.; Woźniak, E.; Gorzelak-Pabiś, P. Neuroprotective Effect of SGLT2 Inhibitors. *Molecules* **2021**, *26*, 7213. [CrossRef]
144. Angelopoulou, E.; Piperi, C. DPP-4 inhibitors: A promising therapeutic approach against Alzheimer's disease. *Ann. Transl. Med.* **2018**, *6*, 255. [CrossRef]
145. Castro, M.M.; Kandasamy, A.D.; Youssef, N.; Schulz, R. Matrix metalloproteinase inhibitor properties of tetracyclines: Therapeutic potential in cardiovascular diseases. *Pharmacol. Res.* **2011**, *64*, 551–560. [CrossRef] [PubMed]
146. Jiang, Y.; Zhu, J.; Wu, L.; Xu, G.; Dai, J.; Liu, X. Tetracycline inhibits local inflammation induced by cerebral ischemia via modulating autophagy. *PLoS ONE* **2012**, *7*, e48672. [CrossRef] [PubMed]
147. Griffin, M.O.; Ceballos, G.; Villarreal, F.J. Tetracycline compounds with non-antimicrobial organ protective properties: Possible mechanisms of action. *Pharmacol. Res.* **2011**, *63*, 102–107. [CrossRef] [PubMed]
148. Robinson, B.D.; Isbell, C.L.; Melge, A.R.; Lomas, A.M.; Shaji, C.A.; Mohan, C.G.; Huang, J.H.; Tharakan, B. Doxycycline prevents blood–brain barrier dysfunction and microvascular hyperpermeability after traumatic brain injury. *Sci. Rep.* **2022**, *12*, 5415. [CrossRef]
149. Salimiaghdam, N.; Singh, L.; Schneider, K.; Nalbandian, A.; Chwa, M.; Atilano, S.R.; Bao, A.; Kenney, M.C. Potential adverse effects of ciprofloxacin and tetracycline on ARPE-19 cell lines. *BMJ Open Ophthalmol.* **2020**, *5*, e000458. [CrossRef]
150. Tikka, T.; Fiebich, B.L.; Goldsteins, G.; Keinanen, R.; Koistinaho, J. Minocycline, a tetracycline derivative, is neuroprotective against excitotoxicity by inhibiting activation and proliferation of microglia. *J. Neurosci.* **2001**, *21*, 2580–2588. [CrossRef]
151. Elewa, H.F.; Hilali, H.; Hess, D.C.; Machado, L.S.; Fagan, S.C. Minocycline for short-term neuroprotection. *Pharmacotherapy* **2006**, *26*, 515–521. [CrossRef]
152. Bradley, C.P.; Berry, C. Microvascular arterial disease of the brain and the heart: A shared pathogenesis. *QJM Int. J. Med.* **2023**, *116*, 829–834. [CrossRef]
153. Peters, C.H.; Sokolov, S.; Rajamani, S.; Ruben, P.C. Effects of the antianginal drug, ranolazine, on the brain sodium channel Na(V)1.2 and its modulation by extracellular protons. *Br. J. Pharmacol.* **2013**, *169*, 704–716. [CrossRef]
154. Litman, M.; Spratt, N.J.; Beard, D.J. The effect of nitroglycerin treatment on cerebral ischaemia: A systematic review and meta-analysis of animal studies. *Nitric Oxide* **2024**, *151*, 10–16. [CrossRef]
155. Etter, E.F.; Eto, M.; Wardle, R.L.; Brautigan, D.L.; Murphy, R.A. Activation of myosin light chain phosphatase in intact arterial smooth muscle during nitric oxide-induced relaxation. *J. Biol. Chem.* **2001**, *276*, 34681–34685. [CrossRef] [PubMed]
156. Zhang, F.; White, J.G.; Iadecola, C. Nitric oxide donors increase blood flow and reduce brain damage in focal ischemia: Evidence that nitric oxide is beneficial in the early stages of cerebral ischemia. *J. Cereb. Blood Flow. Metab.* **1994**, *14*, 217–226. [CrossRef]
157. Zhu, H.Y.; Hong, F.F.; Yang, S.L. The Roles of Nitric Oxide Synthase/Nitric Oxide Pathway in the Pathology of Vascular Dementia and Related Therapeutic Approaches. *Int. J. Mol. Sci.* **2021**, *22*, 4540. [CrossRef] [PubMed]
158. Alborch, E.; Salom, J.B.; Torregrosa, G. Calcium channels in cerebral arteries. *Pharmacol. Ther.* **1995**, *68*, 1–34. [CrossRef]
159. Lee, E.M. Calcium channel blockers for hypertension: Old, but still useful. *Cardiovasc. Prev. Pharmacother.* **2023**, *5*, 113–125. [CrossRef]

160. Lou, J.S.; Su, M.; Wang, J.; Do, H.N.; Miao, Y.; Huang, X.Y. Distinct binding conformations of epinephrine with α- and β-adrenergic receptors. *Exp. Mol. Med.* **2024**, *56*, 1952–1966. [CrossRef]
161. Iaccarion, G.; Sorriento, D. Novel Insights in β-Adrenergic Receptor Signaling. In *Encyclopedia of Endocrine Diseases*; Hayward, L.F., Mueller, P.J., Hasser, E.M., Eds.; Academic Press: Oxford, UK, 2004; Volume 3, pp. 112–115.
162. Asano, N.; Hishiyama, S.; Ishiyama, T.; Kotoda, M.; Matsukawa, T. Effects of β1-adrenergic receptor blockade on the cerebral microcirculation in the normal state and during global brain ischemia/reperfusion injury in rabbits. *BMC Pharmacol. Toxicol.* **2020**, *21*, 13. [CrossRef]
163. Gasier, H.G.; Demchenko, I.T.; Zhilyaev, S.Y.; Moskvin, A.N.; Krivchenko, A.I.; Piantadosi, C.A. Adrenoceptor blockade modifies regional cerebral blood flow responses to hyperbaric hyperoxia: Protection against CNS oxygen toxicity. *J. Appl. Physiol.* **2018**, *125*, 1296–1304. [CrossRef] [PubMed]
164. Aldakkak, M.; Camara, A.K.; Heisner, J.S.; Yang, M.; Stowe, D.F. Ranolazine reduces Ca^{2+} overload and oxidative stress and improves mitochondrial integrity to protect against ischemia reperfusion injury in isolated hearts. *Pharmacol. Res.* **2011**, *64*, 381–392. [CrossRef]
165. Müller-Werdan, U.; Stöckl, G.; Werdan, K. Advances in the management of heart failure: The role of ivabradine. *Vasc. Health Risk Manag.* **2016**, *12*, 453–470. [CrossRef]
166. Kukovetz, W.R.; Holzmann, S.; Pöch, G. Molecular mechanism of action of nicorandil. *J. Cardiovasc. Pharmacol.* **1992**, *20* (Suppl. S3), S1–S7. [CrossRef]
167. Guo, Y.; Li, Y.; Liu, X.; Cui, Y.; Zhao, Y.; Sun, S.; Jia, Q.; Chai, Q.; Gong, G.; Zhang, H.; et al. Assessing the effectiveness of statin therapy for alleviating cerebral small vessel disease progression in people ≥75 years of age. *BMC Geriatr.* **2020**, *20*, 292. [CrossRef]
168. Dichgans, M.; Markus, H.S.; Salloway, S.; Verkkoniemi, A.; Moline, M.; Wang, Q.; Posner, H.; Chabriat, H.S. Donepezil in patients with subcortical vascular cognitive impairment: A randomised double-blind trial in CADASIL. *Lancet Neurol.* **2008**, *7*, 310–318. [CrossRef] [PubMed]
169. Lavallée, P.C.; Labreuche, J.; Gongora-Rivera, F.; Jaramillo, A.; Brenner, D.; Klein, I.F.; Touboul, P.J.; Vicaut, E.; Amarenco, P. Placebo-controlled trial of high-dose atorvastatin in patients with severe cerebral small vessel disease. *Stroke* **2009**, *40*, 1721–1728. [CrossRef]
170. Webb, A.; Werring, D.; Dawson, J.; Rothman, A.; Lawson, A.; Wartolowska, K. Design of a randomised, double-blind, crossover, placebo-controlled trial of effects of sildenafil on cerebrovascular function in small vessel disease: Oxford haemodynamic adaptation to reduce pulsatility trial (OxHARP). *Eur. Stroke J.* **2021**, *6*, 283–290. [CrossRef]
171. Wardlaw, J.M.; Woodhouse, L.J.; Mhlanga, I.I.; Oatey, K.; Heye, A.K.; Bamford, J.; Cvoro, V.; Doubal, F.N.; England, T.; Hassan, A.; et al. Lacunar Intervention Trial-2 (LACI-2) Investigator Group. Isosorbide Mononitrate and Cilostazol Treatment in Patients With Symptomatic Cerebral Small Vessel Disease: The Lacunar Intervention Trial-2 (LACI-2) Randomized Clinical Trial. *JAMA Neurol.* **2023**, *80*, 682–692. [CrossRef] [PubMed]
172. Blair, G.W.; Janssen, E.; Stringer, M.S.; Thrippleton, M.J.; Chappell, F.; Shi, Y.; Hamilton, I.; Flaherty, K.; Appleton, J.P.; Doubal, F.N.; et al. Effects of Cilostazol and Isosorbide Mononitrate on Cerebral Hemodynamics in the LACI-1 Randomized Controlled Trial. *Stroke* **2022**, *53*, 29–33. [CrossRef]
173. Han, S.W.; Lee, S.S.; Kim, S.H.; Lee, J.H.; Kim, G.S.; Kim, O.J.; Koh, I.S.; Lee, J.Y.; Suk, S.H.; Lee, S.I.; et al. Effect of cilostazol in acute lacunar infarction based on pulsatility index of transcranial Doppler (ECLIPse): A multicenter, randomized, double-blind, placebo-controlled trial. *Eur. Neurol.* **2013**, *69*, 33–40. [CrossRef]
174. Rudilosso, S.; Rodríguez-Vázquez, A.; Urra, X.; Arboix, A. The Potential Impact of Neuroimaging and Translational Research on the Clinical Management of Lacunar Stroke. *Int. J. Mol. Sci.* **2022**, *23*, 1497. [CrossRef]

current issues in molecular biology

MDPI

Review

The Role of HSP47 in Thrombotic Disorders: Molecular Mechanism and Therapeutic Potential

Minodora Teodoru [1,2], Oana-Maria Stoia [1,2], Maria-Gabriela Vladoiu [1,2,*] and Alexandra-Kristine Tonch-Cerbu [1,2,*]

1 Medical Clinical Department, Faculty of Medicine, "Lucian Blaga" University, 550024 Sibiu, Romania; minodora.teodoru@ulbsibiu.ro (M.T.); oana.stoia@ulbsibiu.ro (O.-M.S.)
2 County Clinical Emergency Hospital of Sibiu, 550245 Sibiu, Romania
* Correspondence: catanamariagabriela@gmail.com (M.-G.V.); alexandra.kristine@yahoo.com (A.-K.T.-C.)

Abstract: This review aims to analyze the role of heat shock protein 47 (HSP47) in thrombosis and evaluate its potential as a therapeutic target for deep vein thrombosis (DVT). A systematic literature review was conducted using PubMed, Scopus, and Web of Science to identify studies on HSP47, thrombosis, and collagen, selecting only relevant and methodologically rigorous articles. HSP47 regulates platelet function and collagen interaction, playing a key role in deep vein thrombosis (DVT). HSP47, known for stabilizing collagen, also improves platelet–collagen binding and thrombus formation. In addition, reduced HSP47 levels reduce platelet adhesion, resulting in reduced thrombus formation, while inhibitors that target HSP47 decrease platelet aggregation in animal models. Naturally low levels of HSP47 during prolonged immobility are also found in hibernating mammals, such as bears, and are associated with reduced formation of thrombi, indicating a possible natural mechanism of thrombo-protection. This observation could inform new therapeutic approaches. Current studies use in vitro platelet aggregation assays, flow chamber assays, and collagen binding studies to investigate the role of HSP47 in clotting. This review aims to synthesize existing evidence to better understand HSP47's role in clot formation and explore its potential as a target for novel DVT therapies.

Keywords: HSP47; thrombosis; platelet aggregation; collagen stability; therapy

1. Introduction

Heat shock proteins (HSPs) are crucial molecular chaperones that facilitate proper protein folding and prevent aggregation under stress conditions, thereby maintaining cellular proteostasis. Among these, heat shock protein 47 (HSP47) is specifically involved in the folding and secretion of collagen, binding to procollagen to ensure its stability during the maturation process in the endoplasmic reticulum (ER). While its fundamental role in collagen biology is well-documented, emerging evidence suggests that HSP47 may also significantly influence thrombosis, particularly in the context of deep vein thrombosis [1–4].

HSP47, or SERPINH1, is a collagen-specific molecular chaperone of the serpin superfamily, which uniquely does not possess protease inhibitory function. The serpin fold is further characterized with three β-sheets bestowing stability, nine α-helices providing flexibility, and a modified reactive center loop with a strong penchant for binding collagen (RDEL sequence at the sulfonated C-terminus), which help keep it within the endoplasmic reticulum. HSP47 is synthesized in the endoplasmic reticulum of collagen-producing cells, including fibroblasts and myofibroblasts, and is encoded by the SERPINH1 gene located on chromosome 11q13.5 to transcribe in the nucleus and translate on ER-bound ribosomes. It stabilizes procollagen's triple-helix structure and is expressed in a heat-shock and collagen demand-dependent manner, rendering it important in normal physiology and when diseases develop, such as fibrosis [5].

Although it has beneficial roles, opposing evidence exists in the context of HSP47 in disease pathology. HSP47 dysregulation has been shown in some studies to aggravate the inflammatory response, thereby resulting in the progression of disease rather than resolution thereof. Similarly, high activity levels of HSP47 can result in the aggregation of misfolded proteins, causing ER stress and promoting inflammation pathways. Such inflammatory response can establish a pro-thrombotic state, which aggravates the risk of diseases such as DVT [1,5].

HSP47, thus, has distinct roles, mediating collagen stability but also, when dysregulated, inflammatory potential. Understanding these dynamics will be important for evaluating HSP47 as a therapeutic target, as interventions must navigate the entangled history of these multifunctional proteins and will therefore need to augment its protective roles while averting any pernicious effects. Therefore, further studies are necessary to delineate the numerous functions of HSP47 in maintaining cellular homeostasis and its potential roles in pathologies [6–9].

This review aims to explore the molecular mechanisms underlying the role of HSP47 not only in collagen dynamics but also in thrombus formation and stability. It will highlight key findings and studies that link HSP47 to coagulation processes and potential therapeutic targets for preventing thrombosis. By examining both aspects, this review seeks to provide insights into the dual roles of HSP47 in connective tissue disorders and its implications for vascular health.

To provide a concise overview of the key themes and findings discussed in this text, the following table summarizes the critical roles, therapeutic potential, and broader implications of HSP47 across various biological contexts and disease mechanisms (Table 1).

Table 1. Summary of HSP47 Functions Across Biological Contexts and Disease Mechanisms.

Category	Key Points
HSP47 in Collagen Dynamics	- Acts as a chaperone for procollagen, ensuring proper folding and stability in the ER. - Prevents misfolding and aggregation of collagen molecules.
Role in Platelet Function	- Facilitates platelet adhesion and aggregation by stabilizing platelet-collagen interactions. - Key player in thrombus formation and stability.
Therapeutic Potential	- Targeting HSP47 offers potential for managing DVT, thrombosis, and connective tissue disorders. - Promising applications in wound healing and fibrosis.
Hibernation and Clot Resistance	- Decreased HSP47 levels in hibernating bears reduce clot formation. - Provides a model for developing antithrombotic drugs for immobilized patients.
Research Techniques	- Proteomics, RNA sequencing, and in vitro/in vivo models elucidate HSP47's role. - Functional assays confirm its influence on platelet binding to collagen.
Adipose Tissue Function	- Influences collagen turnover and ECM integrity in adipose tissue. - Variability in HSP47 levels linked to obesity and fat distribution.
Atherosclerosis and Plaque Stability	- Modulates collagen stability in plaques. - Targeting HSP47 could enhance plaque stability and reduce cardiovascular risks.
Applications in Vascular Health	- Potential for novel therapies targeting HSP47 in clotting disorders, obesity, and inflammation. - Supports tissue repair and vascular remodeling.
Future Research Directions	- Explore HSP47's role in ECM interactions, immune response, and inflammation. - Develop specific drugs targeting HSP47 pathways for disease treatment.

2. Background

A systematic literature search was performed in PubMed, Scopus, and Web of Science to retrieve studies related to HSP47, thrombosis, and collagen. The search terms were "HSP47", "deep vein thrombosis", "platelet aggregation", "collagen stability", "thrombosis therapy". Articles were selected based on their topical relevance, recency, and methodological rigor.

Included studies provided experimental or clinical data on the role of HSP47 in thrombosis. Exclusion criteria were applied to remove studies that had no direct evidence of HSP47 involvement or had incomplete methodologies.

Data were extracted from eligible studies with regard to experimental models, effects on platelet function, effects on thrombus formation, and/protocols on HSP47 transduction. Physio-pathological effects were compared by analyzing animal and human studies.

Relevant in vitro and in vivo studies examining platelet aggregation, collagen–HSP47 interactions, and thrombus formation were summarized. Mouse models and hibernating animal studies were examined to understand natural thrombus resistance mechanisms.

3. Main Text

3.1. Introduction to Heat Shock Proteins

Heat shock proteins (HSPs) play an essential role as molecular chaperones in the cellular stress response, helping to maintain correct protein folding and preventing aggregation under stress conditions like heat, inflammation, or hypoxia. HSP47, a collagen-specific molecular chaperone, is particularly associated with collagen stability and vascular integrity. Abnormal HSP47 expression leads to pathological collagen synthesis and extracellular matrix (ECM) disruption associated with DVT pathogenesis [6,10].

HSP47 is an endoplasmic reticulum (ER)-resident molecular chaperone for collagens with the vital role of stabilizing procollagen during synthesis through binding of polypeptide chains to avoid misfolding and ensure proper triple-helix formation. It does this in all cells that synthesize collagen such as fibroblasts and endothelial cells but is tissue-specific for various types of collagens. In blood vessels, HSP47 helps vascular fibroblasts to synthesize types I and III of collagen—found in the adventitia and media of vessels, respectively—providing them with strength and elasticity as well as helping endothelial cells and pericytes to synthesize type IV of collagen for the basement membrane. Types I and III of collagen, expressed in vessel subendothelial injury, cause aggregation of platelets for hemostasis with HSP47's role in their synthesis indirectly contributing to the process [4,11,12]. Procollagen is then channeled to the Golgi where HSP47 is released and recycled back to the ER through its RDEL motif for mature collagen formation vital for vascular health and blood clotting [13–15].

3.1.1. Ensuring Protein Integrity: The Functions of HSP47 and Chaperones in Endoplasmic Reticulum Stress

HSP47 is a specialized chaperon for procollagen and is indispensable for correct folding and assembly. Such specificity reduces endoplasmic reticulum (ER) stress due to the accumulation of misfolded proteins. On the other hand, an imbalance of procollagen synthesis overwhelms HSP47, which causes excessive ER stress and inflammation, which lead to diseases like cardiovascular disease and pulmonary fibrosis [1,4,11].

Previous research has shown that HSP47 is capable of binding newly synthesized procollagen chains, thus helping align them properly for triple-helix formation in the ER. In the event of misfolding, HSP47 pools misfolded proteins in the ER, activating quality control mechanisms that are frequently composing obstinate ER stress and inflammation. Most studies of HSP47's role in the heart are conducted in animal models, which if small in scope, may not fully account for potential compensatory pathways through the HSP47 pathways [1,12,13].

3.1.2. The Folding Process of Procollagen

Procollagen exemplifies how it engages with ER enzymes and chaperones during folding. It consists of trimeric regions separated by a triple-helical forming region. If folding is hindered, the C-propeptides associate with BiP, leading to degradation. Folded C-propeptides then form trimers or associate with PDI. After trimerization, the triple-helical domain folds, depending on proline hydroxylation. Trimeric non-triple-helical molecules interact with P4H. Finally, the protein is transported to the Golgi apparatus, forming higher-order aggregates. HSP47 is a crucial mediator in the procollagen folding process within the endoplasmic reticulum (ER), necessary for the formation and stability of the collagen triple helix, which is vital for tissue mechanical strength and function. HSP47 specifically binds to the C-propeptides of nascent procollagen chains during the late synthesis phase in the ER, promoting the optimal alignment of proα1 and proα2 chains to ensure stable triple helical formation. This chaperone function prevents premature aggregation of procollagen molecules [5,16,17].

Once procollagen is properly folded, HSP47 facilitates its transport from the ER to the Golgi apparatus. In cases of improper folding, HSP47 retains misfolded procollagen in the ER, initiating quality control pathways to manage these proteins. Deregulation of HSP47 activity due to genetic mutations, overexpression under stress, or inadequate cellular responses can lead to misfolded procollagen. This misfolding may overwhelm the ER's capacity, resulting in chronic ER stress [1,14].

HSP47, the ER chaperone, stabilizes the triple-helix structure of procollagen and prevents misfolding during synthesis but does not directly facilitate transport to the Golgi. Following proper folding, procollagen is transported to the Golgi via vesicle-mediated transport (COPII vesicles), where HSP47 is released typically due to pH change and recycles back to the ER via its RDEL motif, while procollagen is further processed. This is repeated in all tissues that synthesize collagen, including cells like fibroblasts, myofibroblasts, endothelial cells, osteoblasts, and chondrocytes, although different collagens are synthesized, for example, types I and III from fibroblasts and type IV from endothelial cells in blood vessels. Though universal, the relevance and types of collagen vary with each tissue's function, such as vascular hemostasis or skin structure [13,14,18].

The consequences of elevated ER stress include activation of the unfolded protein response (UPR), which can trigger inflammation, cell apoptosis, or fibrosis—factors critical in various disease processes [19,20].

Abnormalities in collagen synthesis and assembly caused by HSP47 dysregulation can lead to alterations in the vascular extracellular matrix (ECM) and blood vessel dysfunction. Such irregular collagen turnover compromises vascular stability and increases the risk of thrombus formation. Furthermore, inflammation induced by ER stress may create a pro-thrombotic state, heightening the risk of developing DVT [4,6].

3.1.3. The Role of HSP47 in Procollagen Interaction

The role of HSP47 in maintaining cellular homeostasis and endoplasmic reticulum (ER) quality control is pivotal, particularly in the context of procollagen maturation and the secretory pathway. HSP47 acts as a specific chaperone for procollagen, ensuring proper protein folding and preventing aggregation and toxicity that arise from misfolded proteins, similar to general chaperones in protein homeostasis. It collaborates with other ER-resident proteins, such as Prolyl 4-Hydroxylase and protein disulfide isomerase, to facilitate collagen maturation through oxidative folding and essential post-translational modifications. HSP47 also plays a crucial role in retaining misfolded procollagen in the ER, preventing the secretion of non-functional proteins, thereby safeguarding the quality of proteins entering the extracellular environment [1,16,21].

Additionally, HSP47 is integral to the unfolded protein response (UPR), a protective mechanism activated when the ER is overwhelmed by misfolded proteins. This chaperone helps in upregulating other chaperones and managing the protein load through degradation pathways, reinforcing its crucial position in quality control and stress response. The functionality of HSP47 is directly linked to cellular health; its dysfunction can lead to chronic ER stress, contributing to various pathologies such as fibrosis, connective tissue disorders, and cardiovascular diseases. Therefore, HSP47 not only supports protein folding and quality control within the ER but also plays a significant role in maintaining cellular homeostasis and preventing disease through effective protein management [1,4,22].

3.1.4. Procollagen Stability and the Role of HSP47

HSP47 serves as a critical collagen-specific chaperone in the endoplasmic reticulum (ER), crucial for ensuring the proper folding and stabilization of procollagen. Unlike other chaperones such as Protein Disulfide Isomerase (PDI) and Prolyl 4-Hydroxylase (P4H), which assist in oxidative folding and post-translational modifications, HSP47 directly binds procollagen during its folding process. This interaction is essential for preventing premature aggregation and facilitates the correct formation of the stable triple-helix structure of collagen, thereby influencing the quality of collagen that cells secrete [23–25].

Collagen, being a fundamental component of connective tissues, plays a key role in maintaining the structural integrity and strength of organs, skin, cartilage, tendons, and blood vessels. Dysfunction in HSP47 can lead to misfolded or unstable procollagen, adversely affecting collagen formation and consequently weakening connective tissue architecture. This dysfunction is linked to various connective tissue disorders (cTDs), such as osteogenesis imperfecta, Ehlers–Danlos syndrome, and fibrosis. These disorders often result from mutations in collagen or HSP47, leading to clinical symptoms such as fragile bones, hyperelastic skin, and impaired wound healing due to improperly assembled collagen structures [26–28].

The critical role of HSP47 in maintaining procollagen stability underscores its potential pathophysiological significance in connective tissue disorders. It highlights the promise of targeting HSP47 in therapeutic strategies aimed at optimizing collagen stability and addressing extreme connective tissue pathway disorders [1,4,29].

3.1.5. Collagen and Connective Tissue Disorders

Collagens are essential components of the extracellular matrix, playing a critical role in tissue structure and function. The collagen-specific chaperone HSP47 is integral to collagen biosynthesis, particularly in stabilizing and assembling various collagen types, which has significant implications for human collagenopathies [1,25,30].

Notably, HSP47 directly affects collagen biosynthesis and is central in connective tissue disorders (osteogenesis imperfecta and Ehlers–Danlos syndrome). HSP47 expression levels vary in diseases where collagen structures are mutated or collagen disposal is deregulatory. In addition, HSP47 has distinct interactions with different collagen subtypes, including transmembrane collagens such as Collagen XVII and IX. Overexpression of these collagens aids in providing stability to the tissue and repair mechanisms including wound healing and fibrosis-related remodeling. HSP47 interacts differently with various collagen subtypes, essential for proper tissue function. It primarily associates with fibrillar collagens (types I, II, III), crucial for the structural stability of bone, cartilage, and skin. Its role in these interactions ensures the correct triple-helix formation needed for mechanical strength. Conversely, HSP47's interaction with basement membrane collagens (types IV, VII) involves different binding kinetics, crucial for the structural integrity of tissues like skin and blood vessels [1,28,31].

As a chaperone, HSP47 aids in procollagen folding, identifying, and retaining misfolded procollagen within the ER to maintain protein quality. It also facilitates the transport of correctly folded procollagen to the Golgi apparatus for further processing before secretion into the extracellular matrix [1,14,32] (Table 2).

Table 2. Stages of Collagen Biosynthesis and HSP47's Role in Intracellular Processing.

Stage	Location	Key Events	Key Components
1. Transcription	Nucleus	Collagen genes (e.g., COL1A1) are transcribed into mRNA.	Collagen genes, RNA polymerase
2. Translation	Rough ER ribosomes	mRNA is translated into alpha chains, which enter the ER lumen.	Ribosomes, mRNA, translocon
3. ER Processing	Endoplasmic Reticulum	Hydroxylation of proline and lysine (post-translational modification). Glycosylation of hydroxylysine. Three alpha chains form a triple helix (procollagen), stabilized by HSP47 to prevent misfolding.	Prolyl 4-hydroxylase, lysyl hydroxylase, vitamin C, HSP47
4. ER-to-Golgi Transport	ER to cis-Golgi	Folded procollagen is packaged into elongated COPII vesicles and transported along microtubules to the cis-Golgi; HSP47 remains bound.	COPII vesicles, TANGO1, microtubules, kinesin
5. Golgi Processing	Golgi Apparatus	HSP47 dissociates due to pH shift and recycles to ER via COPI vesicles. Minor glycosylation may occur. Procollagen is sorted into secretory vesicles.	COPI vesicles, secretory vesicles
6. Secretion	Plasma Membrane to ECM	Secretory vesicles fuse with the plasma membrane, releasing procollagen into the ECM.	SNARE proteins, calcium
7. ECM Maturation	Extracellular Matrix	Propeptides are cleaved by peptidases, forming tropocollagen. Tropocollagen self-assembles into fibrils. Lysyl oxidase cross-links fibrils into stable collagen fibers.	ADAMTS-2, BMP-1, lysyl oxidase

In multiple studies evaluating the function of HSP47, knockout models have been used to explore the effects of the loss of HSP47 on collagen stability. For example, one prominent study analyzed collagen composition in HSP47-deficient mice by mass spectrometry, revealing impaired collagen integrity.

HSP47 is a key player in collagen biosynthesis, directly affecting the stability and assembly of collagen subtypes. Its role in maintaining collagen integrity has profound implications for connective tissue disorders, offering novel insights for therapeutic targets in treating such pathologies [27,33].

3.1.6. Binding Characteristics of HSP47

HSP47, a collagen-specific chaperone localized in the endoplasmic reticulum (ER), plays a critical role in the synthesis and folding of procollagen, significantly influencing structural integrity and disease prevention in connective tissues. The specific binding interactions between HSP47 and various collagen types are crucial to understanding tissue health and the pathophysiology of collagen-related diseases [1,34,35].

As a structural protein, HSP47 promotes the stable formation of the triple-helix structure in procollagen, which is essential for maintaining the mechanical strength and elasticity of collagen matrices in tissues such as skin, bone, and cartilage. By preventing the aggregation and misfolding of procollagen, HSP47 serves a protective function against diseases caused by collagen dysregulation, such as osteogenesis imperfecta and Ehlers–Danlos syndrome. Efficient binding of HSP47 minimizes the production of incorrectly synthesized collagen precursors, thereby mitigating the onset of associated pathologies [27,28,36].

3.1.7. The Importance of Collagen in Atherosclerotic Plaques

HSP47's role in collagen dynamics modulates stability of atherosclerotic plaque. Research indicates the potential role of HSP47 modulation in promoting fibrin plaque stability, thereby mitigating the risk of rupture and ultimately decreasing the incidence of cardiovascular events [4,37,38].

The relationship between collagen thickness in atherosclerotic plaques and the likelihood of plaque rupture is a significant topic in cardiovascular health research. Understanding how collagen content and structure contribute to plaque stability offers valuable insights into risk assessment and therapeutic strategies [39,40].

Early research established a crucial link between collagen content and plaque stability. The study demonstrated that plaques with thicker fibrous caps, which contained a higher amount of collagen, exhibited greater structural stability. Specifically, collagen-dominant caps were associated with a reduction in rupture risk compared to caps consisting primarily of thin collagen layers. Further investigation reported an inverse correlation between collagen content in the fibrous cap and the risk of rupture. The study measured collagen thickness in human carotid plaques, revealing that plaques with a thickness of less than 65 μm were significantly more prone to rupture [41–43].

Another study reinforced the significance of collagen dynamics by showing that lower collagen content in atherosclerotic plaques was associated with heightened inflammatory signals. This inflammation correlated with increased propensity for rupture. The study emphasized that a favorable collagen-to-fat ratio was linked to fewer rupture events, suggesting that maintaining a robust collagen matrix is vital for plaque stability [44–46].

Advanced imaging modalities, such as magnetic resonance imaging (MRI) and computed tomography (CT), are crucial for non-invasive assessments of plaque structure, including collagen content and thickness. These technologies enable precise categorization of patients based on their risk for cardiovascular events, facilitating early intervention and management strategies [47].

Current research also explores biomarkers associated with collagen turnover, specifically degradation products like C-telopeptides. These biomarkers have the potential to predict plaque stability and risk of rupture, thereby enhancing risk stratification in clinical settings [48,49].

Emerging therapeutic strategies aim to target collagen metabolism to stabilize plaques. Experimental approaches include increasing collagen production via specific growth factors or anti-inflammatory agents to promote a stable collagenous matrix. Additionally, investigational studies are evaluating therapies that enhance collagen crosslinking or inhibit collagenase activity with the intent to strengthen plaque integrity and reduce the risk of rupture. Collectively, these studies and perspectives provide a cohesive framework linking measurable changes in collagen thickness to plaque rupture risk. Understanding the dynamics of collagen in atherosclerotic plaques not only aids in diagnosing cardiovascular risk but also opens avenues for targeted therapeutic interventions aimed at stabilizing plaques and preventing cardiovascular events. As research advances, focusing on collagen dynamics may yield transformative strategies for managing cardiovascular health [50,51].

3.1.8. The Backbone of Atherosclerotic Plaques

The prospect of targeting heat shock protein 47 presents significant therapeutic potential for modifying plaque stability and progression in atherosclerosis. This review examines the potential advantages of HSP47 targeting and its implications for cardiovascular health. HSP47 functions as a critical molecular chaperone responsible for collagen biosynthesis and stability. By modulating HSP47 activity, it may be possible to enhance the maturation of collagen within existing atherosclerotic plaques. Increased collagen stability could lead to thicker fibrous caps, thereby reducing the likelihood of plaque rupture. Evidence suggests that strengthening collagen networks correlates with more stable plaques, which in turn is associated with a lower incidence of acute cardiovascular events [4,5,37].

The inhibition of HSP47 may influence collagen dynamics within plaques by promoting collagen degradation and reducing aberrant collagen accumulation. This process could lead to a decrease in plaque lumen narrowing and contribute to the formation of more quiescent, less inflammatory lesions over time. HSP47 plays a role in cellular responses to stress and is implicated in inflammatory processes within plaques. Targeting HSP47 may alleviate inflammation, which is known to contribute to plaque instability. By reducing inflammatory cell infiltration and cytokine release, HSP47 modulation may create a more favorable microenvironment for plaque stabilization.

There is potential for developing small-molecule inhibitors or monoclonal antibodies aimed at HSP47 as direct therapeutic agents. These agents could be designed to specifically alter HSP47's function, thereby influencing extracellular matrix (ECM) remodeling pathways during the progression of atherosclerosis.

Further research is essential to elucidate the biological role of HSP47 in atherosclerotic plaques and to establish the safety and efficacy of HSP47-targeted therapies in large-scale clinical trials. While preclinical studies show promise in modulating HSP47 expression and its effects on plaque composition, human clinical trials are necessary to validate these findings.

Targeting HSP47 represents a promising strategy for enhancing plaque stability and mitigating progression in atherosclerosis. By addressing these alterations to plaque stability, therapies aimed at HSP47 could offer new avenues for cardiovascular protection and reduce the risk of acute events such as myocardial infarctions (MI) and strokes. Continued investigation into the therapeutic potential of HSP47 will be crucial for translating these findings into clinical practice [4,52,53].

3.1.9. HSP47: Targeting Plaque Stability in Atherosclerosis

Recent research highlights the potential of targeting heat shock protein 47 (HSP47) in managing atherosclerosis and enhancing plaque stability. Experimental approaches, including histological analyses, immunohistochemistry, and mouse models, reveal the significant role of HSP47 in collagen metabolism and plaque integrity. Studies demonstrate that HSP47 expression correlates with collagen organization in atherosclerotic plaques, while genetic modifications in mouse models illustrate impaired collagen handling and decreased plaque stability when HSP47 is deficient.

Furthermore, in vitro assays using vascular smooth muscle cells and macrophages support the influence of HSP47 on collagen synthesis and inflammatory processes. The implications of HSP47 research extend to therapeutic strategies, as understanding its modulation of collagen turnover may guide the development of targeted treatments aimed at stabilizing plaques and preventing rupture.

Incorporating HSP47 targeting into personalized medicine may enhance cardiovascular outcomes by tailoring interventions based on individual plaque characteristics. Additionally, combining HSP47-targeted therapies with established treatments like statins could yield synergistic benefits in reducing cardiovascular risk.

Overall, targeting HSP47 offers promising avenues for therapeutic development in atherosclerosis treatment, warranting further exploration to solidify its role as a critical target in reducing the burden of atherosclerotic disease [4,37,53,54].

3.2. Expression Patterns of HSP47 in Atherosclerosis

Transforming Growth Factor Beta 1 (TGF-β1) and Fibroblast Growth Factor 2 (FGF-2) are key cytokines that regulate heat shock protein 47 (HSP47) and collagen production, essential for extracellular matrix (ECM) maintenance in various tissues, including atherosclerotic plaques. TGF-β1 increases HSP47 expression and collagen synthesis through the SMAD signaling pathway, whereas FGF-2 enhances HSP47 and type I procollagen expression via the MAPK/ERK and PI3K/Akt pathways [55,56].

However, the interplay between HSP47 and type I procollagen can be disrupted under inflammatory conditions, oxidative stress, and hypoxia, leading to compromised collagen synthesis and plaque instability. For instance, high levels of TNF-α can upregulate HSP47 while downregulating type I procollagen expression, resulting in weak fibrous caps prone to rupture.

Therapeutically, modulating HSP47 expression presents a promising approach for stabilizing plaques. Activating HSP47 may enhance collagen deposition and plaque integrity, while inhibition may be beneficial in cases of excessive collagen accumulation. Additionally, HSP47 levels could serve as biomarkers for plaque stability, aiding in cardiovascular risk stratification. Combining HSP47-targeted therapies with existing treatments, such as statins and anti-inflammatory agents, could optimize outcomes by synergistically promoting collagen production and plaque stability.

Overall, understanding the regulatory role of TGF-β1 and FGF-2 on HSP47 highlights the potential for targeted therapeutic strategies in the prevention and management of atherosclerosis and its related cardiovascular events [4,37,55,56].

3.2.1. Overview of Collagen Types and Their Functions

The externalization of phospholipids, particularly phosphatidylserine (PS), on activated platelets is essential for promoting coagulation, a critical step in hemostasis to prevent excessive bleeding following vascular injury. Upon injury, platelets adhere to collagen and extracellular matrix components, initiating activation and morphology changes-from disk-like shapes to sphere-like forms with filopodia extensions. This transformation increases the surface area available for interactions and exposes PS on the outer membrane leaflet, providing a pro-coagulant surface [57–59].

Exposed PS enhances the binding affinities of key coagulation factors, such as factor X and prothrombin, facilitating thrombin generation and promoting further platelet aggregation. This interaction orchestrates the formation of catalytic complexes necessary for converting prothrombin into thrombin and fibrinogen into fibrin, effectively driving the coagulation cascade [57,59,60].

Tissue factor, expressed at the site of injury, interacts with factor VIIa, forming a complex that activates factor X and further enhances coagulation through its localized activity on phospholipid-rich platelets. This localized concentration of active factors amplifies the coagulation response, increasing thrombin production and fibrin generation, with calcium ions (Ca^{2+}) playing a crucial role in facilitating factor binding to the phospholipid surfaces [61,62].

Overall, the exposure of phospholipids on activated platelets significantly enhances the binding and activity of various coagulation factors, promotes catalytic complex formation, and amplifies the coagulation cascade. Understanding these mechanisms is vital for developing therapeutic strategies to manage conditions of excessive hemorrhage or thrombosis effectively [57,59,63].

3.2.2. Role of Collagen in Hemostasis

Type IV collagen plays a pivotal role in hemostatic dynamics by regulating coagulation factors such as factor IX (FIX) and von Willebrand factor (VWF). FIX, a vitamin K-dependent serine protease, directly binds to type IV collagen through its propeptide region, which is essential for localizing FIX to vascular injury sites and facilitating its activation by factor XIIa or the tissue factor VIIa complex. Studies indicate that immobilized collagen type IV enhances FIX's binding affinity, significantly improving its activation rate compared to other surfaces. VWF, an essential mediator of hemostasis, binds to collagen type IV, which helps anchor VWF at injury sites for platelet adhesion. This interaction is crucial under high shear stress conditions, promoting the formation of a primary hemostatic plug by enabling VWF to bind platelet glycoprotein Ib (GPIb). The presence of type IV collagen stabilizes thrombus formation, as it enhances platelet activation and aggregation through VWF interactions. In vitro studies reveal that the binding of FIX to collagen type IV markedly increases FIX activation rates, while animal models show that collagen IV supplementation can partially restore platelet function in VWF knockout mice, highlighting its stabilizing effects in hemostasis. Understanding the interactions between type IV collagen, FIX, and VWF is vital for developing therapeutic strategies aimed at enhancing hemostatic potential in patients with bleeding disorders. Overall, these mechanisms underscore the importance of type IV collagen in promoting effective hemostasis, particularly following vascular injury [64–66].

3.2.3. Platelet Adhesion and Shear Rate Dynamics

Single-layer adhesion and aggregation are two critical processes in platelet function during hemostasis. Single-layer adhesion refers to the initial attachment of platelets to exposed extracellular matrix (ECM) components, such as collagen and von Willebrand factor (VWF), at sites of vascular injury. This process involves interactions between specific platelet surface receptors (e.g., glycoprotein Ib/IX/V with VWF and integrin $\alpha 2\beta 1$ with collagen), resulting in the formation of a monolayer of platelets that aids in blocking blood loss and establishing the initial hemostatic plug.

Aggregation, on the other hand, occurs after initial adhesion, where activated platelets recruit and aggregate into a larger mass or thrombus. This process is driven by the release of soluble agonists (e.g., ADP and thromboxane A2) and conformational changes in integrins (e.g., $\alpha IIb\beta 3$), enabling platelet–platelet interactions facilitated by fibrinogen bridging. The resulting platelet-rich thrombus covers tissue damage and prevents bleeding. The dynamics of blood flow shear rates are crucial in influencing these processes. At high shear rates, such as those in arterioles, platelet adhesion occurs rapidly, but aggregation may be limited due to strong binding forces that can disperse platelets before full aggregation. Understanding shear dynamics is essential for assessing thrombus formation efficiency and risk stratification for cardiovascular events, helping guide therapeutic interventions such as antithrombotic treatments.

In summary, single-layer adhesion represents the initial tethering of platelets to injury sites, while aggregation involves the clustering of platelets into a stable thrombus. Investigating shear rate dynamics is vital for optimizing thrombus formation and developing appropriate clinical strategies to mitigate the risks of thrombosis and bleeding [67–69].

3.2.4. Characterization of Collagen Fragments and Platelet Response

Research on linear and triple-helical peptides presents promising avenues for therapeutic strategies targeting coagulation and collagen-related disorders.

Targeting platelet function: linear peptides derived from collagen sequences have been shown to influence platelet adhesion and activation. These peptides can inhibit platelet aggregation by blocking binding sites on glycoprotein receptors, paving the way for the development of specific antiplatelet therapies aimed at preventing thrombus formation in conditions like atherosclerosis and minimizing re-thrombosis during surgical procedures.

Increased stability of collagen: triple-helical peptides that mimic the structural stability of native collagen may enhance vascular collagen stabilization, potentially preventing aneurysms. Their application can support tissue repair and reinforce collagen networks within atherosclerotic plaques, reducing rupture risk and lowering the incidence of cardiovascular events.

Drug delivery systems: the distinctive structural features of linear and triple-helical peptides can be harnessed for drug delivery systems. Peptide carriers designed to target collagens or adhesion molecules could facilitate the localized delivery of therapeutic agents to vascular injury sites or areas with abnormal collagen deposition.

Biomarker development: investigating collagen-derived peptides may help identify biomarkers indicative of pathophysiological conditions related to disturbed collagen metabolism, such as fibrotic diseases or vascular complications. Monitoring circulating peptide concentrations could provide valuable diagnostic and prognostic insights [70].

Responsive personalized medicine: insights into the interactions between linear and triple-helical peptides and platelet receptors can inform personalized medicine approaches. Understanding polymorphisms related to peptide interactions can guide the selection of optimal treatment regimens, maximizing therapeutic efficacy while minimizing side effects.

Findings from studies on linear and triple-helical peptides highlight new therapeutic strategies for modulating platelet function, enhancing tissue repair, and optimizing vascular health. These developments lay the groundwork for targeted therapies addressing various vascular and coagulation abnormalities, advancing the potential for personalized medicine in this field [71–74].

3.2.5. Interaction Between Integrins and Collagen

Integrin $\alpha2\beta1$ acts as a crucial collagen receptor on platelets, significantly impacting platelet adhesion and aggregation during hemostasis. Its modulation presents several therapeutic opportunities, particularly in cardiovascular diseases and thrombotic disorders.

Antithrombotic strategies: inhibition of $\alpha2\beta1$ may serve as a novel antithrombotic approach, especially for patients at high risk for thrombus formation due to atherosclerosis or vascular injury. By reducing platelet adhesion to collagen, this inhibition could decrease thrombus growth and lower the incidence of acute cardiovascular events such as myocardial infarction and stroke.

Enhancing platelet function: conversely, activating $\alpha2\beta1$ is beneficial in contexts like bleeding disorders or surgical settings, as it promotes effective platelet adhesion to collagen-rich tissues, enhancing hemostatic efficiency.

Collagen dynamics regulation: modulating $\alpha2\beta1$ can also affect platelet adhesion in collagen-rich ECM, helping to restore proper collagen dynamics during disease states associated with excessive or insufficient remodeling, contributing to improved healing post-injury.

Personalized therapies: patient-specific variations in $\alpha2\beta1$ expression may influence therapeutic responses, enabling the development of customized treatment strategies tailored to individual platelet receptor profiles or genetic backgrounds.

PDI's role in $\alpha2\beta1$ regulation: protein disulfide isomerase (PDI) enhances $\alpha2\beta1$ activation by facilitating conformational changes that shift the integrin to a high-affinity state for collagen. PDI's regulation intersects with other platelet activation pathways, including GPVI and thrombin signaling, which may amplify platelet responses and increase the risk of thrombosis in pathological conditions like atherosclerosis.

The ability to modulate $\alpha2\beta1$ activity presents significant therapeutic potential in managing hemostatic responses and thrombotic risks. Understanding the regulatory role of PDI and the interplay of various activation pathways will be crucial for developing targeted therapies that balance effective hemostasis without triggering excessive thrombosis [75–77].

3.2.6. The Implications of GPVI and Integrin Interplay

Glycoprotein VI (GPVI) and integrin α2β1 are critical receptors on platelets that mediate their activation and adhesion to collagen during hemostasis. These receptors exhibit both redundancy and synergy in their functions, ensuring effective platelet responses to vascular injury.

Redundancy in function: both GPVI and α2β1 bind collagen, a key ligand present in the extracellular matrix upon vascular damage. Their overlapping functions allow compensation if either receptor pathway is impaired, enabling continued platelet adhesion and activation. GPVI primarily activates platelets through immunoreceptor tyrosine-based activation motif (ITAM) signaling, leading to calcium mobilization, while α2β1 enhances activation via integrin signaling pathways.

Synergy in function: at intermediate collagen concentrations, GPVI and α2β1 work cooperatively. Initial activation through GPVI enhances the affinity of α2β1 to collagen, promoting stable platelet attachment and larger thrombus formation, particularly in high-shear environments like arterioles.

Clinical significance: understanding the interactions between GPVI and α2β1 is crucial for developing antiplatelet therapies. Targeting GPVI can reduce platelet reactivity; however, α2β1 may still function independently, necessitating a dual inhibition strategy for more effective antithrombotic effects. Moreover, inhibiting these pathways can come with the risk of increased bleeding, emphasizing the need for careful dosing and individualized treatment strategies.

The redundancy and synergy between GPVI and α2β1 underpin the complexity of hemostatic responses and have important implications for antiplatelet therapy development. Insights into these interactions can lead to improved therapeutic strategies for managing thrombotic diseases while minimizing side effects [70,78].

3.2.7. Novel Investigations into Platelet Collagen Receptors

Single-chain antibodies (scFvs) are engineered antibodies formed by linking the variable regions of heavy and light chains with a flexible peptide, facilitating access to conformational epitopes. Their small size, ease of production, and specificity make scFvs valuable tools for research and therapeutic applications, particularly in targeting receptors involved in coagulation, such as glycoprotein VI (GPVI).

Development and validation: the development of scFvs commonly employs phage display technology, allowing the creation of diverse libraries and the selection of high-affinity binders against target antigens. Validation of scFvs includes techniques like ELISA and flow cytometry to confirm binding and biological activity.

Species-specific differences in collagen receptors: research highlights the functional implications of species-specific variations in collagen receptors like GPVI and integrin α2β1. Studies indicate that human GPVI shows higher binding affinity to type I collagen compared to murine GPVI, affecting therapeutic efficacy and safety when translating findings from animal models to human applications.

Impact of GPVI polymorphisms on cardiovascular risk: genetic studies have linked polymorphisms in the GP6 gene, which encodes GPVI, to altered platelet function and increased cardiovascular disease risk. Specifically, certain GPVI polymorphisms correlate with heightened platelet reactivity to collagen, leading to an increased incidence of myocardial infarction. These genetic variations may influence platelet activation and inflammatory cytokine release, thereby contributing to atherogenesis.

ScFvs offer a promising platform for therapeutic development, while understanding interspecies differences in collagen receptor function and the implications of GPVI polymorphisms can enhance personalized medicine approaches in cardiovascular care.

Genotyping GPVI variations may aid in defining individual cardiovascular risk and in tailoring effective anti-platelet therapies [79–81].

3.2.8. The Importance of Adhesion Motifs and Cytoskeletal Extension

Platelet adhesion is a crucial process involving interactions between platelet receptors, such as integrin α2β1, GPVI, and CD36, with collagen fibers, significantly influenced by shear stress environments. Understanding these interactions is vital for addressing thrombotic diseases, including arterial stenosis and venous thrombosis.

Effective platelet adhesion and cytoskeletal extension are essential for thrombus formation. Under high shear conditions, robust receptor engagement, particularly integrin α2β1 and the von Willebrand factor (VWF) axis, is necessary for stable adhesion. Consequently, antiplatelet therapies may need to be tailored based on the hemodynamic environment-different approaches may be required for patients experiencing high shear stress versus those in slower flow conditions.

The hypothesis of a two-site, two-step adhesion mechanism reveals that initial weak binding via GPVI is followed by stronger adhesion through integrin α2β1. This model highlights the varied roles of platelet receptors in different thrombotic contexts, where GPVI and CD36 dominate in venous thrombus formation, promoting aggregation in low-flow conditions.

In arterial stenosis, elevated shear rates create a pro-thrombotic environment that increases platelet activation and underscores the need for aggressive antithrombotic therapies to prevent cardiovascular events. In the case of venous thrombosis, pronounced receptor interactions can lead to large, fibrin-rich thrombi obstructing venous flow [82].

Additionally, the presence of long, insoluble collagen fibers in pathological conditions can enhance thrombus stability, indicating that disrupting receptor–collagen interactions might be a viable therapeutic strategy.

Insights into platelet receptor interactions with collagen under varying shear conditions are critical for developing effective clinical approaches to manage thrombotic diseases. A deeper understanding of these dynamics will aid in the design of targeted antithrombotic therapies, ultimately improving patient outcomes in conditions characterized by abnormal thrombus formation [68–70,76,83].

3.2.9. The Role of Collagen Type I in Thrombus Formation

Perfusion chambers are essential experimental tools that simulate physiological and pathophysiological conditions relevant to hemostasis and thrombosis, allowing researchers to investigate platelet behavior under controlled flow rates and collagen preparations.

Flow rates and shear stress: perfusion systems can replicate low and high shear rates, simulating venous and arterial blood flow conditions, respectively. This enables the study of the effects of shear stress on platelet adhesion and thrombus formation, providing insights into the dynamics of thrombus stability during vascular injury, such as in atherosclerosis or following angioplasty.

Collagen preparation: collagen is derived from human or animal tissues, providing control over the types and concentrations of collagen that mimic pathological conditions. The incorporation of other extracellular matrix components like fibronectin and von Willebrand factor (vWF) enhances the relevance of the models to real-world biological interactions.

Therapeutic targeting of vWF: vWF plays a critical role in platelet adhesion to collagen and stabilizes factor VIII in circulation, thus facilitating effective clotting. Under high shear conditions, vWF adopts an active form that promotes thrombus formation. Targeting vWF with antibodies or small molecules can reduce platelet aggregation, offering potential strategies for managing acute coronary syndromes and preventing thrombus formation following angioplasty.

Relevance to atherosclerosis and restenosis: understanding the interplay between vWF, platelets, and collagen is crucial in the context of atherosclerosis and post-angioplasty restenosis, where platelet activation can lead to adverse thrombotic events. Inhibiting vWF may represent a therapeutic avenue to prevent thrombus formation on ruptured plaques and reduce re-occlusion risks after stenting procedures.

Perfusion chamber studies provide valuable in vitro insights into platelet dynamics across varying shear conditions, highlighting their significance in developing antithrombotic therapies aimed at mitigating thrombotic complications associated with vascular diseases. This research trajectory promises to enhance clinical management and improve outcomes for patients at risk for thrombotic events [84–87].

3.3. Impact of HSP47 on Platelet Binding to Collagen

HSP47 plays a prominent role in platelet activity with impact on thrombus generation in hemostasis. It helps create stable collagen in the ECM and enables proper platelet–collagen interaction necessary for stable thrombus. HSP47 has been implicated in the development of both arterial and venous thrombi and has also been shown to inhibit thrombus formation during in vitro assays; as such, it represents a promising target for development of therapeutics in the setting of thrombotic events.

Targeting heat shock protein 47 (HSP47) emerges as a critical strategy for understanding collagen-mediated platelet interactions, particularly regarding the selective impact of HSP47 inhibitors. These inhibitors effectively reduce collagen-associated platelet adhesion and aggregation while leaving thrombin-induced activation unaffected, indicating HSP47's specific role in stabilizing collagen–platelet interactions.

HSP47 inhibitors specifically influence collagen-mediated interactions by stabilizing collagen fibers and facilitating receptor engagement on platelets, primarily through receptors such as GPVI and integrin $\alpha 2\beta 1$. In contrast, thrombin activates platelets through distinct receptors like PAR-1 and PAR-4, which do not rely on collagen binding or HSP47 stabilization, thereby remaining unaffected by HSP47 blockade.

The ability of HSP47 inhibitors to diminish platelet adhesion exclusively in the presence of collagen highlights HSP47's role as a major modulator in the initial interactions between platelets and collagen. This direct involvement underscores HSP47's importance in mediating platelet adhesion and the subsequent signaling necessary for thrombus formation.

Given its targeted influence on collagen-dependent platelet interactions, HSP47 presents a promising therapeutic target for preventing thrombosis without the side effects associated with broader antiplatelet therapies that disrupt ADP or thrombin signaling. Overall, HSP47 is best characterized as a primary mediator of collagen–platelet interactions, making it an attractive candidate for selective blockade in thrombotic disease management [6,52,88].

3.3.1. The Effect of HSP47 Inhibition on Platelet Aggregation

The use of 96-well plate assays to examine the effects of heat shock protein 47 (HSP47) inhibition on platelet aggregation provides valuable insights into therapeutic strategies aimed at modulating thrombotic diseases.

These assays demonstrate high sensitivity in detecting collagen-mediated changes in platelet aggregation, allowing for both partial and complete inhibition depending on the collagen concentration. Assay reproducibility is ensured through multiple runs and the inclusion of appropriate controls, such as baseline wells devoid of inhibitors and standardized platelet preparations, to mitigate variability.

The results reveal a dose-dependent relationship in the inhibition of platelet aggregation by HSP47 inhibitors, indicating that complete inhibition occurs at lower collagen concentrations, while higher concentrations lead to variable inhibition levels. This suggests that competition between collagen and inhibitors at binding sites informs the effectiveness of HSP47 stabilization on platelet binding.

Given the critical role of HSP47 in platelet–collagen interactions, targeting this pathway may provide new strategies for managing thrombotic conditions, such as deep vein thrombosis and myocardial infarction. Optimizing inhibitor dosages can facilitate tailored therapeutic approaches that effectively reduce pathological platelet activity while preserving essential hemostatic functions.

The 96-well assay methodology offers a robust framework for assessing the potential of HSP47 inhibitors in platelet aggregation, contributing to the development of refined therapeutic strategies to combat thrombotic diseases through precise dose management [6,52].

3.3.2. HSP47's Role in Calcium Mobilization and Platelet Function

Inhibition of heat shock protein 47 (HSP47) has been shown to reduce calcium ion (Ca^{2+}) mobilization, a critical process that influences platelet activation and thrombus stabilization. Ca^{2+} signaling is essential for platelet shape change, granule release, and aggregation, all of which are necessary for effective hemostasis and stable thrombus formation. HSP47 inhibition decreases Ca^{2+} rise in response to collagen stimulation, thereby preventing platelet activation and potentially destabilizing thrombi during thromboembolic events.

Reliable data on Ca^{2+} mobilization necessitate proper experimental controls, employing isotype control antibodies to establish the specificity of HSP47's effects. Additional stimulants like thrombin and ADP help confirm that HSP47 inhibition specifically impacts collagen signaling rather than universally compromising platelet responses. Notably, while the response to CRP-XL (a GPVI-activating agent) is delayed with HSP47 inhibition, it is not entirely abolished, indicating that HSP47 is significant for collagen-mediated interactions but not essential for all platelet activation pathways.

Targeting HSP47 presents beneficial therapeutic implications for thrombotic disorders, such as deep vein thrombosis and acute coronary syndrome. Selective inhibition of platelet aggregation could be achieved while maintaining necessary hemostatic functions. Furthermore, investigating individual variability in responses to HSP47 inhibition may lead to personalized treatments that enhance efficacy and minimize bleeding risks.

HSP47 plays a crucial role in Ca^{2+} mobilization vital for platelet function and thrombus stability. The findings underscore the potential for developing tailored therapies for thrombotic conditions, facilitated by robust experimental controls in research [6,52].

3.3.3. Influences of HSP47 on Thrombus Formation and Aggregation

To investigate HSP47's role in platelet reactivity under physiological conditions, DiOC6-labeled blood from mice lacking platelet HSP47 and control mice was perfused over collagen. Thrombus formation, measured by fluorescence intensity, was reduced by about half in HSP47-deficient mice compared to controls. Similarly, human blood treated with HSP47 inhibitors showed a reduction in thrombus volume by roughly one-third with SMIH and a slightly smaller reduction with anti-HSP47 compared to controls.

Importantly the role of HSP47 in collagen stability has been demonstrated in murine models and corroborated in humans. While HSP47 deficiency leads to severe phenotypes in mice, including embryonic lethality because of unstable collagen type I, issues of collagen stability directly relate to HSP47. Interestingly, high HSP47 levels in adipose tissues are linked with obesity and increased collagen deposition in humans, implying a similar functional significance for mammals. The conservation of this function throughout species highlights the importance of HSP47 as a key player in the interplay between collagen and health [6,33].

3.3.4. Evaluating the Specificity of HSP47 in Platelet Interactions

Heat shock protein 47 (HSP47) is a molecular chaperone with a high affinity for collagen, playing a crucial role in the synthesis and functionality of different collagen types. Its specificity in binding to procollagen and mature collagen fibrils is determined by the unique structural and biochemical properties of the binding site, particularly interactions with proline and hydroxyproline residues essential for collagen's triple-helical conformation.

Unlike most serpins, HSP47 does not serve as a protease inhibitor; instead, it stabilizes collagen by engaging with its hydrophobic zones and forming salt bridges with specific amino acids, such as arginine and aspartate. HSP47 exhibits promiscuous binding to various collagen types, including I, II, III, and IV, and preferentially interacts with higher-order collagen structures containing interrupted triple helices, ensuring proper collagen folding and assembly.

Moreover, HSP47 is critical for platelet adhesion and aggregation by stabilizing collagen in the extracellular matrix. Inhibiting HSP47 leads to decreased platelet aggregation and thrombus formation specifically in response to collagen while not affecting pathways activated by thrombin or GPVI.

The binding specificity of HSP47 highlights its essential role in collagen dynamics and platelet function, suggesting it as a promising therapeutic target for disorders related to connective tissue and thrombotic diseases. Understanding these mechanisms could pave the way for the development of new treatments aimed at modulating collagen interactions and platelet recruitment [5,6,15].

3.3.5. Exploring Early Signaling Events Following GPVI Ligation

HSP47 plays a critical role in binding distinct types of collagen, which significantly influences the regulation of platelet activation and aggregation. This relationship offers valuable insights into the mechanisms that maintain hemostatic balance and may lead to the development of novel therapeutic modalities.

During vascular injury, exposed collagen serves as a potent inducer of platelet activation, while HSP47 stabilizes collagen structure within the extracellular matrix. The interaction between platelets and collagen is vital for activating signaling pathways via GPVI and integrin $\alpha 2\beta 1$, resulting in platelet adhesion, shape change, and degranulation. HSP47's role in structurally stabilizing collagen enhances platelet aggregation by facilitating the optimal clustering of adhered platelets to form a stable thrombus [80].

In contrast, the inhibition of HSP47 significantly reduces platelet aggregation and impairs crucial downstream signaling pathways, including calcium mobilization, which is essential for optimal platelet function. This understanding of HSP47's functional role reveals potential therapeutic targets for ameliorating hemostatic disorders characterized by either excessive or insufficient platelet aggregation. Specifically, HSP47 modulation could provide a new therapeutic avenue for preventing abnormal platelet coalescence during acute events, such as myocardial infarction.

Moreover, the SH2 domain of HSP47 is integral to its binding with collagen and subsequent platelet activation. These insights may help inform novel therapeutic strategies aimed at enhancing patient outcomes in bleeding disorders and vascular diseases. Overall, targeting HSP47 could be instrumental in developing treatments that balance hemostatic function and mitigate thrombotic risk [6,52,88].

3.3.6. Intravital Microscopy: Examining Platelet Recruitment Post-Injury

HSP47's specificity for binding various collagen types positions it as a critical regulator in platelet activation and aggregation while maintaining a delicate balance between hemostasis and thrombosis. This has significant implications for therapeutic strategies targeting cardiovascular diseases.

Platelet activation is primarily initiated by the exposure of collagen following vascular injury. HSP47 enhances the structural stability of collagen in the extracellular matrix, facilitating effective receptor binding and subsequent platelet signaling through receptors such as glycoprotein VI (GPVI) and integrin α2β1. These interactions govern platelet adhesion and trigger essential processes, including shape change and granule secretion. HSP47's stabilization of collagen allows for effective platelet aggregation, promoting thrombus formation.

Inhibition of HSP47 markedly decreases platelet aggregation and disrupts downstream signaling, particularly calcium mobilization, which is vital for platelet function. An established model for studying vascular responses to collagen enables real-time imaging of thrombus dynamics in responses to injury, showcasing how HSP47's role affects thrombus growth compared to other injury models.

While early thrombus kinetics remain similar across models, HSP47 inhibition leads to reduced growth phases, highlighting its significance in thrombus development. Targeting HSP47 may present a novel therapeutic approach to selectively modulate thrombus stability and size, potentially preventing excessive aggregation during acute conditions like myocardial infarction while still allowing necessary platelet–collagen adhesion for hemostasis.

HSP47's binding specificity to collagen is crucial for efficient platelet activation and aggregation. Understanding its role opens avenues for developing targeted therapies that can better manage hemostatic disorders associated with dysfunctional platelet aggregation, offering potential solutions for improving vascular health and controlling thrombotic conditions [6,52,88].

3.3.7. Impact of HSP47 on Hemostasis in Mice

Inhibition of heat shock protein 47 (HSP47) using the small molecule inhibitor SMIH, alongside data from platelet-specific HSP47-deficient mice, highlights HSP47 as a critical regulator of hemostasis. The administration of SMIH to C57Bl/6 mice resulted in an increase in tail bleeding time, indicative of hemostatic dysfunction and impaired platelet function. Further analysis revealed that HSP47-deficient mice exhibited a 1.8-fold increase in bleeding time compared to controls, confirming the essential role of HSP47 in maintaining normal platelet activity.

These findings correlate with clinical observations in humans, where patients with functional platelet disorders, such as Glanzmann thrombasthenia (characterized by defective integrin αIIbβ3), experience prolonged bleeding times due to impaired platelet aggregation. The parallels between enhanced bleeding resulting from HSP47 inhibition in experimental models and the challenges faced by patients with intrinsic platelet defects underscore HSP47's vital function in preserving platelet interactions with collagen, primarily mediated by integrins.

While HSP47 presents a promising target for therapeutic interventions aimed at managing pathological thrombus formation, it is crucial to balance the need for sufficient hemostatic responses against the risk of excessive clotting. The extended bleeding times associated with HSP47 inhibition further emphasize the importance of platelet function in hemostatic control, providing a rationale for targeting HSP47 in the treatment of hemostatic disorders while minimizing the risk of bleeding complications. Overall, HSP47 stands out as a significant player in the regulation of platelet function and hemostasis [6].

3.3.8. The Role of Surface Receptors in Platelet Activation

Platelet adhesion to collagen at sites of vascular injury is a vital step in hemostasis, primarily mediated by two key receptors: glycoprotein VI (GPVI) and integrin α2β1. The significant overlap between these receptors has profound implications for platelet physiology and the development of therapeutic strategies.

Glycoprotein VI (GPVI): this receptor is crucial for initiating signaling pathways that activate platelets and facilitate thrombus formation in response to collagen.

Integrin α2β1: serving as the primary collagen-binding site, α2β1 is essential for the firm adhesion of platelets to the extracellular matrix, enhancing platelet–collagen interactions.

In addition to GPVI and integrin α2β1, other receptors such as integrin αIIbβ3 and CD36 also contribute to collagen adhesion, increasing redundancy in platelet response mechanisms.

The redundancy among collagen receptors allows for compensatory signaling pathways when one receptor's function is impaired. For instance, if GPVI signaling is diminished, integrin α2β1 can help rescue platelet adhesion, thus preserving hemostatic function.

Differences in receptor expression levels among individuals can influence platelet responses to collagen, indicating the need for a multifaceted approach in therapeutic targeting. Recognizing the redundancy in receptor functions is essential for developing effective treatment strategies for thromboembolic diseases.

Focusing solely on GPVI as a therapeutic target may lead to suboptimal outcomes due to compensatory mechanisms from other receptors. Therefore, a strategy that targets multiple receptors may enhance therapeutic efficacy while mitigating risks associated with thrombotic disorders.

Vascular conditions and shear stress can alter thrombus formation dynamics, necessitating the involvement of multiple receptors to maintain proper function under high-shear environments.

Overall, the redundancy of receptors mediating platelet–collagen adhesion is critical for maintaining hemostatic balance. This understanding provides a rationale for exploring receptor interactions further, paving the way for novel receptor-based therapies aimed at improving the management of bleeding and thrombotic disorders [78,82,89].

3.3.9. Interaction of HSP47 with Collagen in Platelet Function

HSP47, a collagen-specific molecular chaperone, plays a critical role in the interactions between platelets and collagen during vascular injury, essential for thrombus stability and successful hemostasis. Following vascular damage and collagen exposure, platelets adhere through various receptors, including glycoprotein VI (GPVI) and integrin α2β1, forming a vital platelet plug to prevent bleeding. HSP47 enhances the binding between platelets and collagen after this initial adhesion, promoting stability in thrombus formation.

Recent studies utilizing flow cytometry and imaging have shown that HSP47 facilitates the recruitment of additional platelets to the thrombus, thereby strengthening platelet–collagen interactions. Inhibition of HSP47 leads to less stable platelet aggregates, underscoring its importance in maintaining thrombus integrity. Furthermore, HSP47's ability to bind to phosphorylated forms of the extracellular matrix (ECM) and interact with various ECM proteins, including fibronectin, laminin, and proteoglycans, indicates its broader role in structural support and platelet activation.

These multifaceted functions of HSP47 highlight its significance in hemostasis by stabilizing platelet–collagen interactions and facilitating dense platelet aggregation. Understanding HSP47's interactions with diverse ECM components may provide new insights into thrombus biology and uncover potential therapeutic strategies for managing hemostatic disorders and enhancing vascular health [6,90].

3.4. Understanding HSP47's Modulatory Effects on Platelet Responses

Heat shock protein 47 (HSP47) is recognized for its essential role in thrombosis, particularly in platelet–collagen interactions that strengthen thrombus structure and facilitate efficient hemostasis. However, its relevance extends beyond thrombotic conditions to include critical functions in wound healing and tissue repair, underscoring its multifactorial importance in vascular biology.

Following vascular injury, HSP47 stabilizes the binding between platelets and collagen, enhancing platelet activation and aggregation while recruiting additional platelets to the injury site. In the wound healing process, HSP47 assists in the proper folding and secretion of procollagen, promoting healthy collagen deposition within the extracellular matrix (ECM).

Furthermore, HSP47 facilitates cell migration by binding to ECM components like fibronectin and laminin, which enhances granulation and vascularization. It also modulates interactions between collagens and immune cells, influencing the recruitment and activation of inflammatory cells essential for effective tissue regeneration.

Understanding the dual contributions of HSP47 in both thrombosis and wound healing can inform therapeutic strategies. Inhibiting HSP47 might exacerbate thrombus formation without hindering tissue repair, while enhancing HSP47 functionality could benefit both thrombotic events and the healing process.

HSP47 represents a promising target for drug development aimed at addressing bleeding disorders and improving recovery following vascular injuries. Ongoing studies of HSP47's interactions within the ECM may reveal further therapeutic opportunities for managing disorders related to hemostasis and tissue repair [6,11,55].

3.4.1. Unraveling HSP47's Role in Collagen Dynamics: Future Research and Therapeutic Implications

Despite the insights gained from HSP47 knockout models, a complete understanding of the dependence of different collagen types on HSP47 remains elusive. More comprehensive research is needed to elucidate the mechanisms underlying HSP47 binding to various collagen types and the consequences of its absence in physiological contexts.

Future research directions include investigating the client-specific dependence of HSP47 on different collagen types, as some may require HSP47 more critically for proper folding and stability than others. Additionally, characterizing less understood collagen types, such as FACIT and MACIT subtypes, may identify new HSP47 clients and their physiological relevance. Employing advanced genetic and biochemical techniques, such as CRISPR/Cas9 and proteomic analyses, could provide deeper insights into the functional roles of HSP47 across various tissues and conditions [91].

Understanding HSP47's interactions with collagen is essential for developing therapeutic strategies aimed at enhancing collagen stability in connective tissue diseases. Identifying small molecules that can promote HSP47 activity or strengthen its binding to specific collagen types could lead to new treatment approaches. Collectively, the binding properties of HSP47 are vital for maintaining tissue homeostasis and preventing disease, highlighting the need for further research to explore these complex interactions and their therapeutic potential in connective tissue disorders [1,31,91].

3.4.2. Heat Shock Protein 47: A Dual Regulator of Platelet–Collagen Interactions and Immune Response in Plaque Thrombosis and VTE

Heat shock protein 47+ is a collagen-specific chaperone that plays a critical role in plaque thrombosis by enhancing platelet–collagen interactions and modulating immune cell activity. By maintaining collagen integrity, HSP47 ensures proper platelet adhesion and aggregation. Studies using knockout models and pharmacological inhibition have demonstrated that downregulation of HSP47 leads to impaired thrombus formation, longer bleeding times, and reduced platelet aggregation.

Additionally, decreased HSP47 expression can alter immune cell activities by modifying cytokine secretion and preventing immune cell influx, which delays inflammation. In the context of venous thromboembolism (VTE), chronic low-grade inflammation may lower HSP47 levels, impairing collagen remodeling and thrombus stability.

Alterations in HSP47 expression in patients with VTE correlate with risk factors such as immobility and obesity, emphasizing its potential as a therapeutic target for VTE management. Manipulating HSP47 levels may influence thrombus stability and aid in inflammation resolution. These findings suggest that HSP47 plays a dual role in hemostasis and inflammation, providing novel insights for pharmacological targeting in thrombotic disorders [6,91].

3.4.3. Investigative Journey of Manuela Thienel

Manuela Thienel, a cardiologist, usually works in a hospital but ventured into a snowy Swedish forest in 2019 to study hibernating bears. She sought answers to why bears do not develop blood clots during hibernation, unlike immobile humans. After years of research involving various animals and human volunteers, her team identified a protein, HSP47, which decreases in bears during hibernation and may reduce clotting risk in humans. During hibernation, bears conserve energy by lowering their body temperature and slowing their metabolism. This likely reduces the need for HSP47, a protein that helps with collagen production, since tissue repair slows down. Hormonal changes, like lower insulin or growth factor levels, may also decrease HSP47 production. With less HSP47, blood clotting could decrease due to reduced collagen-related platelet activity, potentially protecting against clots during long periods of inactivity. Studying this could help find ways to lower clotting risks in humans, such as in deep vein thrombosis or stroke. This discovery suggests a potential new target for clot-preventing drugs, although further research and clinical trials are needed. Coagulation experts are intrigued by the unconventional approach taken by Thienel's team in addressing a major medical challenge.

Dr. Manuela Thienel's research explores the unique physiological adaptations of hibernating bears, particularly their ability to enter hibernation without developing blood clots. This investigation employs several methodologies to study protein expression during hibernation.

Proteomic analysis: mass spectrometry-based proteomic analysis of muscle tissue samples was conducted to identify differences in protein expression between active and hibernating bears.

Comparative studies: the research team compared the function and expression of HSP47 across various animal species and in human volunteers.

Functional analysis of blood coagulation: blood assays revealed that HSP47 levels correlate with coagulation efficiency, suggesting its role in preventing clot formation during hibernation.

Challenges in drug development: while HSP47 presents an exciting target for drugs aimed at preventing blood clotting, challenges remain, including ensuring specificity and safety of the drugs, addressing species variability, designing appropriate clinical trials, and navigating regulatory requirements for safety and efficacy testing.

Dr. Thienel's research lays the foundation for developing novel strategies to manage clotting disorders, necessitating further investigation into HSP47 as a potential therapeutic target in clinical settings [2,92].

4. Discussion

This review underscores the emerging significance of heat shock protein 47 (HSP47) in thrombotic disorders, particularly deep vein thrombosis (DVT) and atherosclerosis. Traditionally recognized for its role as a collagen-specific chaperone in the endoplasmic reticulum (ER), HSP47 also facilitates platelet–collagen interactions critical for thrombus formation and vascular stability [1,5,6].

Recent studies demonstrate that inhibition or genetic deletion of HSP47 specifically reduces collagen-mediated platelet aggregation, calcium mobilization, and thrombus volume, without affecting thrombin-induced pathways [6,52,88]. This specificity suggests a therapeutic window for HSP47 modulation that could minimize thrombosis risk while preserving overall hemostasis.

Findings from hibernating mammals further support HSP47's role in thromboprotection. Bears, which experience prolonged immobility during hibernation, show decreased HSP47 expression and reduced thrombus formation, potentially due to lower collagen-dependent platelet activation and inflammation [2,6,93]. These natural adaptations provide a foundation for new antithrombotic strategies in immobile or critically ill patients.

In atherosclerosis, HSP47 influences collagen content and fibrous cap stability in plaques, thereby modulating rupture risk and cardiovascular outcomes [4,37]. Depending on disease context, both inhibition and enhancement of HSP47 activity may have therapeutic value—either by preventing excessive collagen accumulation in fibrosis or by reinforcing ECM stability in vulnerable plaques.

Despite promising evidence, challenges remain. Redundancy in platelet receptors (e.g., GPVI, integrin $\alpha 2\beta 1$) and species-specific differences in HSP47 function necessitate careful therapeutic design [78,79]. Moreover, long-term systemic modulation of HSP47 could impact wound healing or connective tissue integrity.

Future studies should focus on clarifying collagen-type specificity, shear-dependent activity, and the interplay between HSP47 and immune signaling. The development of selective inhibitors or enhancers, combined with biomarker-based risk stratification, may unlock the full potential of HSP47-targeted therapies for thrombotic and connective tissue disorders.

To further contextualize the role of HSP47 in thrombotic disorders, Table 3 presents a comparative overview of HSP47 expression patterns and functional consequences across various physiological and pathological states. This includes data from both animal models such as hibernating bears and knockout mice and human conditions, including obesity, immobility, and atherosclerosis. Such comparisons underscore the translational relevance of HSP47, highlighting its potential as both a therapeutic target and a biomarker in diverse clinical settings.

Table 3. Comparative Analysis of HSP47 Expression and Functional Outcomes in Physiological and Pathological States.

Species/Model	Physiological State	HSP47 Expression Profile	Functional Consequence	Relevance to Human Disease
Hibernating bears	Prolonged immobility	Downregulated	Decreased clot formation; lower platelet-collagen reactivity	Model for thrombo-protection during inactivity
Healthy humans	Normal activity	Homeostatic expression	Maintains collagen turnover and vascular integrity	Baseline for therapeutic modulation
Immobilized patients	Post-surgery, ICU, etc.	Variable; often dysregulated	Increased clotting risk due to inflammation and stasis	Target for clot-prevention therapies
Obese humans	Chronic inflammation	Upregulated in adipose tissue	Promotes fibrosis, collagen accumulation	Linked to metabolic syndrome and cardiovascular risk
Atherosclerotic plaques	Chronic vascular damage	Heterogeneous (↑ in fibrous cap, ↓ in necrotic core)	Affects plaque stability; promotes collagen cross-linking	Target for plaque stabilization or regression therapy
Mouse knockout models	HSP47-deficient	Absent	Impaired collagen synthesis, lethal phenotypes	Confirms essential role in connective tissue integrity

↑ high, ↓ low.

5. Conclusions

In conclusion, HSP47, a key collagen-specific molecular chaperone, plays a vital role in collagen maturation, folding, and stabilization in humans. Its activity is crucial for maintaining the stability and assembly of collagen types essential to connective tissue health. While primarily functioning within the ER, HSP47's influence extends beyond cellular protein synthesis, impacting adiposity regulation and cardiovascular health.

Research has highlighted its involvement in conditions such as DVT, atherosclerosis, and platelet-induced blood clotting. Particularly intriguing is the discovery of HSP47's decreased expression in hibernating bears, potentially informing human therapeutic approaches for blood clot prevention in immobilized patients.

These studies suggest that modulating HSP47 could offer novel treatment pathways for connective tissue disorders and contribute to therapeutic innovations targeting clotting irregularities. Future research is essential to fully unearth HSP47's mechanisms and therapeutic potential across different biological contexts.

Mirta Schattner at CONICET, Buenos Aires, described a breakthrough study on hibernating bear resistance to blood clotting, headed up by Ole Frobert, as "amazingly new". It is one part of a growing trend toward seeking medical breakthroughs from non-model organisms, facilitated by next-generation sequencing and molecular tools. The research isolated heat shock protein 47 as responsible for preventing clotting, contrary to earlier assumptions regarding its role. Using techniques including whole transcriptome sequencing, proteomics, imaging (FRET, PET), single-cell RNA sequencing, and bioinformatics, scientists discovered that reduced HSP47 levels in hibernating bears inhibit clotting neutrophil extracellular traps. The mechanism appears translatable to human beings, as identical HSP47 reductions were observed in immobilized patients, zero-gravity simulations, and postpartum pigs. In hibernating bears, reduced HSP47 levels inhibit clotting by suppressing neutrophil extracellular trap (NET) formation, possibly by limiting inflammation and collagen signaling in their hypometabolic state, aided by unique physiological adaptations like altered blood viscosity. In immobilized patients with similar HSP47 reductions, clots still form due to the absence of these protective adaptations, allowing venous stasis, endothelial injury, and hypercoagulability to drive thrombosis, potentially via HSP47-independent NET pathways triggered by inflammation or damage [92,93].

These findings offer novel therapeutic potential for clotting diseases, such as venous thromboembolism, and perhaps for thrombosis-risk patients with cancer. HSP47's role stabilizing collagen and regulating clotting makes it a potential drug target as well as a thrombotic risk biomarker. Challenges are off-target effects on platelets, regulation of clotting responses, and variability in protein expression. Future research will include validating HSP47 throughout hibernating animals, elucidating its mechanisms, and defining its uses for aging, chronic inflammation, and oncology, where it may regulate drug delivery and tumor microenvironments. Despite financial barriers, this nature-inspired strategy holds potential for advancing human health.

Therapeutic development: strategies might include pharmacological enhancers of HSP47, specific inhibitors in fibrotic conditions, gene therapy to increase HSP47 expression, and biologics targeting tissue repair. Combination therapies with existing anticoagulants or anti-inflammatory treatments could maximize efficacy by addressing multiple pathways involved in clotting and inflammation (Table 4).

Table 4. Therapeutic Strategies Targeting HSP47 Across Clinical Conditions.

Biological Context	Therapeutic Approach	Potential Benefits	Challenges/Considerations	References
Collagen Stability (Connective Tissue Disorders)	Enhance HSP47 activity (small molecules, gene therapy)	Improved collagen folding and stability in disorders like osteogenesis imperfecta and Ehlers–Danlos syndrome	Risk of excessive collagen deposition leading to fibrosis; specificity to target collagen types	[4]
Fibrosis	Inhibit HSP47 (small-molecule inhibitors like SMIH)	Reduce excessive collagen accumulation in fibrotic conditions (e.g., pulmonary fibrosis, liver fibrosis)	Potential disruption of normal collagen synthesis; off-target effects on other tissues	[94]
Thrombosis (DVT, Atherosclerosis)	Inhibit HSP47 (selective inhibitors, antibodies)	Decrease platelet-collagen interactions, reducing thrombus formation and plaque rupture risk	Balancing hemostasis to avoid bleeding risks; specificity to collagen-mediated pathways	[6]
Atherosclerotic Plaque Stability	Modulate HSP47 (enhance or inhibit depending on context)	Enhance collagen content in fibrous caps for plaque stability; reduce inflammation	Need for precise control to avoid over- or under-stabilization; long-term safety unclear	
Wound Healing/Tissue Repair	Enhance HSP47 functionality	Promote proper collagen deposition and ECM remodeling for effective healing	Potential for excessive scarring or fibrosis if overexpressed	[95]
Venous Thromboembolism (VTE)	Modulate HSP47 levels (inhibition or enhancement)	Prevent thrombus formation in immobile patients; improve inflammation resolution	Variability in patient response; need to address underlying inflammation causes	[2]
Cardiovascular Risk Biomarker	Monitor HSP47 levels	Use as a biomarker for plaque stability or thrombotic risk stratification	Requires validation across populations; correlation with clinical outcomes needs study	[1]
Cancer (Tumor Microenvironment)	Target HSP47 for drug delivery or inhibition	Regulate collagen in tumor ECM; enhance drug delivery to tumors	Off-target effects on normal tissues; complexity of tumor-specific collagen dynamics	[96]
Bleeding Disorders	Enhance HSP47 activity	Strengthen platelet-collagen interactions to improve hemostasis	Risk of promoting unwanted thrombosis; individual variability in response	[6]
Combination Therapies	Combine HSP47 modulation with existing treatments (statins, anticoagulants)	Synergistic effects on collagen stability, inflammation, and clotting control	Drug interaction risks; need for clinical trials to optimize dosing and efficacy	[53]

In summary, HSP47 emerges as a significant target for disease-modifying therapies in thrombotic disorders and offers promising avenues for improving vascular health.

Continued research on its mechanisms and therapeutic applications has the potential to enhance clinical outcomes for patients at risk of clotting abnormalities.

Author Contributions: Conceptualization: M.T. and O.-M.S.; methodology: M.-G.V.; investigation: A.-K.T.-C.; writing—original draft: M.T. and O.-M.S.; writing—review and editing: M.-G.V. and A.-K.T.-C. All authors have read and agreed to the published version of the manuscript.

Funding: This research received no external funding.

Data Availability Statement: Data is contained within the article.

Conflicts of Interest: The authors declare no conflicts of interest.

Abbreviations

The following abbreviations are used in this manuscript:

HSP47	Heat Shock Protein 47
DVT	Deep Vein Thrombosis
ER	Endoplasmic Reticulum
ECM	Extracellular Matrix
UPR	Unfolded Protein Response
PDI	Protein Disulfide Isomerase
P4H	Prolyl 4-Hydroxylase
FACIT	Fibril-Associated Collagens with Interrupted Triple Helices
MACIT	Membrane-Associated Collagens with Interrupted Triple Helices
PPARγ	Peroxisome Proliferator-Activated Receptor Gamma
TZDs	Thiazolidinediones
GWAS	Genome-Wide Association Studies
SNPs	Single Nucleotide Polymorphisms
PRS	Polygenic Risk Score
VTE	Venous Thromboembolism
FGF-2	Fibroblast Growth Factor 2
TGF-β1	Transforming Growth Factor Beta 1
MRI	Magnetic Resonance Imaging
CT	Computed Tomography
ITAM	Immunoreceptor Tyrosine-Based Activation Motif
GPVI	Glycoprotein VI
FRET	Fluorescence Resonance Energy Transfer
PET	Positron Emission Tomography
scRNA-seq	Single-Cell RNA Sequencing
HbA1c	Hemoglobin A1c
CRISPR/Cas9	Clustered Regularly Interspaced Short Palindromic Repeats/Cas9
PAR-1 and PAR-4	Protease-Activated Receptors 1 and 4
SMIH	Small Molecule Inhibitor of HSP47
DiOC6	3,3′-Dihexyloxacarbocyanine Iodide
GPIb	Glycoprotein Ib
VWF	Von Willebrand Factor
FIX	Factor IX
ADP	Adenosine Diphosphate
Ca^{2+}	Calcium Ion
scFvs	Single-Chain Variable Fragments
ELISA	Enzyme-Linked Immunosorbent Assay
FTO	Fat Mass and Obesity-Associated Gene
FRET	Förster (or Fluorescence) Resonance Energy Transfer
PET	Positron Emission Tomography
NET	Neutrophil Extracellular Trap

References

1. Ito, S.; Nagata, K. Roles of the endoplasmic reticulum-resident, collagen-specific molecular chaperone Hsp47 in vertebrate cells and human disease. *J. Biol. Chem.* **2019**, *294*, 2133–2141. [CrossRef]
2. Thienel, M.; Müller-Reif, J.B.; Zhang, Z.; Ehreiser, V.; Huth, J.; Shchurovska, K.; Kilani, B.; Schweizer, L.; Geyer, P.E.; Zwiebel, M.; et al. Immobility-associated thromboprotection is conserved across mammalian species from bear to human. *Science* **2023**, *380*, eabo5044. [CrossRef]
3. Grover, S.P.; Mackman, N.; Bendapudi, P.K. Heat shock protein 47 and venous thrombosis: Letting sleeping bears lie. *J. Thromb. Haemost.* **2023**, *21*, 2648–2651. [CrossRef]
4. Khan, E.S.; Däinghaus, T. HSP47 in human diseases: Navigating pathophysiology, diagnosis, and therapeutic strategies. *Clin. Transl. Med.* **2024**, *14*, e1755. [CrossRef]
5. Nagata, K. Hsp47 as a collagen-specific molecular chaperone: Function and expression in normal and pathological conditions. *Histol. Histopathol.* **2003**, *18*, 473–480.
6. Sasikumar, P.; AlOuda, K.S.; Kaiser, W.J.; Holbrook, L.M.; Kriek, N.; Unsworth, A.J.; Bye, A.P.; Sage, T.; Ushioda, R.; Nagata, K.; et al. The chaperone protein HSP47: A platelet collagen binding protein that contributes to thrombosis and hemostasis. *J. Thromb. Haemost.* **2018**, *16*, 946–959. [CrossRef]
7. Mala, J.G.; Rose, C. Interactions of heat shock protein 47 with collagen and the stress response: An unconventional chaperone model? *Life Sci.* **2010**, *87*, 579–586. [CrossRef]
8. Dafforn, T.R.; Della, M.; Miller, A.D. The molecular interactions of heat shock protein 47 (Hsp47) and their implications for collagen biosynthesis. *J. Biol. Chem.* **2001**, *276*, 49310–49319. [CrossRef]
9. Bellaye, P.-S.; Burgy, O.; Kolb, M. HSP47: A potential target for fibrotic diseases and implications for therapy. *Expert Opin. Ther. Targets* **2021**, *25*, 49–62. [CrossRef]
10. Akkız, H.; Gieseler, R.K.; Canbay, A. Liver Fibrosis: From Basic Science towards Clinical Progress, Focusing on the Central Role of Hepatic Stellate Cells. *Int. J. Mol. Sci.* **2024**, *25*, 7873. [CrossRef]
11. Sakamoto, N.; Okuno, D.; Tokito, T.; Yura, H.; Kido, T.; Ishimoto, H.; Tanaka, Y.; Mukae, H. HSP47: A therapeutic target in pulmonary fibrosis. *Biomedicines* **2023**, *11*, 2387. [CrossRef] [PubMed]
12. Hagiwara, S.; Iwasaka, H.; Shingu, C.; Matumoto, S.; Hasegawa, A.; Noguchi, T. Heat shock protein 47 (HSP47) antisense oligonucleotides reduce cardiac remodeling and improve cardiac function in a rat model of myocardial infarction. *Thorac. Cardiovasc. Surg.* **2011**, *59*, 386–392. [CrossRef] [PubMed]
13. Khalil, H.; Kanisicak, O.; Vagnozzi, R.J.; Johansen, A.K.; Maliken, B.D.; Prasad, V.; Boyer, J.G.; Brody, M.J.; Schips, T.; Kilian, K.K.; et al. Cell-specific ablation of Hsp47 defines the collagen-producing cells in the injured heart. *JCI Insight* **2019**, *4*, e128722. [CrossRef] [PubMed]
14. Omari, S.; Makareeva, E.; Gorrell, L.; Jarnik, M.; Lippincott-Schwartz, J.; Leikin, S. Mechanisms of procollagen and HSP47 sorting during ER-to-Golgi transport in living cells. *J. Cell Biol.* **2016**, *215*, 629–640.
15. Widmer, C.; Gebauer, J.M.; Baumann, U. Molecular basis for the action of the collagen-specific chaperone Hsp47/SERPINH1 and its structure-specific client recognition. *Proc. Natl. Acad. Sci. USA* **2012**, *109*, 13243–13247. [CrossRef]
16. Wilson, R.; Lees, J.F.; Bulleid, N.J. Protein disulfide isomerase acts as a molecular chaperone during the assembly of procollagen. *J. Biol. Chem.* **1998**, *273*, 9637–9643. [CrossRef]
17. Koivu, J.; Myllylä, R. Protein disulfide isomerase is an essential component of prolyl 4-hydroxylase. *EMBO J.* **1987**, *6*, 3497–3503.
18. Ito, S.; Nagata, K. Biology of Hsp47 (Serpin H1), a collagen-specific molecular chaperone. *Semin. Cell Dev. Biol.* **2017**, *62*, 142–151. [CrossRef]
19. Malhotra, J.D.; Kaufman, R.J. Endoplasmic reticulum stress and oxidative stress: A vicious cycle or a double-edged sword? *Antioxid. Redox Signal.* **2007**, *9*, 2277–2293. [CrossRef]
20. Hetz, C.; Papa, F.R. The Unfolded Protein Response and Cell Fate Control. *Mol. Cell* **2018**, *69*, 169–181. [CrossRef]
21. Koide, T.; Aso, A.; Yorihuzi, T.; Nagata, K. Conformational requirements of collagenous peptides for recognition by the chaperone protein HSP47. *J. Biol. Chem.* **2000**, *275*, 8934–8940. [CrossRef] [PubMed]
22. Sepulveda, D.; Rojas-Rivera, D.; Rodríguez, D.A.; Groenendyk, J.; Köhler, A.; Lebeaupin, C.; Ito, S.; Urra, H.; Carreras-Sureda, A.; Hazari, Y.; et al. Interactome Screening Identifies the ER Luminal Chaperone Hsp47 as a Regulator of the Unfolded Protein Response Transducer IRE1α. *Mol. Cell* **2018**, *69*, 238–252.e5. [CrossRef] [PubMed]
23. Makareeva, E.; Leikin, S. Collagen chaperones in health and disease. *Front. Biosci. (Landmark Ed.)* **2014**, *19*, 1047–1060.
24. Lamandé, S.R.; Bateman, J.F. Procollagen folding and assembly: The role of endoplasmic reticulum enzymes and molecular chaperones. *Semin. Cell Dev. Biol.* **1999**, *10*, 455–464. [CrossRef]
25. Hendershot, L.M.; Bulleid, N.J. Protein-specific chaperones: The role of hsp47 begins to gel. *Curr. Biol.* **2000**, *10*, R912–R915. [CrossRef]

26. Morello, R.; Bertin, T.K.; Chen, Y.; Hicks, J.; Tonachini, L.; Monticone, M.; Castagnola, P.; Rauch, F.; Glorieux, F.H.; Vranka, J.; et al. CRTAP is required for prolyl 3-hydroxylation and mutations cause recessive osteogenesis imperfecta. *Cell* **2006**, *127*, 291–304. [CrossRef]
27. Syx, D.; Ishikawa, Y.; Gebauer, J.; Boudko, S.P.; Guillemyn, B.; Van Damme, T.; D'hondt, S.; Symoens, S.; Nampoothiri, S.; Gould, D.B.; et al. Aberrant binding of mutant HSP47 affects posttranslational modification of type I collagen and leads to osteogenesis imperfecta. *PLoS Genet.* **2021**, *17*, e1009339. [CrossRef]
28. Christiansen, H.E.; Schwarze, U.; Pyott, S.M.; AlSwaid, A.; Al Balwi, M.; Alrasheed, S.; Pepin, M.G.; Weis, M.A.; Eyre, D.R.; Byers, P.H. Homozygosity for a missense mutation in SERPINH1, which encodes the collagen chaperone protein HSP47, results in severe recessive osteogenesis imperfecta. *Am. J. Hum. Genet.* **2010**, *86*, 389–398. [CrossRef]
29. Sauk, J.J.; Nikitakis, N.; Siavash, H. Hsp47: A novel collagen binding serpin chaperone, autoantigen, and therapeutic target. *Front. Biosci.* **2005**, *10*, 107–118. [CrossRef]
30. Koide, T.; Asada, S.; Takahara, Y.; Nishikawa, Y.; Nagata, K.; Kitagawa, K. Specific recognition of the collagen triple helix by chaperone HSP47: Minimal structural requirement and spatial molecular orientation. *J. Biol. Chem.* **2006**, *281*, 3432–3438. [CrossRef]
31. Köhler, A.; Mörgelin, M.; Gebauer, J.M.; Öcal, S.; Imhof, T.; Koch, M.; Nagata, K.; Paulsson, M.; Aumailley, M.; Baumann, U.; et al. New specific HSP47 functions in collagen subfamily chaperoning. *FASEB J.* **2020**, *34*, 12519–12531. [CrossRef] [PubMed]
32. Satoh, M.; Hirayoshi, K.; Yokota, S.I.; Hosokawa, N.; Nagata, K. Intracellular interaction of collagen-specific stress protein HSP47 with newly synthesized procollagen. *J. Cell Biol.* **1996**, *133*, 469–483. [CrossRef] [PubMed]
33. Nagai, N.; Hosokawa, M.; Itohara, S.; Adachi, E.; Matsushita, T.; Hosokawa, N.; Nagata, K. Embryonic lethality of molecular chaperone Hsp47 knockout mice is associated with defects in collagen biosynthesis. *J. Cell Biol.* **2000**, *150*, 1499–1506. [CrossRef]
34. Ishida, Y.; Nagata, K. Hsp47 as a collagen-specific molecular chaperone. *Methods Enzymol.* **2011**, *499*, 167–182.
35. Nagata, K. Hsp47: A collagen-specific molecular chaperone. *Trends Biochem. Sci.* **1996**, *21*, 22–26. [CrossRef]
36. Besio, R.; Garibaldi, N.; Sala, A.; Tonelli, F.; Aresi, C.; Maffioli, E.; Casali, C.; Torriani, C.; Biggiogera, M.; Villani, S.; et al. The administration of exogenous HSP47 as a collagen-specific chaperone improves cellular homeostasis in osteogenesis imperfecta. *J. Clin. Investig. Insight* **2025**, *10*, e181570.
37. Rocnik, E.F.; van der Veer, E.; Cao, H.; Hegele, R.A.; Pickering, J.G. Functional linkage between the endoplasmic reticulum protein Hsp47 and procollagen expression in human vascular smooth muscle cells. *J. Biol. Chem.* **2002**, *277*, 38571–38578. [CrossRef]
38. Xu, Q.; Metzler, B.; Jahangiri, M.; Mandal, K. Molecular chaperones and heat shock proteins in atherosclerosis. *Cell Stress Chaperones* **2012**, *17*, 123–131. [CrossRef]
39. Hansson, G.K.; Hermansson, A. The immune system in atherosclerosis. *Nat. Immunol.* **2011**, *12*, 204–212. [CrossRef]
40. Virmani, R.; Burke, A.P.; Kolodgie, F.D.; Farb, A. Vulnerable plaque: The pathology of unstable coronary lesions. *J. Interv. Cardiol.* **2002**, *15*, 439–446. [CrossRef]
41. Finn, A.V.; Nakano, M.; Narula, J.; Kolodgie, F.D.; Virmani, R. Concept of vulnerable/unstable plaque. *Arterioscler. Thromb. Vasc. Biol.* **2010**, *30*, 1282–1292. [CrossRef] [PubMed]
42. Virmani, R.; Kolodgie, F.D.; Burke, A.P.; Farb, A.; Schwartz, S.M. Lessons from sudden coronary death: A comprehensive morphological classification scheme for atherosclerotic lesions. *Arterioscler. Thromb. Vasc. Biol.* **2000**, *20*, 1262–1275. [CrossRef] [PubMed]
43. Burke, A.P.; Kolodgie, F.D.; Farb, A.; Virmani, R. Morphological predictors of arterial remodeling in coronary atherosclerosis. *Circulation* **2002**, *105*, 297–303. [CrossRef] [PubMed]
44. Neumeister, V.; Scheibe, M.; Lattke, P.; Jaross, W. Determination of the cholesterol–collagen ratio of arterial atherosclerotic plaques using near-infrared spectroscopy as a possible measure of plaque stability. *Atherosclerosis* **2002**, *165*, 251–257. [CrossRef]
45. Nadkarni, S.K.; Bouma, B.E.; de Boer, J.; Tearney, G.J. Evaluation of collagen in atherosclerotic plaques: The use of two coherent laser-based imaging methods. *Lasers Med. Sci.* **2009**, *24*, 439–445. [CrossRef]
46. Holm Nielsen, S.; Tengryd, C.; Edsfeldt, A.; Brix, S.; Genovese, F.; Bengtsson, E.; Karsdal, M.; Leeming, D.J.; Nilsson, J.; Goncalves, I. A biomarker of collagen type I degradation is associated with cardiovascular events and mortality in patients with atherosclerosis. *J. Intern. Med.* **2018**, *285*, 118–123. [CrossRef]
47. Silvera, S.S.; El Aidi, H.; Rudd, J.H.; Mani, V.; Yang, L.; Farkouh, M.; Fuster, V.; Fayad, Z.A. Multimodality imaging of atherosclerotic plaque activity and composition using FDG-PET/CT and MRI in carotid and femoral arteries. *Atherosclerosis* **2009**, *207*, 139–143. [CrossRef]
48. Barascuk, N.; Skjøt-Arkil, H.; Register, T.C.; Larsen, L.; Byrjalsen, I.; Christiansen, C.; Karsdal, M.A. Human macrophage foam cells degrade atherosclerotic plaques through cathepsin K mediated processes. *BMC Cardiovasc. Disord.* **2010**, *10*, 19. [CrossRef]
49. Si, S.C.; Yang, W.; Luo, H.Y.; Ma, Y.X.; Zhao, H.; Liu, J. Association of bone turnover biomarkers with severe intracranial and extracranial artery stenosis in type 2 diabetes mellitus patients. *World J. Diabetes* **2023**, *14*, 857–870. [CrossRef]
50. Pandit, R.; Yurdagul, A. The Atherosclerotic Plaque Microenvironment as a Therapeutic Target. *Curr. Atheroscler. Rep.* **2025**, *27*, 47. [CrossRef]

51. Newby, A.C. Metalloproteinase Expression in Monocytes and Macrophages and its Relationship to Atherosclerotic Plaque Instability. *Arterioscler. Thromb. Vasc. Biol.* **2008**, *19*, 1707–1711. [CrossRef] [PubMed]
52. Wu, S.; Liang, C.; Xie, X.; Huang, H.; Fu, J.; Wang, C.; Su, Z.; Wang, Y.; Qu, X.; Li, J.; et al. Hsp47 Inhibitor Col003 Attenuates Collagen-Induced Platelet Activation and Cerebral Ischemic–Reperfusion Injury in Rats. *Front. Pharmacol.* **2021**, *12*, 792263. [CrossRef] [PubMed]
53. Abd El-Fattah, E.E.; Zakaria, A.Y. Targeting HSP47 and HSP70: Promising therapeutic approaches in liver fibrosis management. *J. Transl. Med.* **2022**, *20*, 544. [CrossRef] [PubMed]
54. Ham, S.Y.; Pyo, M.J.; Kang, M.; Kim, Y.S.; Lee, D.H.; Chung, J.H.; Lee, S.T. HSP47 Increases the Expression of Type I Collagen in Fibroblasts through IRE1α Activation, XBP1 Splicing, and Nuclear Translocation of β-Catenin. *Cells* **2024**, *13*, 527. [CrossRef]
55. Kim, H.-J.; Park, J.-H.; Shin, J.-M.; Yang, H.-W.; Lee, H.-M. TGF-β1-induced HSP47 regulates extracellular matrix accumulation via Smad2/3 signaling pathways in nasal fibroblasts. *Sci. Rep.* **2019**, *9*, 15644. [CrossRef]
56. Sasaki, H.; Sato, T.; Yamauchi, N.; Okamoto, T.; Kobayashi, D.; Iyama, S.; Kato, J.; Matsunaga, T.; Takimoto, R.; Takayama, T.; et al. Induction of heat shock protein 47 synthesis by TGF-β and IL-1β in murine peritoneal macrophages. *J. Immunol.* **2002**, *168*, 5178–5183. [CrossRef]
57. Agbani, E.O.; Poole, A.W. Procoagulant platelets: Generation, function, and therapeutic targeting in thrombosis. *Blood* **2017**, *130*, 2171–2179. [CrossRef]
58. Aliotta, A.; Alberio, L. Thrombocytopathies: Not Just Aggregation Defects—The Clinical Relevance of Procoagulant Platelets. *J. Clin. Med.* **2021**, *10*, 894. [CrossRef]
59. Zwaal, R.F.A.; Comfurius, P.; Bevers, E.M. Scott syndrome, a bleeding disorder caused by defective scrambling of membrane phospholipids. *Biochim. Biophys. Acta (BBA)-Mol. Cell Biol. Lipids* **2004**, *1636*, 119–128. [CrossRef]
60. Reddy, E.C.; Rand, M.L. Procoagulant Phosphatidylserine-Exposing Platelets in vitro and in vivo. *Front. Cardiovasc. Med.* **2020**, *7*, 15. [CrossRef]
61. Mann, K.G.; Butenas, S.; Brummel, K. The dynamics of thrombin formation. *Arterioscler. Thromb. Vasc. Biol.* **2003**, *23*, 17–25. [CrossRef] [PubMed]
62. Hoffman, M.; Monroe, D.M. A cell-based model of hemostasis. *Thromb. Haemost.* **2001**, *85*, 958–965. [PubMed]
63. Monroe, D.M.; Hoffman, M.; Roberts, H.R. Platelets and Thrombin Generation. *Arterioscler. Thromb. Vasc. Biol.* **2002**, *22*, 1381–1389. [CrossRef]
64. Alisa, S. Wolberg, et. al. Human Factor IX Binds to Specific Sites on the Collagenous Domain of Collagen IV. *J. Biol. Chem.* **1997**, *271*, 26136–26141.
65. Gui, T.; Reheman, A.; Ni, H.; Gross, P.L.; Yin, F.; Monroe, D.; Monahan, P.E.; Stafford, D.W. Abnormal hemostasis in a knock-in mouse carrying a variant of factor IX with impaired binding to collagen type IV. *J. Thromb. Haemost.* **2009**, *7*, 1843–1851. [CrossRef]
66. Savage, B.; Saldivar, E.; Ruggeri, Z.M. Initiation of platelet adhesion by arrest onto fibrinogen or translocation on von Willebrand factor. *Cell* **1996**, *84*, 289–297. [CrossRef]
67. Jackson, S.P. The growing complexity of platelet aggregation. *Blood* **2007**, *109*, 5087–5095. [CrossRef]
68. Nesbitt, W.S.; Westein, E.; Tovar-Lopez, F.J.; Tolouei, E.; Mitchell, A.; Fu, J.; Carberry, J.; Fouras, A.; Jackson, S.P. A shear gradient–dependent platelet aggregation mechanism drives thrombus formation. *Nat. Med.* **2009**, *15*, 665–673. [CrossRef]
69. Ruggeri, Z.M.; Mendolicchio, G.L. Adhesion mechanisms in platelet function. *Circ. Res.* **2007**, *100*, 1673–1685. [CrossRef]
70. Siljander, P.R.M.; Munnix, I.C.; Smethurst, P.A.; Deckmyn, H.; Lindhout, T.; Ouwehand, W.H.; Farndale, R.W.; Heemskerk, J.W. Platelet receptor interplay regulates collagen-induced thrombus formation in flowing human blood. *Blood* **2004**, *103*, 1333–1341. [CrossRef]
71. Farndale, R.W.; Sixma, J.J.; Barnes, M.J.; De Groot, P.G. The role of collagen in thrombosis and hemostasis. *J. Thromb. Haemost.* **2004**, *2*, 561–573. [CrossRef] [PubMed]
72. Hwang, J.; Huang, Y.; Burwell, T.J.; Peterson, N.C.; Connor, J.; Weiss, S.J.; Yu, S.M.; Li, Y. In situ imaging of tissue remodeling with collagen hybridizing peptides. *ACS Nano* **2017**, *11*, 9825–9835. [CrossRef] [PubMed]
73. Shoulders, M.D.; Raines, R.T. Collagen structure and stability. *Annu. Rev. Biochem.* **2009**, *78*, 929–958. [CrossRef] [PubMed]
74. Li, Y.; Foss, C.A.; Summerfield, D.D.; Doyle, J.J.; Torok, C.M.; Dietz, H.C.; Pomper, M.G.; Yu, S.M. Targeting collagen strands by photo-triggered triple-helix hybridization. *Proc. Natl. Acad. Sci. USA* **2012**, *109*, 14767–14772. [CrossRef]
75. Ni, H.; Freedman, J. Platelets in hemostasis and thrombosis: Role of integrins and their ligands. *Transfus. Apher. Sci.* **2003**, *28*, 257–264. [CrossRef]
76. Andrews, R.K.; Shen, Y.; Gardiner, E.E.; Berndt, M.C. Platelet adhesion receptors and (patho)physiological thrombus formation. *Histol. Histopathol.* **2001**, *16*, 969–980. [CrossRef]
77. Estevez, B.; Du, X. New concepts and mechanisms of platelet activation signaling. *Physiology* **2017**, *32*, 162–177. [CrossRef]
78. Nieswandt, B.; Watson, S.P. Platelet-collagen interaction: Is GPVI the central receptor? *Blood* **2003**, *102*, 449–461. [CrossRef]

79. Jandrot-Perrus, M.; Busfield, S.; Lagrue, A.H.; Xiong, X.; Debili, N.; Chickering, T.; Couedic, J.P.L.; Goodearl, A.; Dussault, B.; Fraser, C.; et al. Cloning, characterization, and functional studies of human and mouse glycoprotein VI: A platelet-specific collagen receptor from the immunoglobulin superfamily. *Blood* **2000**, *96*, 1798–1807. [CrossRef]
80. Joutsi-Korhonen, L.; Smethurst, P.A.; Rankin, A.; Gray, E.; IJsseldijk, M.; Onley, C.M.; Watkins, N.A.; Williamson, L.M.; Goodall, A.H.; de Groot, P.G.; et al. The low-frequency allele of the platelet collagen signaling receptor glycoprotein VI is associated with reduced functional responses and expression. *Blood* **2003**, *101*, 4372–4379. [CrossRef]
81. Munnix, I.; Gilio, K.; Siljander, P.; Raynal, N.; Feijge, M.; Hackeng, T.; Deckmyn, H.; Smethurst, P.; Farndale, R.; Heemskerk, J. Collagen-mimetic peptides mediate flow-dependent thrombus formation by high- or low-affinity binding of integrin α2β1 and glycoprotein VI. *J. Thromb. Haemost.* **2008**, *6*, 2132–2142. [CrossRef] [PubMed]
82. Auger, J.M.; Kuijpers, M.J.E.; Senis, Y.A.; Watson, S.P.; Heemskerk, J.W.M. Adhesion of human and mouse platelets to collagen under shear: A unifying model. *Blood* **2005**, *104*, 1801–1809. [CrossRef]
83. Jackson, S.P.; Nesbitt, W.S.; Westein, E. Dynamics of platelet thrombus formation. *J. Thromb. Haemost.* **2009**, *7* (Suppl. S1), 17–20. [CrossRef] [PubMed]
84. Reininger, A.J. Mechanism of platelet adhesion to von Willebrand factor and microparticle formation under high shear stress. *Blood* **2006**, *107*, 3537–3545. [CrossRef] [PubMed]
85. Wohner, N.; Kovács, A.; Machovich, R.; Kolev, K. Modulation of the von Willebrand factor-dependent platelet adhesion through alternative proteolytic pathways. *Thromb. Res.* **2012**, *129*, 274–281. [CrossRef]
86. Houdijk, W.P.; Sakariassen, K.S.; Nievelstein, P.F.; Sixma, J.J. Role of factor VIII-von Willebrand factor and fibronectin in the interaction of platelets in flowing blood with monomeric and fibrillar human collagen types I and III. *J. Clin. Investig.* **1985**, *75*, 531–540. [CrossRef]
87. Sakariassen, K.S.; Orvim, U.; Turitto, V.T. The impact of blood shear rate on arterial thrombus formation. *Future Sci. OA* **1993**, *2*, FSO100. [CrossRef]
88. Alouda, S.; Sasikumar, P.; AlThunayan, T.; Gibbins, J.M. Role of heat shock protein 47 in platelet glycoprotein VI dimerization and signaling. *Res. Pract. Thromb. Haemost.* **2023**, *7*, 100229. [CrossRef]
89. Inoue, O.; Suzuki-Inoue, K.; Dean, W.L.; Frampton, J.; Watson, S.P.; Murakami, M. Integrin α2β1 mediates outside-in regulation of platelet spreading on collagen through activation of Src kinases and PLCγ2. *J. Cell Biol.* **2003**, *160*, 769–780. [CrossRef]
90. Kaiser, W.J.; Holbrook, L.M.; Tucker, K.L.; Stanley, R.G.; Gibbins, J.M. A functional proteomic method for the enrichment of peripheral membrane proteins reveals the collagen binding protein Hsp47 is exposed on the surface of activated human platelets. *J. Proteome Res.* **2009**, *8*, 2903–2914. [CrossRef]
91. Ito, S.; Ogawa, K.; Takeuchi, K.; Takagi, M.; Yoshida, M.; Hirokawa, T.; Hirayama, S.; Shin-Ya, K.; Shimada, I.; Doi, T.; et al. A small-molecule compound inhibits a collagen-specific molecular chaperone and could represent a potential remedy for fibrosis. *J. Biol. Chem.* **2017**, *292*, 20076–20085. [CrossRef] [PubMed]
92. Schattner, M. Sleep like a bear. *Science* **2023**, *380*, 133–134. [CrossRef] [PubMed]
93. Hisada, Y.; Grover, S.P.; Maqsood, A.; Houston, R.; Ay, C.; Noubouossie, D.F.; Cooley, B.C.; Wallén, H.; Key, N.S.; Thålin, C.; et al. Neutrophils and neutrophil extracellular traps enhance venous thrombosis in mice bearing human pancreatic tumors. *Haematologica* **2019**, *105*, 218–225. [CrossRef] [PubMed]
94. Miyamura, T.; Sakamoto, N.; Kakugawa, T.; Taniguchi, H.; Akiyama, Y.; Okuno, D.; Moriyama, S.; Hara, A.; Kido, T.; Ishimoto, H.; et al. Small molecule inhibitor of HSP47 prevents pro-fibrotic mechanisms of fibroblasts in vitro. *Biochem. Biophys. Res. Commun.* **2020**, *530*, 561–565. [CrossRef]
95. Patsnap Synapse. What Are HSP47 Inhibitors and How Do They Work? 2023. Available online: https://synapse.patsnap.com/article/what-are-hsp47-inhibitors-and-how-do-they-work?utm_source=chatgpt.com (accessed on 6 April 2025).
96. Xiong, G.; Chen, J.; Zhang, G.; Wang, S.; Kawasaki, K.; Zhu, J.; Zhang, Y.; Nagata, K.; Li, Z.; Zhou, B.P.; et al. Hsp47 promotes cancer metastasis by enhancing collagen-dependent cancer cell-platelet interaction. *Proc. Natl. Acad. Sci. USA* **2020**, *117*, 3748–3758. [CrossRef]

current issues in molecular biology

MDPI

Article

An Integrated Network Pharmacology, Molecular Docking, Molecular Dynamics Simulation, and Experimental Validation Study to Investigate the Potential Mechanism of Isoliquiritigenin in the Treatment of Ischemic Stroke

Hang Yuan [1,2], Yuting Hou [1,2], Yuan Jiao [1,2], Xin Lu [1,2] and Liang Liu [1,2,*]

[1] Institute of Translational Medicine, Medical College, Yangzhou University, Yangzhou 225009, China; yzjsnyx@163.com (H.Y.); houytd@hotmail.com (Y.H.); 19552398538@163.com (Y.J.); luxin94lx@163.com (X.L.)

[2] Key Laboratory of the Jiangsu Higher Education Institutions for Integrated Traditional Chinese and Western Medicine in Senile Diseases Control, Yangzhou University, Yangzhou 225009, China

* Correspondence: liuliang@yzu.edu.cn

Abstract: Isoliquiritigenin (ISL) is a type of chalcone that widely exists in medicinal plants of the Leguminosae family and exhibits a remarkable anti-ischemic stroke (IS) effect. However, the anti-IS mechanisms of ISL remain to be systematically elucidated. In this study, network pharmacology was used to predict potential targets related to the anti-IS effect of ISL. The binding ability of ISL to potential core targets was further analyzed by molecular docking and molecular dynamics (MD) simulations. By establishing an oxygen–glucose deprivation/reoxygenation (OGD/R)-induced HT22 cell model, the anti-IS mechanisms of ISL were investigated via RT-qPCR and Western Blot (WB). As a result, network pharmacology analysis revealed that APP, ESR1, MAO-A, PTGS2, and EGFR may be potential core targets of ISL for anti-IS treatment. Molecular docking and molecular dynamics simulation results revealed that ISL can stably bind to the five potential core targets and form stable complex systems with them. The results of the cell experiments revealed a significant anti-IS effect of ISL. Additionally, mRNA and protein expression levels of APP, MAO-A and PTGS2 or ESR1 in the ISL treatment group were significantly lower or higher than those in the OGD/R group In conclusion, ISL may improve IS by regulating the protein expression levels of APP, ESR1, MAO-A, and PTGS2.

Keywords: isoliquiritigenin; ischemic stroke; network pharmacology; molecular docking; molecular dynamics simulation; experimental validation

1. Introduction

Stroke, which is one of the foremost causes of mortality worldwide, imposes a significant health burden [1]. Stroke can be divided into two primary types: hemorrhagic stroke and ischemic stroke (IS); moreover, IS accounts for more than 80% of all stroke cases [2]. IS is characterized by high incidence rates, substantial complications, elevated mortality, and frequent recurrence [3]. Modern medicine holds that the majority of ischemic strokes originate from thromboembolism. The brain is a high-oxygen-consuming organ. The deprivation of oxygen and glucose in local brain tissue caused by ischemia not only leads to the permanent loss of brain nerve function in a short period of time, but also causes a sharp decline in cellular energy production [4]. Current treatment strategies primarily include thrombolytic therapy, anticoagulation agents, neuroprotective agents, and surgical interventions [5]. However, due to the suddenness of the disease and the limitations of treatment conditions, as well as serious adverse reactions and many nonnegligible flaws, the clinical efficacy of these treatment methods is not satisfactory. Therefore, the search for new anti-IS drugs is extremely important.

Active compounds from medicinal plants have consistently been important resources for the development of innovative drugs [6,7]. In recent years, the research on medicinal plants in the treatment of IS injury has gradually deepened. Many studies have shown that the active components of medicinal plants can act on pathophysiological processes such as neuroinflammation, blood–brain barrier (BBB) injury, oxidative stress, and apoptosis; regulate their development; and effectively alleviate cerebral ischemic injury [8,9]. Many studies have focused on the significant neuroprotective effects of flavonoids [10]. Chalcones are a subclass of flavonoids; however, compared with other flavonoid subclasses, less attention has been given to chalcones [11]. Isoliquiritigenin (ISL) is a type of chalcones that widely exists in many medicinal plants of the Leguminosae family, including *Glycyrrhiza uralensis* Fisch., *Astragalus membranaceus* (Fisch.) Bunge, *Caragana jubata* (Pall.) Poir., and *C. changduensis* Y. X. Liou [12–15]. ISL has shown potential in treating multiple diseases. It can inhibit fatty acid synthesis and regulate the ERK signaling pathway, thereby suppressing the M2 polarization of macrophages and improving pulmonary fibrosis [16]. It ameliorates sarcomere contraction and inflammatory responses by inhibiting the IL-17RA/Act1/p38 signaling pathway in rats with myofascial trigger points [17]. And ISL can improve energy metabolism and antioxidant enzyme activity to relieve ischemic injury in rats [18]. However, the underlying molecular mechanisms of ISL in the treatment of IS remain to be fully investigated.

Network pharmacology is a new discipline that reveals the regulatory network of drugs in the body at the system level, and it aims to predict the mechanism of action of drugs by constructing a complex network of relationships among diseases, targets, and drugs [19]. This method is characterized by a high degree of integrity, systematization, and comprehensiveness; moreover, it is an effective method for revealing the pharmacological mechanisms of natural compounds. It employs big data and artificial intelligence to rapidly identify key targets and signaling pathways in complex biological systems. And it integrates large-scale data, network analysis, and algorithm optimization for studying drug mechanisms, with its research outcomes progressively translating into clinical applications and novel therapeutics [20]. Molecular docking and molecular dynamics (MD) simulations have emerged as characterization and virtual screening tools, due to the fact that they can visualize interaction details that cannot be identified by in vitro experiments and can quickly screen for bioactive compounds from large databases containing millions of molecules [21,22]. In this study, an integrated network pharmacology, molecular docking, MD simulation, and experimental validation study was performed to investigate the potential mechanism of isoliquiritigenin in the treatment of IS, thereby providing a theoretical basis for its further study and clinical application.

2. Materials and Methods

2.1. Materials and Reagents

Isoliquiritigenin (A0463) was purchased from Chengdu Must Biotechnology Co., Ltd. (Chengdu, China). The purities of the standards were above 98%. HT22 mouse hippocampal neurons and FBS were purchased from Pricella Biotechnology Co., Ltd. (Wuhan, China). DMEM (12100), DMSO (D8371) and phosphate-buffered saline (PBS) (P1020) were supplied by Beijing Solarbio Science & Technology Co., Ltd. (Beijing, China). Cell Counting Kit-8 (CCK-8) (KTA1020) was purchased from Abbkine Scientific Co., Ltd. (Wuhan, China). TRIzol reagent was purchased from Ambion (Austin, TX, USA). The PrimeScript™ RT reagent kit was purchased from TaKaRa (Otsu, Japan). The NovoStart® SYBR qPCR SuperMix Plus kit (E096-01A) was purchased from Novoprotein (Shanghai, China). All of the primers were synthesized by Sangon Biotech Co., Ltd. (Shanghai, China). BCA protein assay kit was purchased from Beyotime Biotech Inc (Shanghai, China). APP (25524-1-AP), PTGS2 (12375-1-AP), and EGFR (30847-1-AP) antibodies were purchased from Proteintech Group, Inc. (Wuhan, China). The positive control drug edaravone (89-25-8), MAO-A antibody (YA2036) and ESR1 (YA3629) antibody were purchased from MedChemExpress LLC (Shanghai, China). HRP-conjugated goat anti-rabbit IgG (H + L) (AS014) was purchased from ABclonal Technology Co., Ltd. (Wuhan, China). PVDF membranes (cat. FFP24) and SDS-PAGE gels were purchased from Beyotime (Shanghai, China). The automated chemiluminescence imaging system (Tanon-4600) was manufactured by Tanon Science & Technology Co., Ltd. (Shanghai, China). The CFX96 Touch RealTime PCR Detection System was from Bio-Rad Laboratories, Inc. (Hercules, CA, USA). The microplate reader (Spark®) was from Tecan Group Ltd., (Männedorf, Switzerland).

2.2. Network Pharmacology Analysis

Potential targets of ISL were predicted using three in silico platforms: SwissTargetPrediction (http://www.swisstargetprediction.ch/, accessed on 12 April 2025), PharmMapper (http://www.lilab-ecust.cn/pharmmapper/, accessed on 12 April 2025), and the Similarity Ensemble Approach (SEA, http://sea.bkslab.org/, accessed on 12 April 2025), with species constraints rigorously set to *Homo sapiens*. To improve the reliability of the results, the targets were screened based on probability >0, fit score >3, and Max Tc >0.5. Disease targets were obtained from the GeneCards database (https://www.genecards.org/, accessed on 12 April 2025) and searched by using "Ischemic stroke" as the keyword [23]. The chemical constituent targets and disease targets were imported into Venny 2.1.0; additionally, the overlapping targets between ISL and IS were identified as potential therapeutic targets of ISL against IS, and the corresponding Venn diagram was drawn. The intersecting genes were subsequently uploaded to the STRING database (https://string-db.org/, accessed on 12 April 2025), with the species parameter set to "*Homo sapiens*", in order to construct the protein–protein interaction (PPI) network [24]. The results were imported into Cytoscape 3.9.1 software for network construction and visual analysis [25]. The degree centrality, betweenness centrality, and closeness centrality of each target node were calculated using the cytoHubba plugin as evaluation metrics for the network analysis to identify the potential core anti-IS targets for ISL. Additionally, the intersecting genes were input into the DAVID database (https://david.ncifcrf.gov/, accessed on 12 April 2025) for Gene Ontology (GO) and Kyoto Encyclopedia of Genes and Genomes (KEGG) enrichment analyses [26]. A visual representation of the enrichment results was subsequently generated.

2.3. Molecular Docking

The selected protein crystal structures were sorted according to comprehensive analyses of degree centrality, betweenness centrality, and closeness centrality, and molecular docking with ISL was performed. The two-dimensional structure of ISL was downloaded from the TCMSP database (https://www.tcmsp-e.com, accessed on 12 April 2025). The three-dimensional structure of the core protein was downloaded in PDB format from the Research Collaboratory for Structural Bioinformatics Protein Data Bank (RCSB PDB) (https://www.rcsb.org/, accessed on 12 April 2025). After water molecules and ligands were removed from the protein via PyMOL 2.5.7 software, the 3D structure was used as a receptor for visualization of the molecular docking features. Docking grids were constructed as binding sites for the receptor, and the grid size was determined to be 40 × 40 × 40 A. Molecular docking of ligands and receptors was performed using AutoDock Vina 1.1.2 software.

2.4. MD Simulation

After molecular docking, the stability of the chosen docking poses was confirmed by MD by using Gromacs 2022 software. The charmm36-jul2021 force field with the TIP3P water model was used for the protein simulations, whereas the GAFF force field was applied for small molecules. The total charge of the system was neutralized by adding a suitable number of Na^+ or Cl^- ions. The steepest-descent method was used for energy minimization. After energy minimization, equilibrium simulations were performed using both isothermal isovolumic (NVT) and isothermal isobaric (NPT) ensembles. MD simulations were subsequently conducted for 100 ns, which consisted of 5 million steps with a step length of 2 fs at 300 K and 1 atmosphere of pressure. Upon completion of the simulation, the gmx tools of Gromacs 2022 software were used to analyze the trajectory. Moreover, the Root-Mean-Square Deviation (RMSD), interaction fraction, and Gibbs free energy were calculated.

2.5. Cell Culture and Treatment

HT22 cells were cultured in DMEM supplemented with 10% FBS and 1% penicillin–streptomycin antibiotics (100 U/mL penicillin and 100 μg/mL streptomycin), and maintained at 37 °C in a 5% CO_2 incubator. The following three experimental groups were established: the control group, the oxygen–glucose deprivation/reoxygenation (OGD/R) group, and the ISL group.

2.6. Cytotoxicity Evaluation of ISL

HT22 cells in the logarithmic growth phase were seeded in 96-well plates at a density of 1×10^4 cells/well and cultured at 37 °C in a 5% CO_2 incubator for 24 h. ISL was subsequently added to the cells for another 24 h or 72 h. Afterwards, DMEM culture solution containing 10% CCK-8 was added to the cells, and the absorbance of each cell at 450 nm was measured with a microplate reader after incubation for 1 h. Control group received an equal volume of culture medium containing 0.01% DMSO without ISL. The final concentrations of ISLwere 6.25, 12.5, 25, 50, 100, and 200 μM.

2.7. OGD/R Model Establishment

The in vitro neuroprotective effect of ISL was evaluated by establishing an OGD/R model. HT22 cells in the logarithmic growth phase were seeded in 96-well plates at a density of 1×10^4 cells/well. After the cells were attached to the walls, they were washed twice with PBS, followed by incubation with ISL or edaravone (5 μM) for 24 h. Afterwards, sugar-free DMEM was added, and the plate was placed into an incubator containing 5% CO_2, 1% O_2, and 94% N_2. After being deprived of glucose and oxygen for 4 h, the medium of the cells was replaced with sugar-containing complete medium, and the cells were cultured under normal oxygen conditions (95% O_2 and 5% CO_2) for 12 h. Following the addition of DMEM supplemented with 10% CCK-8 to the cells, the absorbance at 450 nm was measured. The control group was supplemented with an equal volume of culture medium without ISL and cultured at 37 °C in a 5% CO_2 incubator throughout the experiment. In the OGD/R group, an equal volume of culture medium without ISL was added, and the cells were cultured under the same conditions as those in the ISL group. The final concentrations of ISL were 1.25, 2.5, 5, 10, and 20 μM. Edaravone served as a positive control, and its final concentration was 5 μM.

2.8. RT-qPCR Assay

Total RNA was extracted using the TRIzol reagent according to the manufacturer's protocol from the following four groups of HT22 cells: control group, vehicle control group

(0.01% DMSO), OGD/R group, and OGD/R+ISL group. We performed reverse transcription into cDNA using the PrimeScript™ RT reagent kit. Quantitative PCR (qPCR) analyses were performed using the NovoStart® SYBR qPCR SuperMix Plus kit on the CFX96 Touch RealTime PCR Detection System (Bio-Rad, Hercules, CA, USA). The expression levels of the target genes were quantified using the $2^{-\Delta\Delta CT}$ method, and the final results were normalized to those of GAPDH. The utilized primers for the qRT-PCR assays are listed in Table 1.

Table 1. Primers used in RT-qPCR.

Gene	Forward	Reverse
APP	TCCGAGAGGTGTGCTCTGAA	CCACATCCGCCGTAAAAGAATG
ESR1	AAGACGCTCTTGAACCAGCA	AGGCTTTGGTGTGAAGGGTC
MAO-A	GCCCAGTATCACAGGCCAC	CGGGCTTCCAGAACCAAGA
PTGS2	TGCACTATGGTTACAAAAGCTGG	TCAGGAAGCTCCTTATTTCCCTT
EGFR	CCGAAACTACGTGGTGACAGAT	TGCCATTACAAACTTTGCGAC
GAPDH	CACTCACGGCAAATTCAACGGCACA	GACTCCACGACATACTCAGCAC

2.9. Western Blotting (WB)

Protein was extracted from three experimental groups of HT22 cells using RIPA lysis buffer containing protease and phosphatase inhibitors: control group, OGD/R group, and OGD/R+ISL group. Protein concentrations were quantified using a BCA protein assay kit, after which 20 µg of protein from each sample was electrophoresed on SDS-PAGE gels and subsequently transferred to PVDF membranes, which were blocked with 5% nonfat milk in TBST buffer for 1 h at room temperature and subsequently incubated with primary antibodies overnight at 4 °C. The primary antibodies used in this study included APP, ESR1, MAO-A, PTGS2, EGFR, β-tubulin, and GAPDH. The PVDF membranes were washed with TBST three times and incubated with HRP-conjugated secondary antibodies for 1 h at room temperature. The protein bands were then detected by enhanced chemiluminescence (ECL) using an automated chemiluminescence imaging system (Tanon-4600; Tanon Science & Technology Co., Shanghai, China). The intensity of the blots was quantified via ImageJ 1.54g software [27], and the obtained values were statistically analyzed.

2.10. Statistical Analysis

Statistical analyses were performed by using GraphPad Prism 9.0 software. All treatment allocations used computer-generated randomization. Biological replicates ($n \geq 3$ independent cultures) with technical replicates averaged per experiment are specified in the figure legends. The Shapiro–Wilk test verified the normality assumption. Homogeneity of variances was confirmed by Levene's test. Differences between two groups were assessed using the *t*-test, whereas comparisons among multiple groups were performed using one-way analysis of variance followed by Dunnett post hoc test. The data are presented as the means ± SDs. A value of $p < 0.05$ was considered statistically significant.

3. Results

3.1. Potential Targets of ISL in the Treatment of IS

A total of 94 potential targets of ISL were identified through SwissTargetPrediction, PharmMapper, and SEA databases, while 1876 disease targets of IS were obtained from the GeneCards database (Table S1). The overlapping targets between ISL and IS were identified as potential targets. After removing duplicates, 35 intersecting genes were ultimately obtained, and the generated Venn diagram is shown in Figure 1A.

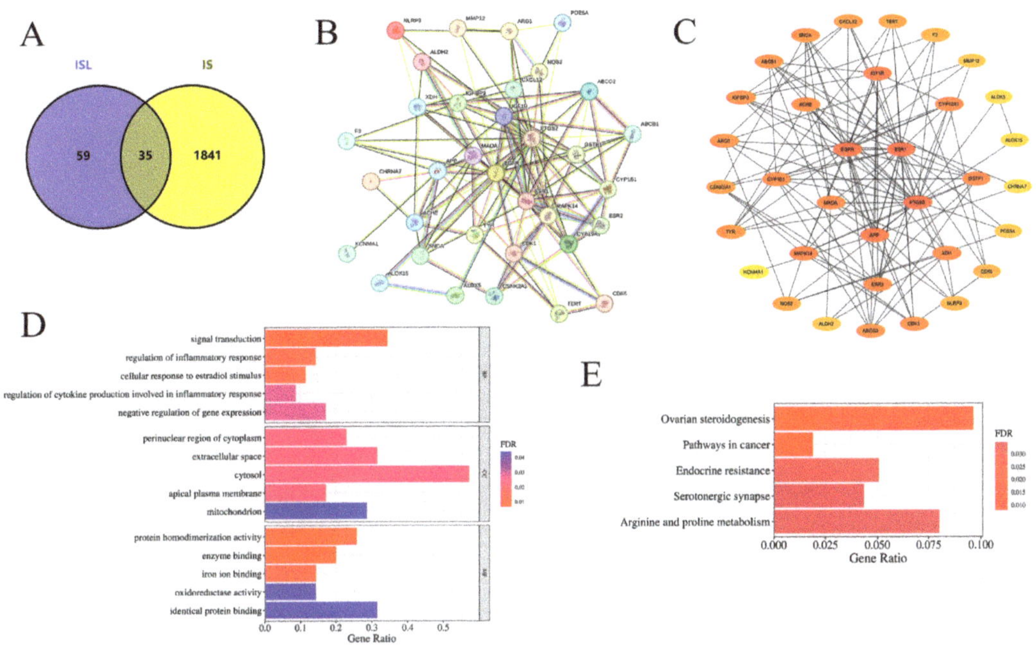

Figure 1. Network pharmacology analysis of the effects of ISL on IS. (**A**) Venn diagram visualization of target acquisition for ISL in the treatment of IS. (**B**) The PPI network of potential targets of ISL in the treatment of IS. (**C**) The key protein targets for ISL in the treatment of IS were ranked according to degree centrality, betweenness centrality, and closeness centrality. (**D**) GO-BP, GO-CC, and GO-MF enrichment analysis of potential targets. (**E**) KEGG enrichment and analysis of the top 5 key signaling pathways associated with potential targets. Bars show gene ratios and colors indicate FDR significance.

3.2. PPI Network Analysis and Screening of Key Targets

The 35 intersecting genes were imported into the STRING database, and the results were imported into Cytoscape 3.9.1 software for network construction and visual analysis, with the unconnected nodes in the network being hidden (Figure 1B). The PPI network comprised 34 nodes and 125 edges, thus reflecting the interactions between these potential ISL genes in the treatment of IS, which were screened via the cytoHubba plugin according to degree centrality, betweenness centrality, and closeness centrality (Table 2). Through the comprehensive considerations of these three parameters, the core targets might be PTGS2, ESR1, APP, EGFR, and MAO-A (Figure 1C).

3.3. GO and KEGG Enrichment Analyses

GO and KEGG functional enrichment analyses of the intersecting genes were conducted using the DAVID database, followed by visualization of the results. Ultimately, we obtained 139 biological process (BP), 27 cellular component (CC), and 63 molecular function (MF) terms based on the GO entry (Table S2). The top 5 terms in each category are shown in Figure 1D. In addition, KEGG enrichment analysis identified 39 pathways (Table S3). The top five enriched pathways suggest that ISL may affect IS via multiple signaling pathways (Figure 1E). KEGG enrichment analysis reveals potential signaling associations, though these pathways remain unvalidated functionally.

Table 2. The corresponding parameters of the top ten targets ranked by degree value.

Gene Name	Degree	Betweenness Centrality	Closeness Centrality
PTGS2	24	320.2112364	28.5
ESR1	22	218.4691729	27.5
EGFR	19	119.9617377	26
APP	11	114.8404762	22
IGF1R	11	18.46190476	21.83333333
MAPK14	9	19.00417711	20.66666667
MAOA	8	26.03433584	20.5
XDH	8	33.59444444	20.33333333
ACHE	8	27.98145363	20.16666667
GSTP1	8	24.95873016	20.16666667

3.4. Molecular Docking Analysis

The top five potential core targets, including APP (PDB: 3SV1), ESR1 (PDB: 7UJM), MAO-A (PDB: 2BXR), PTGS2 (PDB: 5IKR), and EGFR (PDB: 3IKA), were docked with ISL to clarify their possible binding modes and molecular interactions. The docking results were visualized, and the results are shown in Figure 2. ISL was able to effectively bind to the active pocket of the protein, indicating that the compound can undergo hydrogen bond interactions and hydrophobic interactions with the protein binding pocket. These interactions promote the binding of ISL to the active pocket of the protein to form a complex. When the binding affinity is <0 kcal·mol^{-1}, the molecule and protein could be automatically bound; additionally, a lower binding affinity corresponds to a more stable binding conformation [28]. The molecular docking results revealed that the binding affinities of ISL for APP, ESR1, MAO-A, PTGS2, and EGFR were −5.80, −8.60, −8.80, −8.40, and −7.90 kcal·mol^{-1}, respectively. For the targets with available co-crystal structures including ESR1, MAO-A, PTGS2, and EGFR, re-docking validation achieved RMSD values of 1.019 Å, 1.275 Å, 1.799 Å, and 1.677 Å, respectively (Table 3), indicating that ISL could stably bind to the key core targets.

Table 3. Binding affinity and RMSD of ISL docked to potential core targets.

Target Protein	Binding Affinity (kcal·mol^{-1})	RMSD (Å)
APP	−5.80	/
ESR1	−8.60	1.019
MAO-A	−8.80	1.275
PTGS2	−8.40	1.799
EGFR	−7.90	1.677

"/" indicates no RMSD value available due to the absence of a co-crystal structure.

3.5. MD Simulations Results

To further evaluate the influence of ISL on targets, MD simulations were conducted to determine the stability of the ISL-APP (Complex 1), ISL-EGFR (Complex 2), ISL-MAO-A (Complex 3), ISL-ESR1 (Complex 4), and ISL-PTGS2 (Complex 5) complexes obtained from molecular docking in 100 ns. The RMSD is a statistical method used to evaluate the stability of a simulated system. As shown in Figure 3, Complexes 1, 2, 3, 4, and 5 reached equilibrium after 40 ns, 20 ns, 20 ns, 20 ns, and 40 ns, respectively. Meanwhile, the RMSD of protein fluctuated between 3.0 Å and 5.5 Å, 3 Å and 4 Å, 1.5 Å and 2 Å, 2 Å and 2.5 Å, and 2 Å and 2.6 Å, respectively. These results indicated that ISL could generate a stable complex system with APP, EGFR, MAO-A, ESR1, and PTGS2, respectively, which was consistent with the molecular docking results. The MM-GBSA ΔG_{bind} was calculated and is listed in Table 4. The ΔG_{bind} was -41.3 ± 3.4, -40.9 ± 5.1, -46.9 ± 6.2, -31.7 ± 3.7, and

−26.4 ± 5.0, respectively. As shown in Figure 3, for ISL-APP (Complex 1), Glu543, Tyr544, Ile547, and Tyr75 were important residues for binding. For ISL-EGFR (Complex 2), the covalent bond with Cys797 was vital; in addition, the hydrophobic interaction with Phe723 stabilized the complex as well. For ISL-MAO-A (Complex 3), the H-bond with Asn181 and Gly443, as well as the hydrophobic interaction with Tyr407 and Tyr444, contributed to ΔG_{bind}. For ISL-ESR1 (Complex 4), Arg394 formed H-bonds or ionic interactions via the MD. The hydrophobic interactions with Phe404 and Trp383 further stabilized the complex. For ISL-PTGS2 (Complex 5), Tyr355 and Ser540 were most important. Arg120, Leu484, and Trp387 also made a contribution.

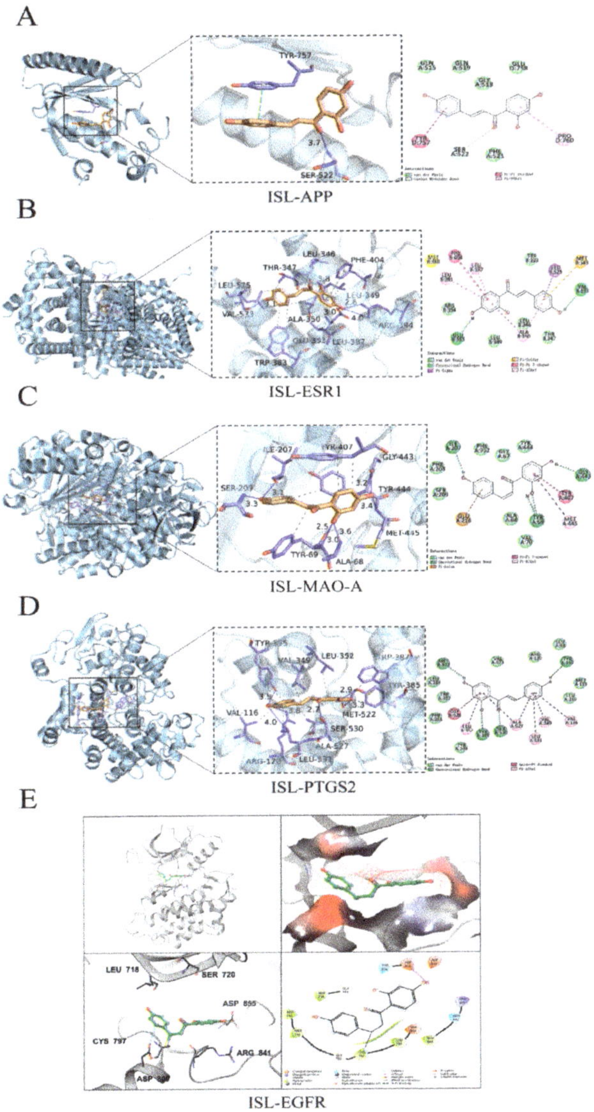

Figure 2. The molecular docking of ISL with the APP, ESR1, MAO-A, PTGS2, and EGFR proteins. The three-dimensional and 2D intermolecular contact between ISL and APP (**A**), ISL and ESR1 (**B**), ISL and MAO-A (**C**), ISL and PTGS2 (**D**), and ISL and EGFR (**E**).

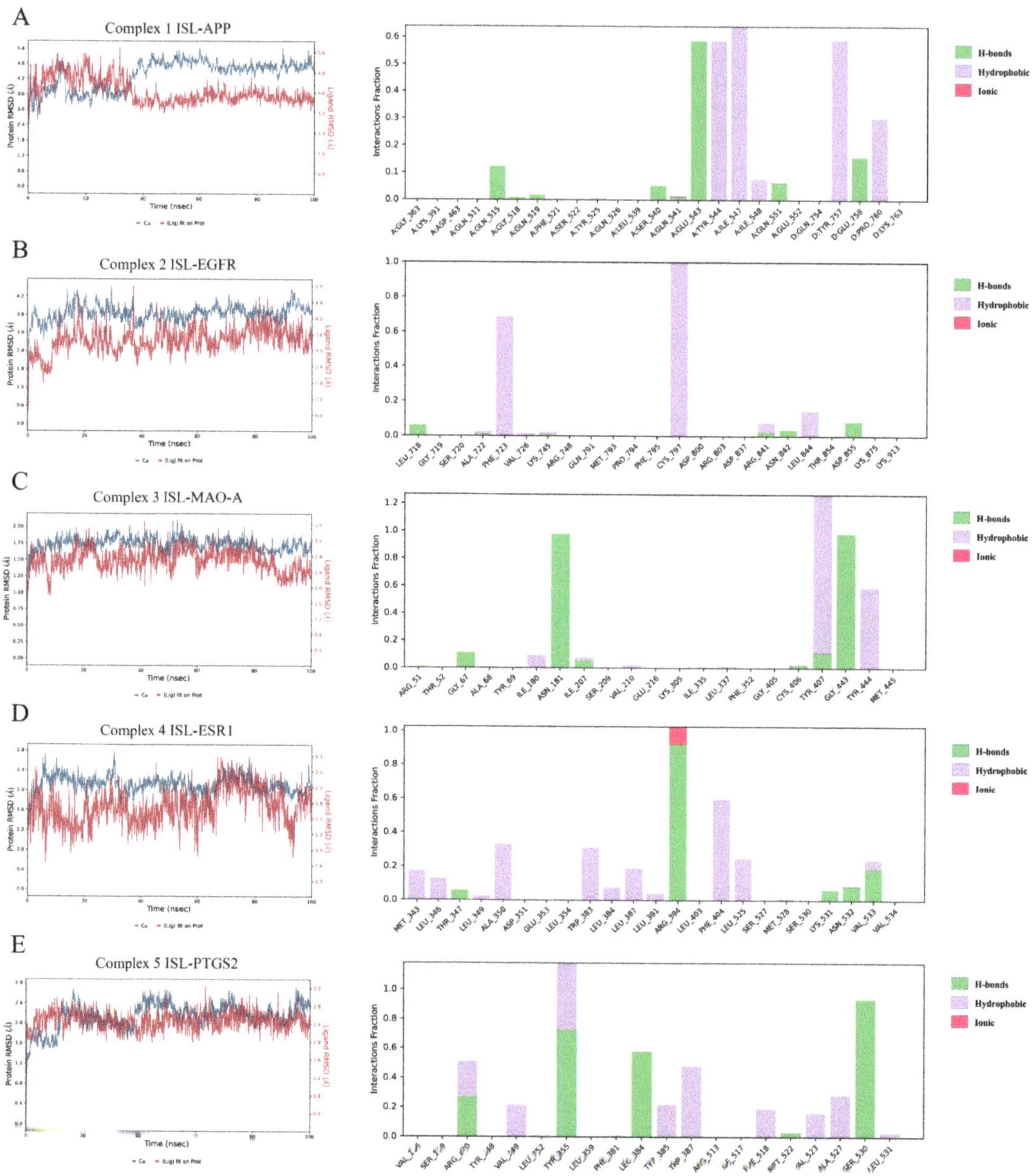

Figure 3. Molecular dynamics (MD) simulations of the ISL-APP (**A**), ISL-EGFR (**B**), ISL-MAO-A (**C**), ISL-ESR1 (**D**), and ISL-PTGS2 (**E**) complexes.

Table 4. The computed MM-GBSA ΔG_{bind} values (n = 3).

Protein	ΔG_{bind} (MM-GBSA)
APP	−41.3 ± 3.4
EGFR	−40.9 ± 5.1
MAO-A	−46.9 ± 6.2
ESR1	−31.7 ± 3.7
PTGS2	−26.4 ± 5.0

3.6. Effects of ISL on OGD/R-Induced HT22 Cells

Within 24 h, ISL showed no obvious damage at the concentrations ranging from 6.25 to 50 μM, while 100 and 200 μM ISL exhibited toxicity ($p < 0.0001$). At 72 h, no obvious damage was observed at the concentrations of 6.25 and 12.5 μM, while ISL exhibited toxicity at the concentrations ranging from 25 to 200 μM ($p < 0.01$, $p < 0.0001$) (Figure 4A). The IC_{50} values at 24 h and 72 h were 127.1 ± 3.98 and 87.6 ± 7.68 μM, respectively (Figure 4B). Afterwards, the effect of ISL on OGD/R-induced HT22 cells was evaluated, and edaravone served as a positive control. As shown in Figure 4C, the cell viability of the OGD/R group was significantly lower than that of the control group ($p < 0.0001$), whereas ISL significantly increased the viability of HT22 cells induced by OGD/R at the concentrations ranging from 2.5 to 20 μM ($p < 0.01$, $p < 0.001$), indicating that ISL exerted an obviously protective effect on OGD/R-induced HT22 cells at the abovementioned concentration range. More importantly, 2.5–20 μM ISL exhibited effects comparable to those of 5 μM edaravone ($p > 0.05$). However, 1.25 μM ISL had no protective effects on OGD/R-induced HT22 cells, likely due to an insufficient concentration to reach the effective therapeutic threshold.

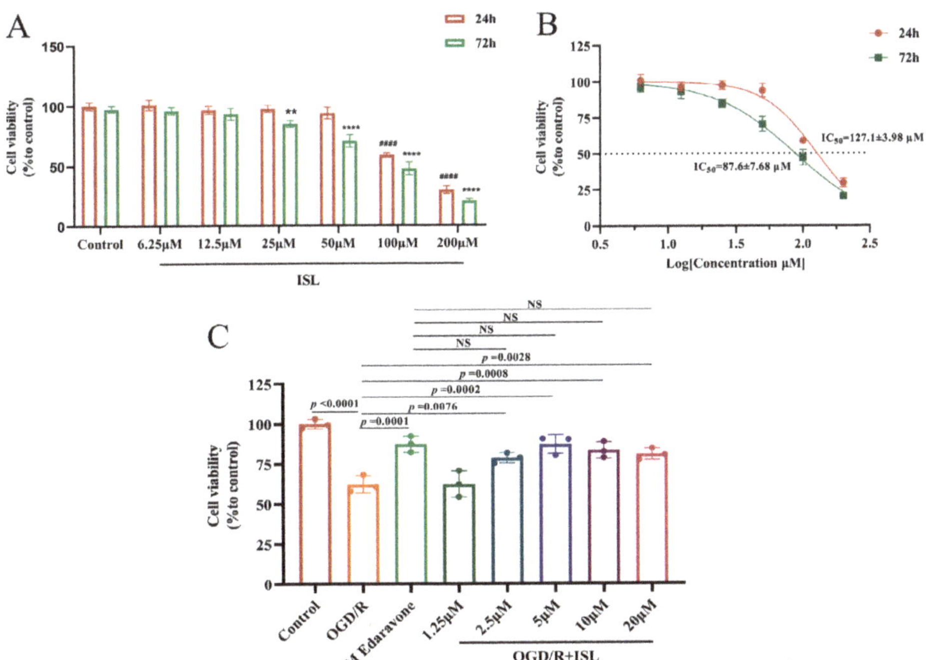

Figure 4. The neuroprotective effect of ISL on OGD/R-induced HT22 cells. Data are presented as the mean ± SD ($n = 3$). (**A**,**B**) The toxic effect of ISL (24 h/72 h) on HT22 cells under normal culture conditions. #### $p < 0.0001$ vs. the control group at 24 h. ** $p < 0.01$ and **** $p < 0.0001$ vs. the control group at 72 h. (**C**) The effect of ISL on OGD/R-induced HT22 cells.

3.7. The Effect of ISL on the mRNA Expression of APP, ESR1, MAO-A, PTGS2, and EGFR

To further confirm the mechanisms of ISL against IS, we subsequently evaluated the mRNA levels of APP, ESR1, MAO-A, PTGS2, and EGFR after ISL treatment via RT-qPCR. As shown in Figure 5, ISL treatment significantly reduced the mRNA level of PTGS2 at the concentrations of 1.25–10 μM ($p < 0.05$, $p < 0.01$, and $p < 0.001$). It also decreased the mRNA levels of APP and MAO-A ($p < 0.05$, $p < 0.01$) and increased that of ESR1 ($p < 0.05$, $p < 0.01$)

at the concentrations ranging from 2.5 to 10 μM, compared with those in the OGD/R group. Notably, when compared to the control group, the 5 μM ISL treatment group showed no significant differences in APP, ESR1, MAO-A, and PTGS2 expression levels ($p > 0.05$), suggesting a trend toward normalization. ISL treatment did not influence the mRNA level of EGFR at the tested concentrations. Additionally, there was no significant difference between the control group and the 0.01% DMSO group, which showed that the usage of 0.01% DMSO was nontoxic.

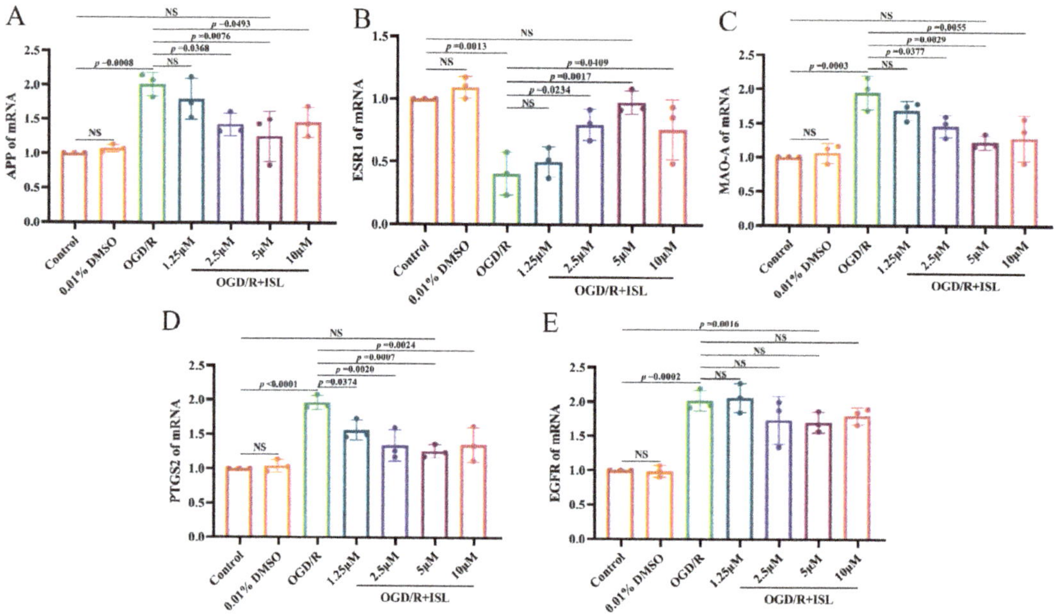

Figure 5. RT-qPCR analysis of the mRNA levels of APP (**A**), ESR1 (**B**), MAO-A (**C**), PTGS2 (**D**), and EGFR (**E**). Data are presented as the means ± SD ($n = 3$).

3.8. The Effect of ISL on the Protein Expression of APP, ESR1, MAO-A, PTGS2, and EGFR

As shown in Figure 6A–E, APP, MAO-A, PTGS2, and EGFR protein expression levels were significantly greater in the OGD/R model group than in the control group ($p < 0.001$ and $p < 0.0001$). And the expression level of ESR1 protein was markedly reduced relative to the control group ($p < 0.001$). Compared with the OGD/R group, ISL was able to significantly reduce the expression level of APP at the concentration of 2.5 μM ($p < 0.05$) and increase that of ESR1 protein at the concentrations of 5 and 10 μM, respectively ($p < 0.05$, $p < 0.01$). In addition, ISL significantly inhibited those of MAO-A and PTGS2 at the concentrations of 2.5–10 μM and 1.25–10 μM, respectively ($p < 0.05$, $p < 0.01$, and $p < 0.001$). Notably, ESR1, MAO-A, and PTGS2 protein expression levels of the 5 μM ISL treatment group were comparable to that of the control group ($p > 0.05$), suggesting a trend toward normalization.

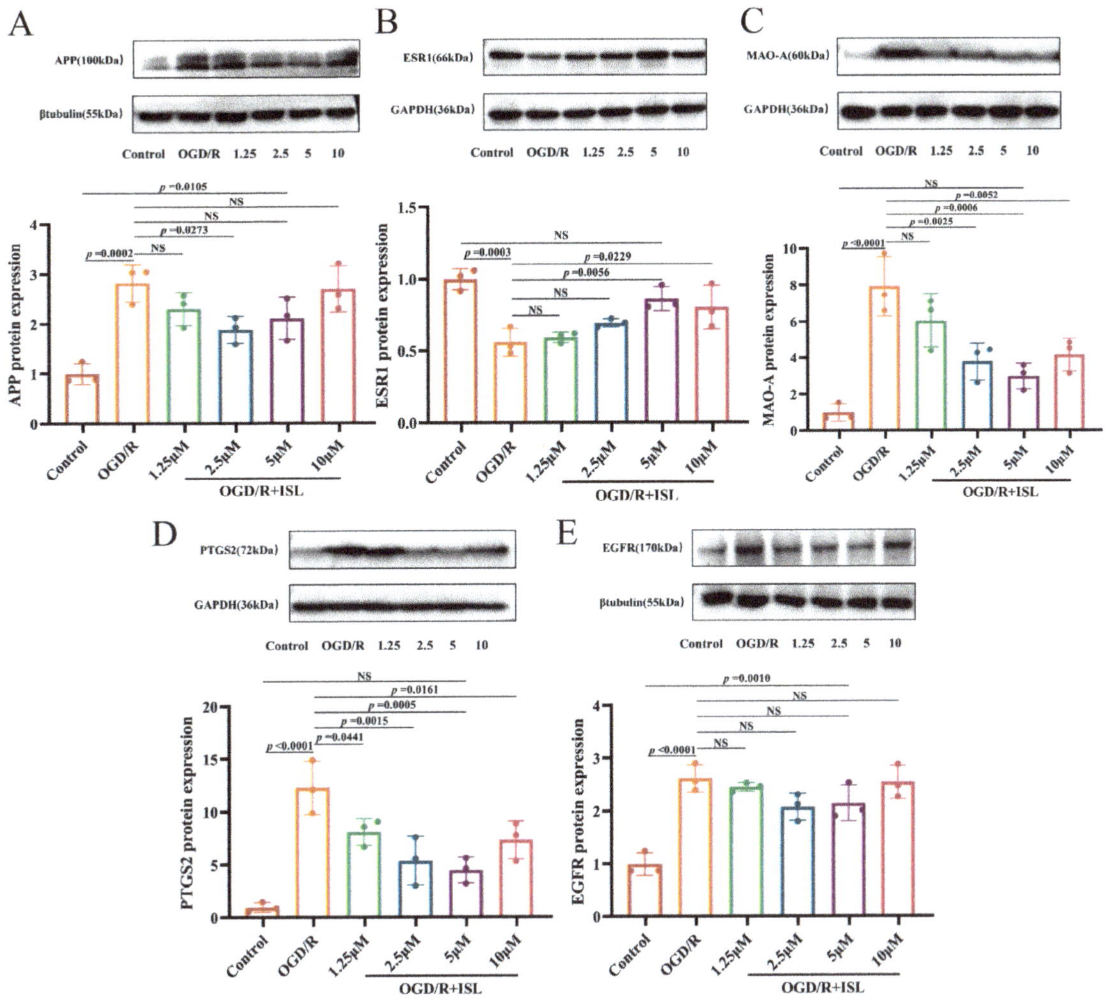

Figure 6. Effects of ISL on the protein expression levels of APP (**A**), ESR1 (**B**), MAO-A (**C**), PTGS2 (**D**), and EGFR (**E**) in HT22 cells. Data are presented as the means ± SD (n = 3).

4. Discussion

The pathogenesis of stroke mainly include cellular excitotoxicity, mitochondrial dysfunctions, neuroinflammation, BBB impairment, and cell death processes [29]. A previous study shows that ISL plays an important role in combating the pathological process of IS [30]. ISL has been proven to alleviate brain injury induced by ischemia in rats by inactivating the p38/MAPK pathway and promoting M2 polarization of microglia, thereby exerting an anti-IS effect. Meanwhile, ISL can improve tMCAO-induced brain injury by inhibiting autophagy and neuronal apoptosis and improving energy metabolism and its antioxidant properties [31,32]. However, there is still a lack of systematic research on its anti-IS characteristics at present. Therefore, we comprehensively investigated the mechanism of the anti-IS action of ISL by network pharmacology, molecular docking, molecular dynamics simulations, and experimental validation.

Based on the results of PPI network topology analysis, we can speculate that APP, ESR1, MAO-A, PTGS2, and EGFR are the main targets of ISL in the treatment of IS. Amyloid precursor protein (APP) is a universally expressed transmembrane protein, and previous studies have demonstrated that an increase in APP production can contribute to a cascade of cellular events that lead to neuronal dysfunction and death [33]. Beta amyloid (Aβ) is produced by the proteolytic cleavage of amino acids 695~770 of APP. A variety of cell types can produce Aβ from APP; moreover, in an acute ischemic model, ischemic brain injury was demonstrated to affect APP processing and lead to increased Aβ production [34]. Studies have demonstrated that the overexpression of APP increases the susceptibility of the brain to ischemic damage. This effect may involve transgene-derived Aβ neurotoxicity and Aβ-induced vascular dysfunction, thereby leading to more severe ischemia in areas at risk of infarction [35,36]. Prostaglandin-endoperoxidase synthase (PTGS2), which is also known as COX-2, is a major response gene involved in maintaining homeostasis in normal tissues and regulating inflammation [37]. The PTGS2 mRNA and protein are highly expressed in IS [38]. Studies have shown that PTGS2 reaction products are harmful to the ischemic brain. Moreover, the inhibition of PTGS2 by highly selective antagonists reduces ischemic damage in both focal and global cerebral ischemia models [39,40]. The genetic deletion of PTGS2 has been described as being an inhibitor of neuroinflammatory events to maintain the integrity of the BBB and mitigate IS-induced damage [41]. Monoamine oxidases (MAOs), including type A (MAO-A) and type B (MAO-B) MAOs in humans and mice, are a family of flavoenzymes associated with the outer mitochondrial membrane in neurons, astrocytes, and many other types of cells [42–44]. MAO produces large amounts of hydrogen peroxide, aldehydes, and ammonia during oxidative deamination, which can damage mitochondria and cause oxidative stress and neuronal death. Therefore, the inhibition of MAO activity exhibits important therapeutic value in the treatment of ischemic stroke-induced brain injury [45]. Estrogen receptor alpha (ESR1), a ligand-activated transcription factor, critically mediates transcriptional activation, hormone binding, and DNA binding, which regulates the expression of estrogen-responsive genes by interacting with estrogen [46]. Studies have shown that ESR1 can downregulate the expression of CYP1A1 and 20-HETE after stroke. And the posterior cortical 20-HETE levels were elevated in stroke patients and MCAO model mice. This indicates that ESR1 plays a crucial role in neuroprotection against hypoxic and ischemic damage [47,48]. The epidermal growth factor receptor (EGFR) serves as a critical therapeutic target for modulating pathological astrocyte overactivation. Following central nervous system (CNS) injury, EGFR expression undergoes rapid upregulation [49]. Evidence shows that p-PLCγ and EGFR expression is significantly upregulated following ischemia/reperfusion (I/R), suggesting a critical role for the EGFR/PLCγ signaling pathway in IS pathogenesis, thereby affecting neurons [50].

Molecular docking studies have shown that ISL has high affinities with the key targets identified by network pharmacology analyses. The results of MD simulations indicate that the bindings of ISL to the core targets are stable. Network pharmacology, molecular docking, and MD simulation rely on data and algorithms. However, due to the limitations of databases and software, it is impossible to fully capture the highly complex and dynamic nature of biological systems [51]. The combination of network pharmacology and in vitro experiments not only verified the accuracy of the prediction, but also provided deeper insights into the multi-target characteristics of natural drugs, thereby enhancing the scientific value and translational potential of the research [52].

Our following experimental results revealed that ISL can exert anti-IS effects by downregulating the mRNA and protein expression levels of APP, MAO-A, and PTGS2, and upregulating ESR1, which confirmed the results of the molecular docking and MD simulations. However, EGFR did not exhibit significant expression differences in the mRNA and protein levels, showing the necessity of verification experiments. These findings indicate that ISL exerts an anti-IS effect via a multitarget and multipathway pattern, which may involve the regulation of the inflammatory response, as well as cell and vascular renewal.

As demonstrated by the in vitro experiments, ISL has significant effects on the mRNA and protein levels of PTGS2 and MAO-A. Subsequent structural optimization of ISL can focus on improving their regulatory roles on PTGS2 and MAO-A. As shown in the results of molecular docking and MD simulations, the hydrogen bonds formed by OH are crucial for bindings. Therefore, the focus of structural optimization may be on the substituents of the benzene ring. In addition, the literature reports that the poor bioavailability and solubility in water of ISL have restricted its clinical application. It is a suitable strategy to make a prodrug to improve in vivo efficacy. Furthermore, proper nanodelivery technology can improve oral bioavailability or increase brain exposure [53]. This indicates that the development of brain-targeted ISL nanocarriers for the treatment of IS is very likely to improve the anti-IS effect of ISL.

5. Conclusions

In conclusion, ISL may exert an anti-IS effect by the regulation of APP, ESR1, MAO-A, and PTGS2, according to the combined results of network pharmacology, molecular docking, MD simulation, and experimental validation. All of the findings of the present study provide a theoretical basis for anti-IS drug development in the future.

Supplementary Materials: The following supporting information can be downloaded at https://www.mdpi.com/article/10.3390/cimb47080627/s1.

Author Contributions: Conceptualization, L.L. and H.Y.; methodology, H.Y. and Y.H.; software, H.Y. and X.L.; formal analysis, H.Y.; investigation, H.Y., Y.H., Y.J., and X.L.; resources, L.L.; data curation, H.Y.; writing—original draft preparation, H.Y., L.L., and X.L.; writing—review and editing, L.L. and H.Y.; supervision, L.L.; project administration, L.L.; funding acquisition, L.L. All authors have read and agreed to the published version of the manuscript.

Funding: This research was supported by the National Natural Science Foundation of China (82374147) and College Students' Innovation and Entrepreneurship Trainings Program of Jiangsu Province (202411117092Z).

Data Availability Statement: The original contributions presented in this study are included in the article/Supplementary Materials. Further inquiries can be directed at the corresponding authors.

Conflicts of Interest: The authors declare no conflicts of interest.

References

1. Zhang, R.; Lin, Y.Q.; Wang, W.S.; Wang, X.Q. Excessive nNOS/NO/AMPK signaling activation mediated by the blockage of the CBS/H2S system contributes to oxygen-glucose deprivation-induced endoplasmic reticulum stress in PC12 cells. *Int. J. Mol. Med.* **2017**, *40*, 549–557. [CrossRef] [PubMed]
2. Paul, S.; Candelario-Jalil, E. Emerging neuroprotective strategies for the treatment of ischemic stroke: An overview of clinical and preclinical studies. *Exp. Neurol.* **2021**, *335*, 113518. [CrossRef] [PubMed]
3. Jin, X.; Wang, R.-H.; Wang, H.; Long, C.-L.; Wang, H. Brain protection against ischemic stroke using choline as a new molecular bypass treatment. *Acta Pharmacol. Sin.* **2015**, *36*, 1416–1425. [CrossRef]
4. Soria, F.N.; Pérez-Samartín, A.; Martin, A.; Gona, K.B.; Llop, J.; Szczupak, B.; Chara, J.C.; Matute, C.; Domercq, M. Extrasynaptic glutamate release through cystine/glutamate antiporter contributes to ischemic damage. *J. Clin. Investig.* **2014**, *124*, 3645–3655. [CrossRef]
5. Bindal, P.; Kumar, V.; Kapil, L.; Singh, C.; Singh, A. Therapeutic management of ischemic stroke. *N-S Arch. Pharmacol.* **2024**, *397*, 2651–2679. [CrossRef]
6. Samkaria, S.; Kumari, P. Wild Edible Flowers of Indian Himalayan Region, Their Traditional Uses and Potential Health Benefits: A Way Forward for Food and Nutritional Security. *Plant Food Hum. Nutr.* **2025**, *80*, 60. [CrossRef]
7. Wang, H.; Lu, J.; Chen, X.; Zhang, K.; Zhao, X.; Zhang, Y. Antidiabetic and antioxidant molecules of *Ficus tikoua* and their combined effect: Bioassay-guided isolation, in vitro and in silico analysis. *Plant Food Hum. Nutr.* **2025**, *80*, 87. [CrossRef]

8. Hong, Z.; Cao, J.; Liu, D.; Liu, M.; Chen, M.; Zeng, F.; Qin, Z.; Wang, J.; Tao, T. Celastrol targeting Nedd4 reduces Nrf2-mediated oxidative stress in astrocytes after ischemic stroke. *J. Pharm. Anal.* **2023**, *13*, 156–169. [CrossRef]
9. Younis, N.S.; Ghanim, A.M.H. The Protective Role of Celastrol in Renal Ischemia-Reperfusion Injury by Activating Nrf2/HO-1, PI3K/AKT Signaling Pathways, Modulating NF-κb Signaling Pathways, and Inhibiting ERK Phosphorylation. *Cell Biochem. Biophys.* **2022**, *80*, 191–202. [CrossRef]
10. Subedi, L.; Gaire, B.P. Phytochemicals as regulators of microglia/macrophages activation in cerebral ischemia. *Pharmacol. Res.* **2021**, *165*, 105419. [CrossRef]
11. Zhou, J.; Sun, F.; Zhang, W.; Feng, Z.; Yang, Y.; Mei, Z. Novel insight into the therapeutical potential of flavonoids from traditional Chinese medicine against cerebral ischemia/reperfusion injury. *Front. Pharmacol.* **2024**, *15*, 1352760. [CrossRef]
12. Lan, X.; Wang, Q.; Liu, Y.; You, Q.; Wei, W.; Zhu, C.; Hai, D.; Cai, Z.; Yu, J.; Zhang, J.; et al. Isoliquiritigenin alleviates cerebral ischemia-reperfusion injury by reducing oxidative stress and ameliorating mitochondrial dysfunction via activating the Nrf2 pathway. *Redox Biol.* **2024**, *77*, 103406. [CrossRef]
13. Cui, L.; Ma, Z.; Li, W.; Ma, H.; Guo, S.; Wang, D.; Niu, Y. Inhibitory activity of flavonoids fraction from *Astragalus membranaceus* Fisch. ex Bunge stems and leaves on *Bacillus cereus* and its separation and purification. *Front. Pharmacol.* **2023**, *14*, 1183393. [CrossRef]
14. Zhao, D.; Chen, X.; Wang, R.; Pang, H.; Wang, J.; Liu, L. Determining the chemical profile of *Caragana jubata* (Pall.) Poir. by UPLC–QTOF–MS analysis and evaluating its anti-ischemic stroke effects. *J. Ethnopharmacol.* **2023**, *309*, 116275. [CrossRef] [PubMed]
15. Fang, S.M. Chemical Constituents of Traditional Tibetan Medicine *Caragana changduensis*. Master's Thesis, Tianjin University, Tianjin, China, 2009.
16. Mei, X.; Zhang, Q.; Shao, W.; Zheng, J.; Zhao, P.; Li, J. Targeted Inhibition of Aldose Reductase by Isoliquiritigenin Suppresses Fatty Acid Synthesis to Inhibit Macrophage M2 Polarization and Alleviate Pulmonary Fibrosis. *J. Agric. Food Chem.* **2025**, *73*, 18252–18267. [CrossRef] [PubMed]
17. Li, X.; Qi, T.; Zhou, L.; Lin, P.; Chen, Q.; Li, X.; He, R.; Yang, S.; Liu, Y.; Qi, F. Isoliquiritigenin alleviates abnormal sarcomere contraction and inflammation in myofascial trigger points via the IL-17RA/Act1/p38 pathway in rats. *Phytomedicine* **2025**, *145*, 156993. [CrossRef] [PubMed]
18. Zhan, C.; Yang, J. Protective effects of isoliquiritigenin in transient middle cerebral artery occlusion-induced focal cerebral ischemia in rats. *Pharmacol. Res.* **2006**, *53*, 303–309. [CrossRef]
19. Liu, X.L.; Zhu, P.Y.; Ma, J.R.; Wang, X.Y.; Wei, Q.; Gao, Y.M.; Zhang, D.Y.; Yuan, W.J. Antioxidant activity and mechanism of *Centranthera grandiflora* Benth roots based on network pharmacology and in vitro experiments. *Nat. Prod. Res. Dev.* **2024**, *36*, 155–166.
20. Tao, L.; Ke, Z.P.; Wang, T.J.; Wang, Z.Z.; Cao, L.; Xiao, W. Frontier technologies and development trends of network pharmacology: A patent bibliometric analysis. *China J. Chin. Mater. Med.* **2025**, *50*, 3070–3078.
21. Surovi, S.; Manobjyoti, B. Molecular Docking: Challenges, Advances and its Use in Drug Discovery Perspective. *Curr. Drug Targets* **2019**, *20*, 501–521. [CrossRef]
22. Yu, Y.; Xu, S.; He, R.; Liang, G. Application of Molecular Simulation Methods in Food Science: Status and Prospects. *J. Agric. Food Chem.* **2023**, *71*, 2684–2703. [CrossRef]
23. Stelzer, G.; Rosen, N.; Plaschkes, I.; Zimmerman, S.; Twik, M.; Fishilevich, S.; Stein, T.I.; Nudel, R.; Lieder, I.; Mazor, Y.; et al. The GeneCards suite: From gene data mining to disease genome sequence analyses. *Curr. Protoc. Bioinform.* **2016**, *54*, 1.30.1–1.30.33. [CrossRef] [PubMed]
24. Szklarczyk, D.; Gable, A.-L.; Nastou, K.-C.; Lyon, D.; Kirsch, R.; Pyysalo, S.; Doncheva, N.-T.; Legeay, M.; Fang, T.; Bork, P.; et al. The STRING database in 2021: Customizable protein-protein networks, and functional characterization of user-uploaded gene/measurement sets. *Nucleic Acids Res.* **2021**, *49*, D605–D612. [CrossRef] [PubMed]
25. Otasek, D.; Morris, J.-H.; Bouças, J.; Pico, A.-R.; Demchak, B. Cytoscape Automation: Empowering workflow-based network analysis. *Genome Biol.* **2019**, *20*, 185. [CrossRef] [PubMed]
26. Dennis, G., Jr.; Sherman, B.-T.; Hosack, D.-A.; Yang, J.; Gao, W.; Lane, H.-C.; Lempicki, R.A. DAVID: Database for annotation, visualization, and integrated discovery. *Genome Biol.* **2003**, *4*, P3. [CrossRef]
27. Schneider, C.A.; Rasband, W.S.; Eliceiri, K.W. NIH Image to ImageJ: 25 years of image analysis. *Nat. Methods* **2012**, *9*, 671–675. [CrossRef]
28. Ouyang, G.; Zhu, Y.; Ouyang, Z. Investigation of *Scutellaria Barbata*'s immunological mechanism against thyroid cancer using network pharmacology and experimental validation. *Sci. Rep.* **2025**, *15*, 2490. [CrossRef]
29. Qin, C.; Yang, S.; Chu, Y.-H.; Zhang, H.; Pang, X.-W.; Chen, L.; Zhou, L.-Q.; Chen, M.; Tian, D.-S.; Wang, W. Signaling pathways involved in ischemic stroke: Molecular mechanisms and therapeutic interventions. *Signal Transduct. Target. Ther.* **2022**, *7*, 215. [CrossRef]

30. He, D.; Wang, P.; Liao, F.; Yu, L.; Gan, B. Cell membrane-coated biomimetic magnetic nanoparticles for the bio-specific extraction of components from Gualou Guizhi decoction exhibiting activities against oxygen-glucose deprivation/reperfusion injury. *J. Pharm. Biomed. Anal.* **2022**, *209*, 114528. [CrossRef]
31. Wang, R.; Zhang, W. Isoliquiritigenin regulates microglial M1/M2 polarisation by mediating the P38/MAPK pathway in cerebral stroke. *J. Pharm. Pharmacol.* **2023**, *75*, 828–836. [CrossRef]
32. Song, W.; Bai, L.; Yang, Y.; Wang, Y.; Xu, P.; Zhao, Y.; Zhou, X.; Li, X.; Xue, M. Long-Circulation and Brain Targeted Isoliquiritigenin Micelle Nanoparticles: Formation, Characterization, Tissue Distribution, Pharmacokinetics and Effects for Ischemic Stroke. *Int. J. Nanomed.* **2022**, *17*, 3655–3670. [CrossRef]
33. Shi, J.; Yang, S.-H.; Stubley, L.; Day, A.-L.; Simpkins, J.-W. Hypoperfusion induces overexpression of β-amyloid precursor protein mRNA in a focal ischemic rodent model. *Brain Res.* **2000**, *853*, 1–4. [CrossRef]
34. Nihashi, T.; Inao, S.; Kajita, Y.; Kawai, T.; Sugimoto, T.; Niwa, M.; Kabeya, R.; Hata, N.; Hayashi, S.; Yoshida, J. Expression and Distribution of Beta Amyloid Precursor Protein and Beta Amyloid Peptide in Reactive Astrocytes After Transient Middle Cerebral Artery Occlusion. *Acta Neurochir.* **2001**, *143*, 287–295. [CrossRef] [PubMed]
35. Huang, K.-L.; Lin, K.-J.; Ho, M.-Y.; Chang, Y.-J.; Chang, C.-H.; Wey, S.-P.; Hsieh, C.-J.; Yen, T.-C.; Hsiao, I.-T.; Lee, T.-H. Amyloid deposition after cerebral hypoperfusion: Evidenced on $[^{18}F]$AV-45 positron emission tomography. *J. Neurol. Sci.* **2012**, *319*, 124–129. [CrossRef] [PubMed]
36. Zhang, F.; Eckman, C.; Younkin, S.; Hsiao, K.-K.; Iadecola, C. Increased susceptibility to ischemic brain damage in transgenic mice overexpressing the amyloid precursor protein. *J. Neurosci.* **1997**, *17*, 7655–7661. [CrossRef] [PubMed]
37. Ma, S.-L.; Tang, N.-L.-S.; Zhang, Y.-P.; Ji, L.-d.; Tam, C.-W.-C.; Lui, V.-W.-C.; Chiu, H.-F.-K.; Lam, L.-C.-W. Association of prostaglandin-endoperoxide synthase 2 (PTGS2) polymorphisms and Alzheimer's disease in Chinese. *Neurobiol. Aging* **2008**, *29*, 856–860. [CrossRef]
38. Cui, Q.; Zhang, Y.-L.; Ma, Y.-H.; Yu, H.-Y.; Zhao, X.-Z.; Zhang, L.-H.; Ge, S.-Q.; Zhang, G.-W.; Qin, X.-D. A network pharmacology approach to investigate the mechanism of Shuxuening injection in the treatment of ischemic stroke. *J. Ethnopharmacol.* **2020**, *257*, 112891. [CrossRef]
39. Nakayama, M.; Uchimura, K.; Zhu, R.-L.; Nagayama, T.; Rose, M.-E.; Stetler, R.-A.; Isakson, P.-C.; Chen, J.; Graham, S.-H. Cyclooxygenase-2 inhibition prevents delayed death of CA1 hippocampal neurons following global ischemia. *Proc. Natl. Acad. Sci. USA* **1998**, *95*, 10954–10959. [CrossRef]
40. Nogawa, S.; Zhang, F.; Ross, M.-E.; Iadecola, C. Cyclooxygenase-2 gene expression in neurons contributes to ischemic brain damage. *J. Neurosci.* **1997**, *17*, 2746–2755. [CrossRef]
41. Yang, C.; Yang, Y.; DeMars, K.-M.; Rosenberg, G.-A.; Candelario-Jalil, E. Genetic Deletion or Pharmacological Inhibition of Cyclooxygenase-2 Reduces Blood-Brain Barrier Damage in Experimental Ischemic Stroke. *Front. Neurol.* **2020**, *11*, 887. [CrossRef]
42. Shih, J.-C.; Chen, K. Regulation of MAO-A and MAO-B Gene Expression. *Curr. Med. Chem.* **2004**, *11*, 1995–2005. [CrossRef]
43. Wang, C.-C.; Borchert, A.; Ugun-Klusek, A.; Tang, L.-Y.; Lui, W.-T.; Chu, C.-Y.; Billett, E.; Kuhn, H.; Ufer, C. Monoamine Oxidase A Expression Is Vital for Embryonic Brain Development by Modulating Developmental Apoptosis. *J. Biol. Chem.* **2011**, *286*, 28322–28330. [CrossRef]
44. Tong, J.; Meyer, J.-H.; Furukawa, Y.; Boileau, I.; Chang, L.-J.; Wilson, A.-A.; Houle, S. Distribution of Monoamine Oxidase Proteins in Human Brain: Implications for Brain Imaging Studies. *J. Cereb. Blood Flow Metab.* **2013**, *33*, 863–871. [CrossRef]
45. Deshwal, S.; Di, S.-M.; Di, L.-F.; Kaludercic, N. Emerging role of monoamine oxidase as a therapeutic target for cardiovascular disease. *Curr. Opin. Pharmacol.* **2017**, *33*, 64–69. [CrossRef]
46. He, P.; Wang, Z.; Yang, J.; Pan, P.; Shi, T.; Xu, S.; Lan, J.; Hao, Z.; Yang, A.; Chen, L.; et al. Mechanism of *Ligusticum wallichii*—Borneol in the Treatment of Cerebral Ischemic Stroke in Rats Based On Network Pharmacology, Molecular Docking, and Experimental Verification. *Chem. Biodivers.* **2025**, *22*, e202401893. [CrossRef]
47. Zhang, Y.; Jiao, X.; Qi, X.; Wang, G.; Ma, Y. Edaravone ameliorates inflammation in ischemic stroke mouse by regulating the CYP1A1 pathway through gut microbiota. *Exp. Neurol.* **2025**, *390*, 115263. [CrossRef] [PubMed]
48. Crago, E.A.; Thampatty, B.P.; Sherwood, P.R.; Kuo, C.-W.J.; Bender, C.; Balzer, J.; Horowitz, M.; Poloyac, S.M. Cerebrospinal Fluid 20-HETE Is Associated with Delayed Cerebral Ischemia and Poor Outcomes After Aneurysmal Subarachnoid Hemorrhage. *Stroke* **2011**, *42*, 1872–1877. [CrossRef] [PubMed]
49. Chen, J.; He, W.; Hu, X.; Shen, Y.; Cao, J.; Wei, Z.; Luan, Y.; He, L.; Jiang, F.; Tao, Y. A role for ErbB signaling in the induction of reactive astrogliosis. *Cell Discov.* **2017**, *3*, 17044. [CrossRef] [PubMed]
50. Ren, Z.-L.; Zhang, C.-X.; Zheng, Y.-X.; Chen, C.-A.; Dan, C.; Lan, X.; Yan, X.; Liu, Y.; He, Y.-H.; Cheng, J.-L.; et al. Refined qingkailing attenuates reactive astrocytes and glial scar formation after ischemia stroke via the EGFR/PLCγ pathway. *Phytomedicine* **2025**, *142*, 156696. [CrossRef]
51. Luo, T.-T.; Lu, Y.; Yan, S.-K.; Xiao, X.; Rong, X.-L.; Guo, J. Network Pharmacology in Research of Chinese Medicine Formula: Methodology, Application and Prospective. *Chin. J. Integr. Med.* **2020**, *26*, 72–80. [CrossRef]

52. Sun, R.; Liu, C.; Liu, J.; Yin, S.; Song, R.; Ma, J.; Cao, G.; Lu, Y.; Zhang, G.; Wu, Z.; et al. Integrated network pharmacology and experimental validation to explore the mechanisms underlying naringenin treatment of chronic wounds. *Sci. Rep.* **2023**, *13*, 132. [CrossRef]
53. Zhang, K.; Wang, Q.; Yang, Q.; Wei, Q.; Man, N.; Adu-Frimpong, M.; Toreniyazov, E.; Ji, H.; Yu, J.; Xu, X. Enhancement of Oral Bioavailability and Anti-hyperuricemic Activity of Isoliquiritigenin via Self-Microemulsifying Drug Delivery System. *AAPS PharmSciTech* **2019**, *20*, 218. [CrossRef]

current issues in molecular biology

MDPI

Review

Pathophysiological Links Between Stroke and Prediabetes: A Systematic Review

Yerushka Naicker and Andile Khathi *

Department of Human Physiology, School of Laboratory Medicine and Medical Sciences, College of Health Sciences, University of KwaZulu-Natal, Durban 3629, South Africa; 220014139@stu.ukzn.ac.za
* Correspondence: khathia@ukzn.ac.za

Abstract: Prediabetes is an intermediate stage between normoglycaemia and type 2 diabetes mellitus (T2DM), affecting over 425 million people globally and contributing to vascular damage and increased stroke risk. Despite the severity of both conditions, their association remains underexplored. This review examines the literature on stroke-related biomarkers in normoglycaemia, prediabetes and T2DM to identify potential links between prediabetes and stroke. This systematic review followed PRISMA-2020 guidelines. PubMed, Google Scholar, Scopus, Web of Science and Science Direct were searched for studies (2003–2023) on stroke biomarkers in prediabetes. Eligible studies were original human research in English, with defined diagnostic criteria (ADA or WHO) for glycaemic status and reported biomarker associations or stroke risk. Studies with major comorbidities were excluded. Data were extracted and bias was assessed using the Newcastle–Ottawa Scale. Meta-analysis was not performed due to limited studies per biomarker. Eight studies (n = 3003) were included. NSE was examined in three studies, all reporting significant elevations in hyperglycaemic individuals. Interleukin-6 (IL-6) was assessed in two studies; one showed a significant increase in diabetes, while the other found a non-significant upward trend. D-dimer and GFAP were each reported in separate single studies, both showing significant elevations in hyperglycaemic individuals with stroke or neurocognitive impairment. S100B was investigated in two studies, with divergent findings: one showed a positive association with glycaemic status, while the other reported lower levels in hyperglycaemia. Findings indicate biomarker alterations in T2DM, suggesting that early changes may occur in prediabetes. Our review suggests that individuals with prediabetes may show alterations in inflammatory (IL-6), coagulation (D-dimer), and neurovascular (S100B, GFAP, NSE) markers, though some findings are inconsistent, reflecting early pathophysiological changes that may increase stroke risk. Further well-designed studies are needed to clarify these associations and establish biomarker-based tools for earlier stroke risk detection and prevention in individuals with prediabetes.

Keywords: systematic review; prediabetes; type 2 diabetes mellitus; stroke; stroke risk

1. Introduction

Type 2 Diabetes Mellitus (T2DM) is a chronic metabolic condition characterised by insulin resistance that results in elevated blood glucose levels [1]. With rapid economic development, dramatic changes in lifestyles and an ageing population, T2DM has become a leading public health problem globally [2]. It is estimated to affect 537 million adults worldwide, with a global prevalence of 10.5% among adults aged 20 to 79 years [3].

Stroke is a well-recognised macrovascular complication associated with T2DM [4]. It is clinically defined as a syndrome of acute, focal neurological deficit attributed to vascular injury, which can either be ischaemic (due to inadequate blood supply) or haemorrhagic (due to bleeding) [5]. Stroke is the second leading cause of death worldwide, accounting for approximately 5.5 million deaths each year [6]. Importantly, individuals with diabetes mellitus and symptomatic cerebral ischaemia have a higher risk of stroke recurrence, and female sex is a demographic factor associated with a worse prognosis [7].

In T2DM, hyperglycaemia leads to oxidative stress, inflammation and endothelial dysfunction, increasing vascular damage and stroke risk [8]. Both T2DM and stroke share common inflammatory pathways; elevated cytokine levels in both conditions accelerate atherosclerosis and enhance thrombogenesis, leading to greater clot formation and ischaemic events [9].

T2DM also impairs cerebral autoregulation and vascular integrity, making the brain more susceptible to ischaemic injury [10]. This can cause reduced cerebral blood flow which contributes to neurodegeneration. Furthermore, insulin resistance affects glucose metabolism in the brain, leaving it more vulnerable to ischaemic events and increasing the risk of stroke [11].

In clinical settings, quick and reliable tests to assess stroke risk and diagnose ongoing stroke events are essential for timely intervention. Current methods, such as CT scans and MRI, although accurate, are not always available in resource-limited settings. Therefore, identifying and using reliable stroke biomarkers is a crucial area of research.

Recent studies have shown that specific biomarkers of inflammation, coagulation and brain injury are effective for evaluating stroke risk and severity in individuals with T2DM [12]. Elevated levels of these biomarkers are linked to higher stroke risk and worse outcomes, indicating their potential role in diagnosis and prognosis.

However, while these associations have been extensively studied in T2DM cohorts, there remains a limited understanding of how such biomarkers change during the earlier stages of dysglycaemia.

Before the onset of T2DM, moderate insulin resistance leads to a gradual increase in glucose levels, known as prediabetes [13]. This condition, also referred to as intermediate hyperglycaemia, is marked by blood glucose levels that are elevated but not high enough to meet the criteria for T2DM [14]. Evidence indicates that complications associated with T2DM often begin during the prediabetes state, with research showing that individuals with prediabetes are at an increased risk of both progression to T2DM and stroke [15].

Yet, no independent synthesis of biomarker data has been undertaken specifically in individuals with prediabetes. Most existing research has inferred stroke risk in prediabetic individuals based on findings from T2DM studies, under the assumption of shared pathophysiological mechanisms. This represents a significant gap in the literature, as it remains unclear whether the same biomarker trends observed in T2DM also apply in earlier dysglycaemic states.

In 2021, the worldwide prevalence of impaired glucose tolerance (IGT) was 9.1%, affecting 464 million people and it is expected to rise to 10.0%, or 638 million individuals, by 2045 [16]. Importantly, prediabetes is increasingly linked to early onset of T2DM-related complications, particularly in low- and middle-income countries where limited awareness and scarce healthcare resources delay diagnosis and management. This contributes to poorer quality of life, increased stroke risk and greater strain on already burdened healthcare systems. Many affected individuals may become unable to work, leading to economic dependency on social grants or family support, which further impacts community and national economies. The rising prevalence of prediabetes among younger populations increases healthcare costs, reduces workforce productivity and elevates morbidity and mortality. Therefore, prioritising early detection, education and prevention is critical to reduce the long-term public health impact of prediabetes and its complications.

Importantly, prediabetes is a reversible condition, and early identification of at-risk individuals could enable interventions to prevent progression to T2DM and reduce stroke risk. This review uniquely synthesises biomarker data reflecting mechanisms underlying both stroke and hyperglycaemia, with a focus on younger adults in populations where prediabetes prevalence is rising. This panel of biomarkers holds promise for independent monitoring of stroke risk in at-risk individuals.

An observational study found that T2DM doubles stroke risk compared to individuals without diabetes [17]. While the association between T2DM and stroke is well-documented, there is a lack of comprehensive research examining stroke in individuals with prediabetes.

There is a need to better understand whether prediabetes itself confers measurable risk in terms of biomarker expression, or if stroke-related pathophysiological changes only emerge at later stages of metabolic dysfunction.

Therefore, this systematic review aims to bridge this gap by examining the pathophysiological mechanisms between prediabetes and stroke in individuals aged 18 and above, addressing the rising prevalence of prediabetes in younger adults and its potential long-term implications for stroke risk. This review will summarise articles that have analysed stroke biomarkers in individuals with type 2 diabetes mellitus and prediabetes.

1.1. Research Question

What evidence exists regarding the pathophysiological mechanisms and stroke biomarkers in individuals with prediabetes?

1.2. Objectives

Primary objective:

- To summarise the evidence on blood biomarker changes associated with stroke risk in individuals with prediabetes.

Secondary objective:

- To evaluate the reliability and clinical relevance of blood biomarkers for assessing stroke risk and outcomes in individuals with prediabetes.

2. Materials and Methods

The protocol for this systematic review has been registered with the International Prospective Registry of Systematic Reviews, PROSPERO; CRD42024530755, as required in the Prepared Reporting Items for Systematic Reviews and Meta-Analysis (PRISMA-P) Protocols 2020 [18].

2.1. Eligibility Criteria for the Study

The review targeted prospective observational studies, case–control, cohort and cross-sectional studies that assessed associations of type 2 diabetes, prediabetes and stroke risk through various stroke biomarkers.

This review included only studies written in English. This restriction was due to limited translation resources and the risk of misinterpretation. As most high-impact studies in this field are published in English, the likelihood of capturing relevant evidence remained high. While this may introduce language bias, it ensured the feasibility and accuracy of the review.

Grey literature was not included due to concerns about inconsistent reporting standards and limited peer review, which may compromise the reliability of biomarker data. However, manual screening of reference lists from relevant articles was conducted to help mitigate potential gaps.

All participants with prediabetes or type 2 diabetes aged above 18 years from all ethnic groups were included in this study. Included studies must have had a clear criterion used to diagnose participants with prediabetes or T2DM such as the ADA or WHO criteria. Also, studies that reported participants with a history of stroke or participants who currently experienced a stroke were included.

Studies reporting participants with a history of traumatic brain injury, hypertension, blood disorders, depression and kidney/heart/liver diseases were excluded.

Studies meeting this criterion were considered eligible for the systematic review.

2.2. Ethics Approval and Consent to Participate

Data was not collected from individuals; instead, the review analysed previously published data. Since no informed consent was required, the systematic review did not need ethics approval.

2.3. Diagnostic Criteria

For this study, eligible participants had to be clinically diagnosed with prediabetes according to the diagnostic criteria set by the ADA or WHO; however, there were some differences in their criteria. For fasting plasma glucose, the ADA defined impaired fasting glucose (IFG) as a level between 5.6 and 6.9 mmol/L, while the WHO specified a range of 6.1 to 6.9 mmol/L. Impaired glucose tolerance (IGT) was defined as a 2-h plasma glucose measurement. Both ADA and WHO used the same criteria for IGT, setting the range at 7.8 to 11.0 mmol/L. Unlike the WHO, the ADA included an HbA1c guideline for assessing glucose levels, specifying a range of 5.7 to 6.4%. Despite the minor differences, both the ADA and WHO criteria provided clear guidelines designed to diagnose prediabetes, as depicted in Table 1. Only studies that used the ADA/WHO criteria were included. Since different countries use different diagnostic standards, including both criteria allowed us to capture the largest number of relevant studies while maintaining clinical accuracy.

Table 1. Difference between the American Diabetes Association and World Health Organization criteria for prediabetes diagnosis.

Criteria	ADA	WHO
Fasting Plasma glucose (IFG)	5.6–6.9 mmol/L	6.1–6.9 mmol/L
2 Hour plasma glucose (IGT)	7.8–11.0 mmol/L	7.8–11.0 mmol/L
HbA1c %	5.7–6.4%	-

In this review, stroke diagnosis was primarily based on clinical criteria as reported in the respective studies, including neurological assessments and imaging techniques such as computed tomography (CT) scans and magnetic resonance imaging (MRI), where available. However, if the diagnostic method for stroke was not explicitly stated, we included studies that met the broader clinical definitions according to the World Health Organization (WHO) or American Heart Association (AHA) guidelines, which define stroke as a sudden onset of neurological symptoms caused by an interruption of blood flow to the brain.

2.4. Search Strategy

The databases screened were PubMed, Google Scholar, Scopus, Web of Science and Science Direct. The Boolean Logical operator "AND" and "OR" was used between search terms. Additional filters were applied to narrow results. To refine the results, additional filters were applied: studies involving humans aged 18 and older, published between 2003 and 2023 and available as free full text. This timeframe was chosen to include the most recent research and avoid outdated information. Some of the following medical subject headings were used in our search strategy: "prediabetes AND stroke risk," "type 2 diabetes mellitus AND stroke" AND "NSE OR D-dimer OR GFAP OR IL-6". Additionally, bibliographies of relevant articles were manually screened for other relevant articles. A comprehensive PubMed search strategy is provided in Table 2.

Table 2. PubMed search strategy for studies.

Search	Search Terms
1	Search ("Prediabetes"[MeSH Terms] OR "Impaired Glucose Tolerance" OR "Type 2 Diabetes Mellitus")
2	Search ("Stroke"[MeSH Terms] OR "Stroke Risk" OR "Cerebrovascular Accident")
3	Search ("NSE" OR "Neuron-Specific Enolase" OR "D-dimer" OR "GFAP" OR "Glial Fibrillary Acidic Protein" OR "IL-6" OR "Interleukin-6" OR "S100B" OR "Fibrinogen" OR "Stroke Biomarkers")
4 (Additional Filters)	Humans, Age 18+, English, Free Full Text, Publication Date: 2003–2023
5	Search (#1 AND #2 AND #3 AND #4)

2.5. Identification of Eligible Studies

Duplicate studies were removed using Endnote X20 reference manager. YN screened the articles and removed irrelevant studies based on the titles and abstracts. Studies were deemed eligible and included in the review if they met the stated inclusion criteria. Thereafter, YN conducted a second screening of all articles to ensure accuracy and consistency. Uncertainty of results was settled by AK. The selection process is represented in the PRISMA flow chart (Figure 1).

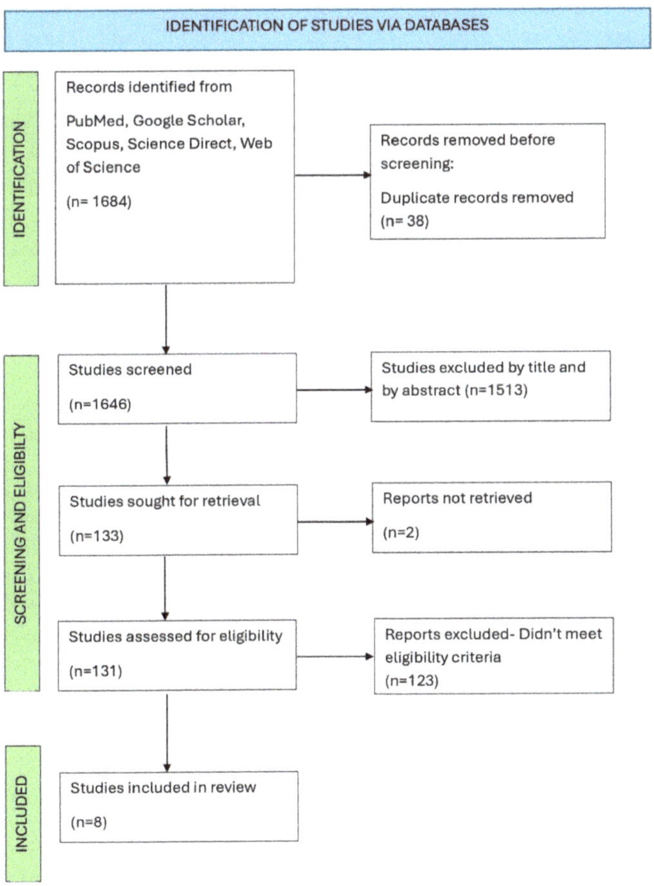

Figure 1. Preferred reporting items for systematic review flow diagram of included studies.

2.6. Data Extraction

Relevant data were extracted from the selected studies and recorded in a Microsoft Excel file. Among the baseline characteristics extracted from the eligible reports included author, publication year, country, study setting, methodology categorised into study design, number and age range of participants. Quantitative data, including biomarker levels and statistical significance, were extracted and organised for comparative analysis.

In addition to biomarkers, qualitative information regarding the pathophysiological mechanisms explaining the link between prediabetes or type 2 diabetes mellitus and stroke was also extracted. Proposed mechanisms, such as insulin resistance, chronic inflammation, endothelial dysfunction and vascular damage induced by hyperglycaemia, are described in detail. Theoretical discussions had been summarised and then categorised into distinct mechanisms to allow a comprehensive understanding of how these processes contribute to the risk of stroke in individuals with prediabetes. Accordingly, these pathophysiological insights have been synthesised and reported in the narrative sections of this review, complementing the quantitative data on biomarkers of stroke.

2.7. Risk of Bias and Quality Appraisal

The potential risk of bias was assessed by YN using a risk of bias checklist. The Newcastle Ottawa quality assessment scale checklist was used as it assessed the quality of cohort, cross-sectional and case–control studies (newcastle—ottawa quality assessment scale. (n.d.). Newcastle—ottawa quality assessment scale. https://cdn-links.lww.com/permalink/eandh/a/eandh_1_1_2021_02_03_meghji_eandh-d-20-00242_sdc3.pdf) (accessed on 15 October 2024). Risk of bias was assessed for all 8 studies included in the final review. Given the limited number of eligible studies, we retained all eight, regardless of their NOS scores. However, studies with lower methodological quality were interpreted with caution during synthesis and their findings were weighed accordingly. This approach allowed for a comprehensive review while maintaining transparency and critical appraisal standards. Disagreements were settled by an independent researcher. The overall quality of the included studies was good with a score of 8/9.

2.8. Data Synthesis and Analysis

While forest plots were initially planned for statistical aggregation, the limited number of studies available for each marker made this approach unfeasible. As a result, the findings were presented without the usual meta-analysis aggregation. Results from the included studies were first collated into a summary table and then synthesised narratively. Studies were grouped according to shared biomarker targets and synthesised based on their associated pathophysiological mechanisms, such as inflammation, endothelial dysfunction and neuronal injury. This thematic organisation allowed for comparison of biomarker trends across populations with T2DM and prediabetes, enabling the identification of patterns and gaps relevant to stroke risk.

3. Results

3.1. Search Report Results and Eligible Reports

The search strategy generated 1684 studies from the following databases: PubMed, Google Scholar, Scopus, Science Direct and Web of Science. A total of 38 duplicates were removed before screening, and 1513 studies were excluded based on review of the title and abstracts. 133 studies were sought for retrieval, of which only 131 were assessed for eligibility. Thereafter, 123 studies were excluded after full text analysis as they did not meet the eligibility criteria or report on stroke biomarkers in groups of individuals with T2DM or prediabetes. Hence, a total of 8 studies were included in this systematic review.

3.2. Study Characteristics

All included studies were of those published in journals between 2003 and 2023. The total number of participants across the included eight studies is 3003. Study characteristics and results have been summarised and presented in Table 3. The results have been grouped according to the investigated pathophysiological mechanisms, namely inflammation, coagulation, neurovascular injury and further categorised by specific biomarker type to facilitate clearer interpretation of patterns, consistencies and contradictions across studies.

Table 3. Overview of characteristics of the included studies.

Biomarker	Author (Year)	Study Design	Country	Population (n)	Avg. Age	Diabetes Status	Stroke Type/Condition	Key Findings	Risk of Bias
Neurovascular Injury									
NSE	Pandey et al. (2011)	Cross-sectional case–control	India	191	60.5	Diabetes Mellitus	Ischaemic Stroke	Stroke: 18.0 ± 4.5 ng/mL Controls: 7.5 ± 1.5 ng/mL ($p = 0.001$) Normoglycemic Ischaemic Stroke Patients: 15.2 ± 2.4 ng/mL Hyperglycaemic Ischaemic Stroke Patients: 19.7 ± 4.7 ng/mL ($p = 0.05$).	Low
NSE	Nayak et al. (2016)	Cohort	India	104	50	Type 2 Diabetes Mellitus	Ischaemic Stroke	NSE < 0.65 µg/L in controls. Significant group differences but specific p-values not given.	High
NSE	Roshdy et al. (2007)	Cross-sectional	Egypt	66	52.3	Type 2 Diabetes Mellitus	Ischaemic Stroke	Diabetic Stroke: 14.42 ± 4.09 µg/L Controls: 6.71 ± 1.29 µg/L ($p < 0.001$) Strong positive correlation with infarct size (r = 0.938).	Low
S100B	Roshdy et al. (2007)	Cross-sectional	Egypt	66	52.3	Type 2 Diabetes Mellitus	Ischaemic Stroke	Diabetic Stroke: 115.18 ± 7.45 µg/L vs. Controls: 28.22 ± 7.9 µg/L ($p < 0.001$) Strong correlation with infarct size (r = 0.9816).	Low
S100B	Yu et al. (2020)	Case–control	Turkey	164	49.67	Type 2 Diabetes Mellitus	Cognitive Dysfunction	Lower levels in T2DM with cognitive dysfunction (0.117 µg/L) vs. controls (0.344 µg/L) ($p < 0.05$).	Moderate
GFAP	Ayala-Guerrero et al. (2022)	Comparative study	Mexico	138	72.3	Type 2 Diabetes Mellitus	Neurocognitive Disorders	GFAP higher in NCD with T2D (2046 pg/mL) vs. control without T2D (583.2 pg/mL) ($p < 0.0001$).	Moderate
Inflammation									
IL-6	Pickup et al. (2000)	Case–control	UK	37	55.4	Type 2 Diabetes	Diabetic Complications	T2D: 1.8 pg/mL vs. Controls: 1.1 pg/mL ($p < 0.01$); No difference between diabetics with and without complications.	Low
IL-6	Wu et al. (2012)	Case–control	China	327	55	Type 2 Diabetes Mellitus	Atherosclerosis	Higher IL-6 in both diabetic and non-diabetic groups with atherosclerosis vs. without ($p = 0.007$ and $p = 0.004$, respectively).	Low
Coagulation									
D-Dimer	Cheng et al. (2022)	Retrospective cohort	China	1976	59.6	Type 2 Diabetes Mellitus	Stroke Events (general)	Higher quartiles of D-dimer levels associated with more stroke events. $p < 0.001$ across all quartile comparisons.	Low

4. Discussion

The global prevalence of prediabetes is significant and, on the rise, particularly in low-income countries, where the most substantial growth is anticipated by 2045. In 2021, the global prevalence was 9.1%, affecting 464 million people, and this is expected to rise to 10.0%, impacting 638 million by 2045 [16]. Additionally, IGT prevalence is projected to grow in adults aged 20–74 years, emphasising the need for intervention [16].

Individuals with prediabetes are at a higher risk of developing T2DM and its associated complications [19]. An investigation conducted in 2019 found that, in comparison to people with normal glucose levels, those with prediabetes had a higher risk of having a new stroke and a worse outcome [20].

While evidence is available that links prediabetes with an increased risk of stroke, the mechanisms through which this happens are not well-defined. Further challenges remain in explaining how elevated blood glucose levels during the prediabetes state may facilitate the development of cerebrovascular disease. Even though the diagnosis of stroke today is optimally performed through neuroimaging studies using CTs or MRIs [21], there is still a dire need for quicker and easier diagnostic tools, especially in resource-deprived settings.

Studies have shown that certain blood biomarkers may be a reliable predictor in assessing the risk of stroke or facilitating its diagnosis [22–24]. With the rising global incidence of stroke and prediabetes, especially among the younger population, this is a concern that should be given even greater emphasis. This systematic review explored the association between prediabetes and stroke through significant biomarkers like IL-6, D-dimer, NSE and GFAP. Further, this review also tried to elucidate the underlying pathophysiological mechanisms in the state of prediabetes that increase susceptibility to stroke or trigger stroke events.

It was, however, hard to determine studies suitable for this review since few studies have been conducted concerning this subject. To maximise inclusivity, we applied both the ADA and WHO criteria for prediabetes. We anticipated that using these broader international criteria would allow us to capture more studies; however, despite this approach, the retrieved studies still did not specifically investigate individuals with prediabetes. Although prediabetes is a prevalent condition worldwide, it often goes undetected due to its asymptomatic nature, hence contributing to the deficiency of research concerning this topic.

This study aimed to summarise the existing evidence on the pathophysiological mechanisms linking stroke and stroke biomarkers in individuals with prediabetes. However, despite a thorough search, no studies were found that directly investigated this specific relationship. Instead, this study identified numerous articles assessing various stroke biomarkers in individuals who had experienced a previous stroke, though these did not specifically address prediabetes as a potential predisposing factor. Several researchers also explored the diagnostic value of certain stroke biomarkers in individuals with established T2DM.

While this focus on T2DM was not the original target of the study, the findings remain highly relevant. Since prediabetes is well established as an intermediary state on the glycaemic spectrum, sharing key metabolic disturbances with T2DM, including insulin resistance, low-grade inflammation and endothelial dysfunction, it is biologically plausible that the same biomarkers elevated in T2DM-associated stroke may also exhibit early alterations in individuals with prediabetes, though likely to a lesser degree.

These early molecular changes may indicate subclinical neurovascular injury or coagulation imbalances that precede overt cerebrovascular events. This supports the concept of a continuum of vascular risk, positioning prediabetes as a critical window for early detection and intervention through sensitive biomarker monitoring—even before structural or clinical signs of stroke emerge.

Pathophysiology Mechanisms Linking Stroke and Type 2 Diabetes Mellitus

In T2DM, both insulin resistance and hyperglycaemia are important components that predispose individuals to cerebrovascular events such as stroke [9]. Insulin resistance results in a hyperglycaemic state due to the disruption of the normal glucose metabolism within the body [25]. Such disruptions cause a cascade of pathological changes that will eventually compromise vascular health.

Inflammation is a basic pathophysiological process that highly contributes to the development of stroke and increased stroke risk [26]. IL-6 is an important pro-inflammatory cytokine, very often present in higher levels in individuals with T2DM and prediabetes [27]. Chronic elevation of IL-6 is associated with enhanced inflammation and endothelial dysfunction, a condition wherein the endothelial cells lining the blood vessels become impaired [27]. These cells generally regulate the flow of blood through the vessel and maintain vascular integrity. However, in a state of chronic inflammation resulting from high levels of IL-6, there is decreased production of nitric oxide, a molecule important for the dilating of blood vessels [28]. Thus, the blood vessels have a reduced capacity to dilate properly, leading to increased stiffness and heightened susceptibility to injury.

Findings from two eligible studies in this systematic review portray conflicting perspectives regarding the role of IL-6 in type 2 diabetes and its complications. The study by [29] reports that high levels of IL-6 are greatly associated with the development of macrovascular complications among individuals with T2DM, where 54.9% of the subjects developed such complications. IL-6 concentrations were significantly elevated in individuals with diabetes and atherosclerosis (1.7 ± 0.5 ng/mL) compared to those without, supporting its role as a pro-inflammatory marker in vascular damage.

This means that the development of atherosclerosis in individuals with diabetes is contributed to by inflammatory factors, including IL-6. The study by [30] found that plasma IL-6 levels in individuals with diabetes were significantly higher than those in individuals without diabetes. However, it also found that IL-6 levels were paradoxically lower in diabetic individuals with complications (0.9 pg/mL) compared to those without complications (7.2 pg/mL), suggesting variability in IL-6 expression depending on disease stage or immune adaptation. It was also noted that there was no significant difference in the level of IL-6 between the individuals with diabetes and the group of individuals with diabetes with microvascular or macrovascular problems.

These findings together emphasise the complexity of inflammation in T2DM. Reference [28] suggests that there is a direct association between raised IL-6 and stroke risk, whereas [30] suggests that other inflammatory pathways may be more important. This strengthens the need for further research to better understand these relationships by having the group of individuals with prediabetes as a testing group in future studies. Understanding these mechanisms in earlier metabolic stages may help determine whether IL-6 elevations precede vascular complications or simply reflect existing pathology.

Interestingly, IL-6 also promotes insulin resistance, where further worsening of hyperglycaemia exacerbates vascular health [31]. This would then form a vicious circle, promoting increased stroke risk in individuals with T2DM.

In this situation, platelet aggregation is one of the very important risk factors. It is so easily regulated by increased IL-6 that it supports platelet activation and fast aggregation of platelets. This might lead to blood clotting, which in turn would cause blockage of blood, thereby increasing the risk of stroke events. Given that prediabetes is marked by low-grade inflammation and early endothelial dysfunction, IL-6 may serve as a marker for detection through sensitive biomarker monitoring in this group, offering a window of opportunity for early intervention before overt macrovascular damage occurs.

Another important biomarker in the assessment of stroke risk is D-dimer. The high level of D-dimer, just like the case of platelet aggregation, expresses increased coagulability, which can lead to thrombus formation and subsequent strokes [32]. A 4-year cohort study in China has associated high levels of D-dimer in individuals with diabetes with increased incidence of cardiovascular events, including ischaemic stroke events, hence the importance of monitoring this biomarker in the prediabetes setting [33]. Although D-dimer was assessed in only a single included study, the large-scale nature of this investigation considerably strengthens the reliability of its findings. This was the only study among the reviewed literature to examine such a large population (n = 1976), significantly enhancing its statistical power relative to other biomarker studies. The study by [33] demonstrated a strong and statistically significant association between elevated D-dimer concentrations and increased risk of cardiovascular events, including ischaemic stroke, in individuals with diabetes.

Despite the strength of its design and low risk of bias, the lack of replication across multiple independent studies limits the broader generalisability of these findings. Further research is needed to validate the diagnostic and prognostic utility of D-dimer, especially in prediabetic populations or in studies that specifically stratify stroke subtypes.

When stratified by D-dimer quartiles, the incidence of cardiovascular events rose progressively with $p < 0.001$ for all comparisons. This dose–response trend highlights a strong and statistically significant link between D-dimer concentrations and vascular risk.

This evidence aligns with broader research indicating that hyperglycaemia exacerbates both inflammatory and prothrombotic pathways, thereby increasing the likelihood of macrovascular complications such as stroke. Consequently, these findings may be extended to help explain intermediary pathophysiological mechanisms observed in individuals with prediabetes. This convergence of evidence across distinct pathophysiological pathways highlights D-dimer's potential role as a complementary marker in multi-biomarker panels aimed at early identification of stroke risk in metabolic disease contexts.

GFAP is an important marker for brain damage in stroke diagnosis. GFAP is released into the bloodstream following brain cell death or injury [34]. GFAP levels can be used as an indicator of the extent of brain damage and can also be used to distinguish between stroke types and a stroke from other causes [35].

This systematic review revealed only one study that examined the relationship between GFAP, cognitive impairments and type 2 diabetes mellitus. A case–control/cohort study by Guerro et al. reported that GFAP levels were considerably higher in patients with neurocognitive disorders compared to healthy controls and the highest concentrations were found in individuals with both neurocognitive disorders and type 2 diabetes [36]. This is concerning since elevated GFAP levels are associated with central nervous system damage and stroke risk is increased by type 2 diabetes, which further exacerbates the problem. Increased GFAP is a sign of an astrocyte's response to brain damage, which is made worse by type 2 diabetes. This neuroinflammation and dysfunction most likely increases the risk of stroke even more. Although GFAP was identified as a potentially useful biomarker, it was assessed in only a single included study, limiting the generalisability and strength of the conclusions. The study by [36] was assessed as low risk of bias, with clearly reported quantitative findings and appropriate adjustment for confounding variables. However, the lack of replication across multiple independent studies highlights the need for further research to validate GFAP's diagnostic and prognostic utility in diabetic populations at risk for stroke or neurovascular complications. This trend mirrors that of D-dimer, where consistent associations with stroke risk have been observed, yet limited study replication in prediabetic or diabetic cohorts hinders broader clinical application. Despite this limitation, the consistent elevation of GFAP in T2DM-related neurocognitive and vascular conditions lends support to its potential role as an early biomarker of astrocytic dysfunction and warrants future exploration in larger, well-controlled studies.

Following brain injury, damaged astrocytes release the calcium binding protein S-100B, which is mostly located in the central nervous system [37]. Serum levels can be indicative of several pathological conditions, such as stroke and diabetes. Following an ischaemic stroke, S100B has been reported to be released into the bloodstream as a result of glial cell activation and breakdown of the blood–brain barrier (BBB) [38].

Another example is Neuron-Specific Enolase (NSE), a biomarker of neuronal injury. NSE is usually elevated in diabetes and this could be related to greater neuronal injury and poorer prognosis in individuals with diabetes who may have experienced a stroke event [39]. Among all biomarkers reviewed, NSE demonstrated the strongest and most consistent evidence as a marker of neurovascular injury. Three independent studies, encompassing a combined sample size of 361 participants, consistently reported statistically significant elevations in NSE levels both in individuals with diabetes compared to healthy controls, and in individuals with diabetes who experienced ischemic stroke compared to diabetic individuals without stroke. This reproducibility across diverse populations strengthens confidence in NSE's reliability as a biomarker of neurovascular injury.

Regarding study quality, two of the studies [40,41] were assessed as having low risk of bias and provided detailed quantitative data with clear statistical comparisons, enhancing the interpretability and validity of their findings. In contrast, one study [42] was rated as high risk of bias due to reliance on self-reported outcomes via telephonic follow-up, a small sample size and lack of randomization. Moreover, this study reported only statistically significant differences ($p < 0.05$) between diabetic and non-diabetic acute ischemic stroke patients without providing specific NSE concentration values, limiting comparability with other studies.

Despite this limitation, the overall consistency of elevated NSE levels in diabetic stroke patients across the low-bias studies supports NSE's clinical utility as a reliable biomarker for neurovascular injury. Its consistent elevation in diabetic stroke groups versus diabetic controls indicates NSE's potential to discriminate neurovascular injury severity within diabetic populations.

Similarly to S-100B, the levels of NSE are accepted as a marker of the severity of brain damage in stroke patients, including those with diabetes, since the damaged BBB allows NSE to leak out into the bloodstream.

Roshdy et al. found that elevated levels of S100B and NSE correlated significantly with infarct size and neurological examinations, indicating their diagnostic validity [40]. Additionally, the study by Pandey et al. highlighted the use of NSE as a reliable prognostic marker of stroke severity and outcome in individuals with elevated blood glucose levels [41].

Interestingly, the study by [43] found that in T2DM, S100B is implicated in cognitive dysfunction, a common complication among patients. The relationship between S100B levels and cognitive impairment suggests that this biomarker may reflect both acute neuronal injury and chronic neurodegenerative processes linked to diabetes. Additionally, the study revealed an association between chronic hyperglycaemia and cognitive function, reporting lower S100B levels in individuals with T2DM compared to healthy controls. This reduction is thought to be due to hyperglycaemia downregulating S100B secretion from astrocytes. However, the findings also indicate that lower S100B levels are associated with worse cognitive function, suggesting that while insulin resistance may lead to reduced S100B, the protein could still play a critical role in neuroprotection and cognitive health.

Interestingly, although the included studies primarily focused on individuals with T2DM and stroke, regardless of the biomarker investigated, most participants were middle-aged or older, typically between 50 and 60 years of age. This highlights a gap in current research, as evidence shows that prediabetes often develops at a younger age. In addition, individuals over 50 years are more likely to have other comorbidities, which can complicate the interpretation of biomarker changes and stroke risk. Therefore, future studies should consider focusing on younger populations, particularly those aged 25 to 45 years, to better isolate the early pathophysiological effects of prediabetes and gain clearer insights into its association with stroke risk.

The exact role of biomarkers S100B, GFAP and NSE in diagnosing stroke is still being defined. While elevated levels of these markers indicate neuronal injury, they are not unique to stroke and can also be present in other neurological conditions [44,45].

The results of this systematic review, along with findings from existing literature, indicate that S100B, GFAP and NSE hold promise as biomarkers for brain injury in stroke, particularly in individuals with T2DM. By combining these markers with others that evaluate different pathophysiological pathways associated with stroke such as IL-6 and D-dimer, we can develop a comprehensive panel of blood biomarkers that may provide a more accurate clinical prognosis for stroke patients with prediabetes and T2DM. Although most included studies focused on individuals with established T2DM, the underlying pathophysiological mechanisms are highly relevant to prediabetes. Insulin resistance, a hallmark of prediabetes, contributes to chronic low-grade inflammation and endothelial dysfunction, which can in turn influence levels of circulating biomarkers such as IL-6, D-dimer, and S100B. Given the continuum of metabolic dysfunction between normoglycaemia and overt T2DM, it is biologically plausible to expect intermediate or subclinical elevations in these biomarkers among individuals with prediabetes. Such elevations may reflect early neurovascular stress or low-grade neuroinflammation that precedes overt clinical stroke, offering a critical window for early intervention. This highlights the importance of exploring biomarker trajectories in prediabetes, not only to better understand the early pathogenesis of stroke, but also to identify at-risk individuals before irreversible damage occurs. Future studies should focus on investigating these biomarkers specifically in individuals with prediabetes and T2DM to further clarify their diagnostic value and enhance clinical outcomes.

5. Limitations

The main limitation of this study was the lack of data on changes in stroke biomarker concentrations in individuals with prediabetes. All 8 studies included in this review assessed various stroke biomarkers in individuals with T2DM. Since prediabetes is an asymptomatic condition, many individuals are unaware that they have it. For individuals who suffer a stroke event but were not diagnosed with prediabetes, prediabetes may not be considered a possible cause of the stroke. These individuals have not been included in studies that investigate the relationship between diabetes and stroke. Additionally, factors like gender and race were not always specified in studies. While we know that prediabetes and stroke are becoming more prevalent in younger adults, most studies only investigated in individuals over the age of 50.

6. Conclusions

It is well-established that T2DM increases the risk of stroke due to mechanisms such as insulin resistance, chronic inflammation, endothelial dysfunction and hyperglycaemia-induced vascular damage. Several studies have confirmed these mechanisms and investigated stroke biomarkers, such as IL-6, S100B, NSE, D-Dimer and GFAP in T2DM populations, demonstrating their relevance to stroke risk. While some articles have also explored these mechanisms in individuals with prediabetes, providing evidence of an elevated stroke risk, there remains a significant gap in research focusing specifically on stroke biomarkers in populations with prediabetes. Further studies are needed to better understand the biomarker profiles associated with stroke in prediabetes, which could help identify early interventions to reduce stroke risk in this group.

Author Contributions: Y.N. and A.K. were responsible for developing and drafting the protocol, as well as reviewing the eligible studies and final draft of the review. No generative AI has been used. All authors have read and agreed to the published version of the manuscript.

Funding: This study was funded by the College of Health Science (CHS); however, the funders had no role in the study design and review preparation.

Institutional Review Board Statement: Not applicable.

Informed Consent Statement: Not applicable.

Data Availability Statement: The raw data supporting the conclusions of this article will be made available by the authors on request.

Acknowledgments: The authors would like to acknowledge the Department of Human Physiology at UKZN, Westville, for their support and academic contributions to this protocol.

Conflicts of Interest: The authors declare no conflicts of interest.

Abbreviations

The following abbreviations are used in this manuscript:

ADA	American Diabetes Association
AHA	American Heart Association
BBB	Blood–brain barrier
CT	Computed tomography
GFAP	Glial fibrillary acidic protein
HbA1c	Glycated Haemoglobin
IFG	Impaired fasting glucose
IGT	Impaired glucose tolerance
IL-6	Interleukin 6
MRI	Magnetic resonance imaging
NSE	Neuron-specific enolase
PRISMA	Prepared Reporting Items for Systematic Reviews and Meta-Analyses
S100B	Calcium-binding protein B
T2DM	Type 2 diabetes mellitus
UKZN	University of KwaZulu-Natal
WHO	World Health Organization

References

1. Smushkin, G.; Vella, A. What is type 2 diabetes? *Medicine* **2010**, *38*, 597–601. [CrossRef] [PubMed]
2. Zheng, Y.; Ley, S.H.; Hu, F.B. Global aetiology and epidemiology of type 2 diabetes mellitus and its complications. *Nat. Rev. Endocrinol.* **2018**, *14*, 88–98. [CrossRef]
3. Sun, H.; Saeedi, P.; Karuranga, S.; Pinkepank, M.; Ogurtsova, K.; Duncan, B.B.; Stein, C.; Basit, A.; Chan, J.C.N.; Mbanya, J.C.; et al. IDF Diabetes Atlas: Global, regional and country-level diabetes prsevalence estimates for 2021 and projections for 2045. *Diabetes Res. Clin. Pract.* **2022**, *183*, 109119. [CrossRef]
4. Schlienger, J.-L. Complications du diabète de type 2. *La Presse Médicale* **2013**, *42*, 839–848. [CrossRef]
5. Murphy, S.J.; Werring, D.J. Stroke: Causes and clinical features. *Medicine* **2020**, *48*, 561–566. [CrossRef]
6. Donkor, E.S. Stroke in the 21st Century: A Snapshot of the Burden, Epidemiology, and Quality of Life. *Stroke Res. Treat.* **2018**, *2018*, 3238165. [CrossRef] [PubMed]

7. Arboix, A.; Milian, M.; Oliveres, M.; García-Eroles, L.; Massons, J. Impact of Female Gender on Prognosis in Type 2 Diabetic Patients with Ischemic Stroke. *Eur. Neurol.* **2006**, *56*, 6–12. [CrossRef]
8. Ménégaut, L.; Laubriet, A.; Crespy, V.; Leleu, D.; Pilot, T.; Van Dongen, K.; de Barros, J.-P.P.; Gautier, T.; Petit, J.-M.; Thomas, C.; et al. Inflammation and oxidative stress markers in type 2 diabetes patients with Advanced Carotid atherosclerosis. *Cardiovasc. Diabetol.* **2023**, *22*, 248. [CrossRef]
9. Galicia-Garcia, U.; Benito-Vicente, A.; Jebari, S.; Larrea-Sebal, A.; Siddiqi, H.; Uribe, K.B.; Ostolaza, H.; Martín, C. Pathophysiology of Type 2 Diabetes Mellitus. *Int. J. Mol. Sci.* **2020**, *21*, 6275. [CrossRef]
10. van Sloten, T.T.; Sedaghat, S.; Carnethon, M.R.; Launer, L.J.; Stehouwer, C.D.A. Cerebral microvascular complications of type 2 diabetes: Stroke, cognitive dysfunction, and depression. *Lancet Diabetes Endocrinol.* **2020**, *8*, 325–336. [CrossRef] [PubMed]
11. Ding, P.F.; Zhang, H.S.; Wang, J.; Gao, Y.Y.; Mao, J.N.; Hang, C.H.; Li, W. Insulin resistance in ischemic stroke: Mechanisms and therapeutic approaches. *Front. Endocrinol.* **2022**, *13*, 1092431. [CrossRef] [PubMed]
12. Kamtchum-Tatuene, J.; Jickling, G.C. Blood Biomarkers for Stroke Diagnosis and Management. *Neuromol. Med.* **2019**, *21*, 344–368. [CrossRef]
13. Alvarez, S.; Coffey, R.; Mathias, P.M.; Algotar, A.M. Prediabetes. In *StatPearls*; StatPearls Publishing: Treasure Island, FL, USA, 2023; Copyright © 2024; StatPearls Publishing LLC.: Treasure Island, FL, USA, 2024.
14. Echouffo-Tcheugui, J.B.; Selvin, E. Prediabetes and What It Means: The Epidemiological Evidence. *Annu. Rev. Public Health* **2021**, *42*, 59–77. [CrossRef]
15. Brannick, B.; Wynn, A.; Dagogo-Jack, S. Prediabetes as a toxic environment for the initiation of microvascular and macrovascular complications. *Exp. Biol. Med.* **2016**, *241*, 1323–1331. [CrossRef]
16. Rooney, M.R.; Fang, M.; Ogurtsova, K.; Ozkan, B.; Echouffo-Tcheugui, J.B.; Boyko, E.J.; Magliano, D.J.; Selvin, E. Global Prevalence of Prediabetes. *Diabetes Care* **2023**, *46*, 1388–1394. [CrossRef]
17. Mosenzon, O.; Cheng, A.Y.; Rabinstein, A.A.; Sacco, S. Diabetes and Stroke: What Are the Connections? *J. Stroke* **2023**, *25*, 26–38. [CrossRef] [PubMed]
18. Page, M.J.; Moher, D.; Bossuyt, P.M.; Boutron, I.; Hoffmann, T.C.; Mulrow, C.D.; Shamseer, L.; Tetzlaff, J.M.; Akl, E.A.; Brennan, S.E.; et al. PRISMA 2020 explanation and elaboration: Updated guidance and exemplars for reporting systematic reviews. *BMJ* **2021**, *372*, n160. [CrossRef]
19. Hostalek, U. Global epidemiology of prediabetes—Present and future perspectives. *Clin. Diabetes Endocrinol.* **2019**, *5*, 5. [CrossRef]
20. Lee, M.; Saver, J.L.; Hong, K.-S.; Song, S.; Chang, K.-H.; Ovbiagele, B. Effect of pre-diabetes on future risk of stroke: Meta-analysis. *BMJ Br. Med. J.* **2012**, *344*, e3564. [CrossRef]
21. Merino, J.G.; Warach, S. Imaging of acute stroke. *Nat. Rev. Neurol.* **2010**, *6*, 560–571. [CrossRef] [PubMed]
22. Waje-Andreassen, U.; Kråkenes, J.; Ulvestad, E.; Thomassen, L.; Myhr, K.-M.; Aarseth, J.; Vedeler, C.A. IL-6: An early marker for outcome in acute ischemic stroke. *Acta Neurol. Scand.* **2005**, *111*, 360–365. [CrossRef] [PubMed]
23. Yao, T.; Tian, B.-L.; Li, G.; Cui, Q.; Wang, C.-f.; Zhang, Q.; Peng, B.; Gao, Y.; Zhan, Y.-Q.; Hu, D.; et al. Elevated plasma D-dimer levels are associated with short-term poor outcome in patients with acute ischemic stroke: A prospective, observational study. *BMC Neurol.* **2019**, *19*, 175. [CrossRef]
24. di Biase, L.; Bonura, A.; Pecoraro, P.M.; Carbone, S.P.; Di Lazzaro, V. Unlocking the Potential of Stroke Blood Biomarkers: Early Diagnosis, Ischemic vs. Haemorrhagic Differentiation and Haemorrhagic Transformation Risk: A Comprehensive Review. *Int. J. Mol. Sci.* **2023**, *24*, 11545. [CrossRef]
25. Sowers, J.R.; Standley, P.R.; Ram, J.L.; Jacober, S.; Simpson, L.; Rose, K. Hyperinsulinemia, insulin resistance, and hyperglycemia: Contributing factors in the pathogenesis of hypertension and atherosclerosis. *Am. J. Hypertens* **1993**, *6*, 260s–270s. [CrossRef]
26. Parikh, N.S.; Merkler, A.E.; Iadecola, C. Inflammation, Autoimmunity, Infection, and Stroke: Epidemiology and Lessons from Therapeutic Intervention. *Stroke* **2020**, *51*, 711–718. [CrossRef]
27. Huang, K.; Liang, Y.; Ma, Y.; Wu, J.; Luo, H.; Yi, B. The Variation and Correlation of Serum Adiponectin, Nesfatin-1, IL-6, and TNF-α Levels in Prediabetes. *Front. Endocrinol.* **2022**, *13*, 774272. [CrossRef]
28. Didion, S.P. Cellular and Oxidative Mechanisms Associated with Interleukin-6 Signaling in the Vasculature. *Int. J. Mol. Sci.* **2017**, *18*, 2563. [CrossRef]
29. Wu, W.; Wang, M.; Sun, Z.; Wang, X.; Miao, J.; Zheng, Z. The predictive value of TNF-α and IL-6 and the incidence of macrovascular complications in patients with type 2 diabetes. *Acta Diabetol.* **2012**, *49*, 3–7. [CrossRef]
30. Pickup, J.C.; Chusney, G.D.; Thomas, S.M.; Burt, D. Plasma interleukin-6, tumour necrosis factor alpha and blood cytokine production in type 2 diabetes. *Life Sci.* **2000**, *67*, 291–300. [CrossRef] [PubMed]
31. Rehman, K.; Akash, M.S.H.; Liaqat, A.; Kamal, S.; Qadir, M.I.; Rasul, A. Role of Interleukin-6 in Development of Insulin Resistance and Type 2 Diabetes Mellitus. *Crit. Rev. Eukaryot. Gene Expr.* **2017**, *27*, 229–236. [CrossRef] [PubMed]
32. Choi, K.-H.; Kim, J.-H.; Kim, J.-M.; Kang, K.-W.; Lee, C.; Kim, J.-T.; Choi, S.-M.; Park, M.-S.; Cho, K.-H. D-dimer Level as a Predictor of Recurrent Stroke in Patients with Embolic Stroke of Undetermined Source. *Stroke* **2021**, *52*, 2292–2301. [CrossRef] [PubMed]

33. Cheng, L.; Fu, Q.; Zhou, L.; Fan, Y.; Liu, F.; Fan, Y.; Zhang, X.; Lin, W.; Wu, X. D-dimer as a predictor of cardiovascular outcomes in patients with diabetes mellitus. *BMC Cardiovasc. Disord.* **2022**, *22*, 82. [CrossRef] [PubMed]
34. Žurek, J. Chapter 33—Biomarkers in Traumatic Brain Injury. In *Essentials of Neuroanesthesia*; Prabhakar, H., Ed.; Academic Press: Waltham, MA, USA, 2017; pp. 587–591.
35. Yang, Z.; Wang, K.K. Glial fibrillary acidic protein: From intermediate filament assembly and gliosis to neurobiomarker. *Trends Neurosci.* **2015**, *38*, 364–374. [CrossRef]
36. Ayala-Guerrero, L.; García-delaTorre, P.; Sánchez-García, S.; Guzmán-Ramos, K. Serum Levels of Glial Fibrillary Acidic Protein Association with Cognitive Impairment and Type 2 Diabetes. *Arch. Med. Res.* **2022**, *53*, 501–507. [CrossRef]
37. Gayger-Dias, V.; Vizuete, A.F.; Rodrigues, L.; Wartchow, K.M.; Bobermin, L.; Leite, M.C.; Quincozes-Santos, A.; Kleindienst, A.; Gonçalves, C.A. How S100B crosses brain barriers and why it is considered a peripheral marker of brain injury. *Exp. Biol. Med.* **2023**, *248*, 2109–2119. [CrossRef]
38. Huang, Y.; Wang, Z.; Huang, Z.X.; Liu, Z. Biomarkers and the outcomes of ischemic stroke. *Front. Mol. Neurosci.* **2023**, *16*, 1171101. [CrossRef] [PubMed]
39. Li, J.; Zhang, H.; Xie, M.; Yan, L.; Chen, J.; Wang, H. NSE, a potential biomarker, is closely connected to diabetic peripheral neuropathy. *Diabetes Care* **2013**, *36*, 3405–3410. [CrossRef]
40. Roshdy, N.; Zakaria, A.; Kahla, H.; Samy, S. Diagnostic value of initial s-100b and neuron- specific enolase levels in diabetic patients with ischemic stroke. *Bull. Egypt. Soc. Physiol. Sci.* **2007**, *27*, 275–290. [CrossRef]
41. Pandey, A.; Saxena, K.; Verma, M.; Bharosay, A. Correlative study between neuron-specific enolase and blood sugar level in ischemic stroke patients. *J. Neurosci. Rural. Pract.* **2011**, *2*, 50–54. [CrossRef]
42. Nayak, A.R.; Badar, S.R.; Lande, N.; Kawle, A.P.; Kabra, D.P.; Chandak, N.H.; Raje, D.V.; Singh, L.R.; Daginawala, H.F.; Kashyap, R.S. Prediction of Outcome in Diabetic Acute Ischemic Stroke Patients: A Hospital-Based Pilot Study Report. *Ann. Neurosci.* **2016**, *23*, 199–208. [CrossRef]
43. Yu, H.; Li, H.; Liu, X.; Du, X.; Deng, B. Levels of serum S100B are associated with cognitive dysfunction in patients with type 2 diabetes. *Aging* **2020**, *12*, 4193–4203. [CrossRef] [PubMed]
44. Mochol, M.; Taubøll, E.; Aukrust, P.; Ueland, T.; Andreassen, O.A.; Svalheim, S. Serum Markers of Neuronal Damage and Astrocyte Activity in Patients with Chronic Epilepsy: Elevated Levels of Glial Fibrillary Acidic Protein. *Acta Neurol. Scand.* **2023**, *2023*, 7246373. [CrossRef]
45. Žurek, J.; Fedora, M. The usefulness of S100B, NSE, GFAP, NF-H, secretagogin and Hsp70 as a predictive biomarker of outcome in children with traumatic brain injury. *Acta Neurochir.* **2012**, *154*, 93–103; discussion 103. [CrossRef] [PubMed]

current issues in molecular biology

MDPI

Article

A Retrospective Assessment of Changes in Stroke Risk-Related Biomarkers in Individuals with Prediabetes from Durban, South Africa: Preliminary Findings

Yerushka Naicker and Andile Khathi *

Department of Human Physiology, School of Laboratory Medicine and Medical Sciences, College of Health Sciences, University of KwaZulu-Natal, Durban 3629, South Africa
* Correspondence: khathia@ukzn.ac.za

Abstract: Type 2 diabetes mellitus (T2DM) is a chronic metabolic disorder that significantly increases the risk of stroke, with prediabetes serving as an intermediate stage marked by similar pathophysiological mechanisms such as inflammation and vascular dysfunction. This study investigated the relationship between prediabetes and stroke-related biomarkers in individuals aged 25–45 years in Durban, South Africa. After obtaining ethical approval, a retrospective analysis was performed on blood samples from 100 participants recruited from King Edward Hospital and Inkosi Albert Luthuli Central Hospital. Participants were classified as non-prediabetic (n = 30), prediabetic (n = 35), or type 2 diabetic (n = 35) according to ADA criteria. Plasma concentrations of C-reactive protein (CRP), interleukin-6 (IL-6), fibrinogen, D-dimer, calcium binding protein (S100B), glial fibrillary acidic protein (GFAP), and neuron-specific enolase (NSE) were measured using enzyme-linked immunosorbent assay (ELISA). It is important to note that none of the participants had confirmed stroke events; these biomarkers were assessed as surrogate indicators of stroke risk. Statistical analyses included one-way ANOVA with Tukey–Kramer tests and Pearson's correlations. Biomarker concentrations were significantly elevated in prediabetic individuals compared to non-prediabetic controls, with levels further increasing in T2DM. Strong positive correlations were observed between S100B and both HbA1c (r = 0.75, $p < 0.0001$) and fasting glucose (r = 0.75, $p < 0.0001$). These findings suggest that inflammatory, coagulation, and neurovascular biomarkers, particularly S100B, may indicate early stroke risk in prediabetes. Further investigation into these biomarkers could improve early detection strategies and stroke prevention efforts in at-risk populations.

Keywords: prediabetes; type 2 diabetes mellitus; stroke; Durban; South Africa

1. Introduction

Type 2 diabetes mellitus (T2DM) is a chronic metabolic condition characterised by insulin resistance that results in elevated blood glucose levels [1]. The prevalence of T2DM is rapidly increasing in South Africa, consequently placing a significant burden on the South African health system [2]. A systematic review published in 2021 estimated that the pooled prevalence of T2DM in South Africans aged 25 years and older was 15.25% [3]. In South Africa, T2DM ranks second among the top 10 leading natural causes of death, thus emphasising the magnitude of the diabetes epidemic [4]. Heart failure and stroke are complications frequently associated with T2DM and may become life-threatening [5–7]. Stroke is not an independent disease but can be triggered by a multitude of risk factors and disease mechanisms [8]. A national burden of disease study in 2000 reported that T2DM accounted for approximately 10% of all stroke cases [2,9]. According to studies, people with T2DM have a higher risk of stroke than those without T2DM [7,10]. Notably, individuals with diabetes mellitus who experience symptomatic cerebral ischaemia have been shown to have a higher risk of recurrent stroke, and the female sex is a demographic factor associated with a worse prognosis [11]. These findings highlight the complexity of stroke pathophysiology and suggest that biological factors, including sex-related differences and pre-existing metabolic disturbances, may influence disease progression and prognosis.

The pathophysiology of stroke involves the abrupt interruption of blood supply to the brain, resulting in the brain being deprived of oxygen and causing damage to brain cells [12]. A stroke can manifest as either ischaemic, which is due to a lack of blood flow, or haemorrhagic, which is due to bleeding in a specific area. These disruptions can lead to a range of neurological impairments, such as paralysis, speech difficulties, and cognitive deficits, depending on the location and extent of the brain damage [8]. In T2DM, hyperglycaemia leads to oxidative stress, inflammation, and endothelial dysfunction, increasing vascular damage and stroke risk [13]. Both T2DM and stroke share common inflammatory pathways [14]. Elevated inflammatory markers contribute to the onset of a stroke, while the stroke event itself also triggers a subsequent inflammatory response, resulting in increased levels of circulating inflammatory mediators [15]. Elevated C-reactive protein (CRP) and interleukin 6 (IL-6) levels can signal inflammation, which plays a role in stroke development and progression [16–18]. Increased fibrinogen and D-dimer levels can indicate inflammation and blood clotting, both of which are associated with stroke risk and damage [19–21]. Neuron-specific enolase (NSE), glial fibrillary acidic protein (GFAP), and calcium-binding protein S100B are biomarkers of brain injury. The concentrations of these biomarkers are increased after stroke and help to determine the severity of brain injury, with NSE being a marker of neuronal death, GFAP for astrocytic injury and S100B for brain cell injury. Together these biomarkers have been shown to help distinguish between stroke types, to assess the severity of the stroke and predict the outcomes including the potential for functional recovery in individuals with diabetes [22–24].

Given that metabolic disturbances such as hyperglycaemia and insulin resistance significantly influence stroke risk and outcomes in patients with diabetes, it is important to consider whether similar processes may already be active before the onset of overt disease.

Before the onset of T2DM, moderate insulin resistance causes a gradual increase in glucose levels referred to as prediabetes [25]. Prediabetes or intermediate hyperglycaemia is a severe health condition characterised by blood glucose concentration higher than normal but below the established threshold for T2DM [26]. The global prevalence of prediabetes is estimated at 318 million [27]. A systematic review and meta-analysis conducted in 2022 found that the prevalence of prediabetes among South Africans aged 25 to 45 was 15.56% [25]. Another retrospective study conducted in Durban revealed a concerningly high prevalence of prediabetes among individuals aged 25–45 years [28]. This is alarming as research has shown that individuals with prediabetes have an increased risk of stroke and progression to T2DM [29]. Prolonged elevated blood glucose in the prediabetic state can upregulate markers of chronic inflammation and contribute to the generation of reactive oxygen species (ROS) [30]. Excessive ROS generated can damage the brain tissue and is an important pathological mechanism of stroke [31]. Despite this growing prevalence, there is a notable absence of research investigating stroke-related biomarkers specifically in South African prediabetic populations. Most existing studies were conducted in Western or Asian contexts, which may not be fully generalisable given the distinct genetic, socioeconomic, and healthcare characteristics of African populations. Although a scoping review by Van Vuuren (2022) identified several blood biomarkers with potential prognostic value for stroke outcomes, and other studies have explored the prevalence of prediabetes in stroke patients, none have specifically assessed stroke-related biomarkers in individuals with prediabetes within the South African context [29,32,33]. Measuring these biomarkers in the prediabetic stage is essential because this phase presents a window of opportunity for early intervention before irreversible vascular and neurological damage occurs. Identifying alterations in inflammation, coagulation, and neurovascular injury markers could enable the development of predictive tools for stroke risk stratification. This highlights the urgent need for further investigation to facilitate early detection strategies and preventive measures for this demographic. Previous studies assume prediabetes is an intermediate metabolic state primarily based on associations with later diabetes or cardiovascular outcomes. In contrast, this study directly investigated early neurovascular, inflammatory and coagulation changes in young South African adults, a population in which prediabetes is increasingly prevalent but remains underrepresented in biomarker research. To our knowledge, this is among the first studies to comprehensively characterise these biological changes in a prediabetic cohort, offering novel insights into stroke risk at an earlier disease stage. Therefore, the aim of this study was to evaluate stroke risk in individuals with prediabetes (aged 25–45 years) in Durban, South Africa, using plasma biomarkers of inflammation, coagulation and neurovascular injury.

2. Materials and Methods

2.1. Ethical Approval

Approval was obtained from the UKZN Biomedical Research Ethics Committee (BREC) BE266/2019. Informed consent for storage of samples and potential studies was obtained from each participant.

2.2. Sample Collection

These samples were retrospectively obtained as leftover clinical specimens from a previous ethically approved study and stored in a biobank for future research. A convenience sampling method was used.

These samples, originally collected as part of the previous study, were collected from individuals between the ages of 25–45 years, from King Edward Hospital and Inkosi Albert Luthuli Central Hospital in Durban, South Africa.

The selection of samples was performed according to predefined inclusion and exclusion criteria. The blood samples were collected from adult individuals of all ethnicities and both males and females, aged between 25 and 45 years. Included were samples obtained from individuals clinically diagnosed as non-prediabetic, prediabetic, or with type 2 diabetes mellitus, based on established diagnostic criteria. All participants were systematically screened, and individuals with a history of traumatic brain injury, hypertension, blood disorders, depression, HIV, and kidney/heart/liver diseases were excluded, regardless of their glycaemic status.

The inclusion criteria focused on individuals aged 25–45 years to target a younger adult population at early stages of prediabetes and T2DM, where early biomarker changes might be detectable. We know that inflammation and coagulation are common pathways involved in many illnesses; therefore, to isolate the specific relationship between prediabetes and stroke risk, individuals with conditions known to affect these pathways, such as HIV infection, hypertension, traumatic brain injury, and chronic organ diseases, were excluded. This exclusion helped minimise confounding factors and strengthen the validity of associations observed between glycaemic status and stroke-related biomarker alterations.

Relevant clinical data, including age, sex, fasting glucose, and HbA1c levels, were measured at the time of sample collection and recorded for the purpose of this study.

Lifestyle factors such as smoking, alcohol consumption, diet, and BMI were not recorded in the retrospective dataset.

2.3. Sample Size Calculation

The following formula was used to determine the appropriate sample size for this study.

$$n = \frac{z^2 \times P \times (1 - P)}{E^2}$$

where

- n is the required sample size.
- Z is the Z-score corresponding to the chosen confidence level.
- P represents the assumed prevalence.
- E is the desired margin of error.

$n = \frac{1.96^2 \times 0.10 \times (1 - 0.10)}{0.05^2}$

$n = 138.2976$

$n = 138$

For this study, a 95% confidence level was used, with a corresponding Z-score of 1.96. The assumed prevalence (P) of stroke cases due to diabetes in South Africa is 10% and a margin of error (E) of 5% was considered [2]. Using these parameters, the required sample size was calculated to be 138.

The G*Power software version 3.1.9.7 was also used to calculate the required sample size needed to achieve a power of 0.95, which was determined to be 45 [34]. The input parameters and the corresponding output are shown below.

Test Family: F tests

Statistical test: ANOVA: Repeated measures, within factors

Type of power Analysis: A priori: Compute required sample size

Input:

Effect size f = 0.25

α err prob = 0.05

Power (1-β err prob) = 0.95

Number of groups = 3

Number of measurements = 3

Corr among rep measures = 0.5

Nonsphericity correction ε = 1

Output:

Noncentrality parameter λ = 16.8750000

Critical F = 3.1051566

Numerator df = 2.0000000

Denominator df = 84.0000000

Total sample size = 45

Actual power = 0.9597015

Although the formula-based calculation suggested a sample size of 138 and the GPower analysis indicated that 45 samples were sufficient for 95% power, a total of 100 samples were available for analysis due to the limited availability of stored samples that met the inclusion criteria. This limitation was primarily due to feasibility constraints related to sample accessibility and quality. This study collected a total of 100 blood samples from various patients. Based on HbA1c parameters, the samples were categorised into three groups: non-prediabetic (NPD, n = 30), prediabetic (PD, n = 35), and type 2 diabetes mellitus (T2DM, n = 35). Due to limited plasma volumes and assay cost constraints, subsets of samples were randomly selected for specific biomarker analyses: CRP, fibrinogen, and S100B (n = 20 per group); IL-6, D-dimer, GFAP, and NSE (NPD n = 10, PD/T2D n = 15). Random selection ensured representative sampling and minimised selection bias.

2.4. Study Design

This study employed a retrospective cross-sectional design using stored blood samples collected from hospital patients. Upon arrival at the university, samples were categorised into groups as per criteria outlined by the American Diabetes Association. Blood glucose levels and glycated haemoglobin levels, recorded at the time of admission to the hospital or at the time of recruitment were used to divide the samples into a non-prediabetic group, a prediabetic group, and a type 2 diabetic group. The ADA criteria provide a clear guideline designed to diagnose prediabetes, as depicted in the table below (see Table 1).

Table 1. American Diabetes Association Criteria.

Fasting Plasma Glucose (IFG)	2 Hour Plasma Glucose (IGT)	HbA1c%
5.6–6.9 mmol/L	7.8–11.0 mmol/L	5.7–6.4%

Thereafter, the blood samples were subjected to centrifugation (Eppendorf centrifuge 5403, Hamburg, Germany) at 4 °C, 503× g for a duration of 15 min to isolate plasma. Following centrifugation, these plasma samples were preserved at −80 °C inside a Bio Ultra freezer (Snijers Scientific, Tilburg, The Netherlands) until biochemical analysis was conducted.

2.5. Biochemical Analysis

CRP, IL-6, Fibrinogen, D-dimer, S100B, NSE and GFAP concentrations were determined using their respective ELISA kits (Elabscience Biotechnology Co., Ltd., Houston, TX, USA) according to manufacturer's specifications. Briefly, the ELISA kits contained micro-ELISA plates that were coated with antibodies. One hundred microliters of samples were placed into designated wells, followed by a 90 min incubation at 37 °C. The liquid was then discarded and one hundred microliters of biotinylated detection antibody working solution was promptly added to each well, with an additional incubation of 60 min at 37 °C, after which the unbound constituents were removed using the wash buffer supplied. Post-washing, 100 µL of horseradish peroxidase (HRP) was introduced into the plates and left to incubate for 30 min at 37 °C. Another wash step was performed to eliminate unbound substances. Next, 90 µL of substrate reagent was added to the wells, followed by an additional 15 min of incubation at 37 °C. Fifty microliters of a stop solution was added to the wells to stop the reaction and allow for measurement. Optical densities were then measured at a 450 nm wavelength using a nano spectrophotometer (BMG Labtech, Passau, Germany). The CRP, IL-6, Fibrinogen, D-dimer, S100B, NSE and GFAP concentrations in the samples were extrapolated from their respective standard curves.

To ensure objectivity during biochemical analysis, all plasma samples were coded and laboratory personnel conducting the ELISA were blinded to participant group assignments. While samples were grouped for batch testing, the laboratory technicians were not informed whether the batch belonged to the NPD, PD or T2D category. Relevant clinical data (including age, sex, name, HbA1c and glucose levels) were stored separately in Excel spreadsheets. Each sample was allocated a unique code, which was labelled directly on the sample tube. Additionally, all assays were performed in duplicates to enhance the accuracy and reliability of the results.

2.6. Statistical Analysis

All the data obtained from the experiment were expressed as means ± SEM. Statistical comparisons were performed using Graph Pad InStat Software (version 5.00, GraphPad Software, Inc., San Diego, CA, USA). The differences between the 3 groups were determined using analysis of variance (ANOVA) followed by the Tukey–Kramer multiple comparison test. A value of $p < 0.05$ was considered statistically significant. Correlation analyses were conducted using Pearson's correlation coefficient to examine the strength and direction of the relationship between variables. A p-value of <0.05 was considered statistically significant.

Sensitivity analyses were not required in this study because key potential confounders, particularly HIV infection, were addressed at the design stage. Individuals with known comorbidities, including HIV infection, were excluded from the sample to reduce bias linked to inflammation and coagulation abnormalities. The study specifically targeted participants aged 25–45 years to control for age-related variability. While factors such as smoking, alcohol use, and BMI were not assessed in the present analysis, they are acknowledged as important variables and are recommended for inclusion in future research to further refine understanding of the association between biomarker levels and stroke risk.

3. Results

3.1. Demographic and Clinical Characteristics of Study Participants

Table 2 summarises the demographic and clinical characteristics of participants according to glycaemic status. Age data were normally distributed (Shapiro–Wilk test) and are presented as mean (95% CI), whereas Impaired Fasting Glucose data were not normally distributed and are presented as median (IQR). An upward trend in age was noted, with a mean of 36.1 years in the non-prediabetic group and 40.3 years in the type 2 diabetic group. This suggests a possible association between increasing age and glycaemic deterioration. Sex distribution varied across groups, with females comprising 60% of the non-prediabetic group and 68.6% of the type 2 diabetic group, while males were more common in the prediabetic group at 57.1%. Statistical comparisons revealed that median fasting glucose levels were significantly higher in the prediabetic group compared to the non-prediabetic group ($p < 0.001$) and further significantly elevated in the type 2 diabetic group compared to both the non-prediabetic and prediabetic groups ($p < 0.001$). Similarly, mean HbA1c levels showed a stepwise increase, with significant differences between all groups ($p < 0.001$). These results indicate a clear progression of glycaemic impairment across the study groups.

Table 2. Demographic and clinical characteristics of the study participants in the non-prediabetic, prediabetic and type 2 diabetic groups. Data are presented as mean (95% CI) or median (IQR), depending on the normality and proportion of variables.

	Non-Prediabetic	**Prediabetic**	**Type 2 Diabetic**
Age (mean)	36.13 (33.62–38.64)	38.91 (37.08–40.75)	40.31 (38.69–41.99)
	Sex		
Male	12	20	11
Female	18	15	24
n	30	35	35
Impaired fasting glucose	5.95 mmol/L	7.1 mmol/L	11.6 mmol/L
median (IQR)	(5.7–6.2)	(6.8–7.5)	(9.4–14.9)
HbA1c% mean	5.35% (5.29–5.41)	6.18% (6.04–6.31)	9.72% (8.80–10.64)

3.2. C-Reactive Protein (CRP) Concentration

Figure 1 shows the plasma high-sensitivity C-Reactive Protein (CRP) concentrations in non-prediabetic (NPD), prediabetic (PD) and type 2 diabetic (T2D) individuals in Durban, South Africa. The mean $\pm$ SEM CRP concentrations were 2.37 $\pm$ 0.019 mg/L in the NPD group, 3.12 $\pm$ 0.031 mg/L in the PD group and 4.35 $\pm$ 0.025 mg/L in the T2D group. Statistical comparison tests revealed that CRP levels in the PD group were significantly higher than those in the NPD group ($p = 0.0008$). Furthermore, CRP concentrations in the T2D group were significantly higher than both the PD ($p = 0.0008$) and NPD ($p < 0.0001$) groups (See Figure 1). These findings indicate a progressive increase in CRP across the glycaemic spectrum. For context, Hs-CRP concentrations < 3 mg/L are generally considered to be within the normal range; thus, individuals in both the PD and T2D groups exhibited elevated CRP levels relative to clinical reference values [35].

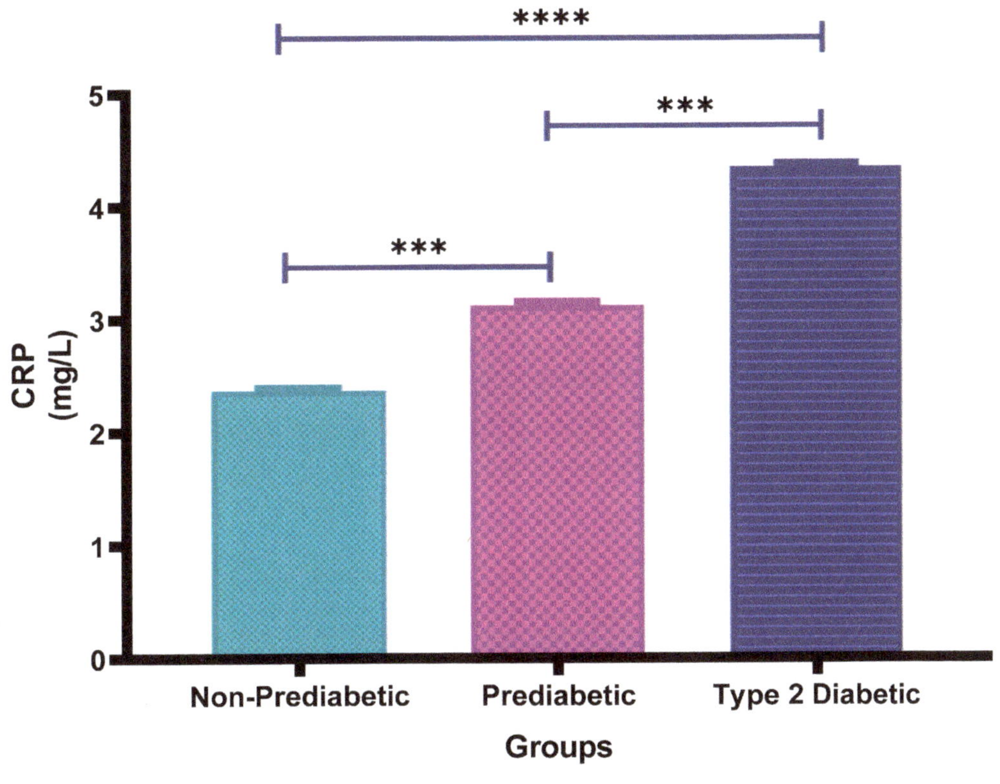

Figure 1. Plasma Hs-C-Reactive Protein (CRP) concentration in the non-prediabetic (NPD) group ($n = 20$), prediabetic (PD) group ($n = 20$) and type 2 diabetic (T2D) group ($n = 20$). Values are presented as means $\pm$ SEM. Asterisk Indicates the significant (*** $p < 0.001$, **** $p < 0.0001$) differences between the groups.

3.3. Interleukin-6 (IL-6) Concentration

Figure 2 shows the plasma Interleukin-6 (IL-6) concentrations in non-prediabetic (NPD), prediabetic (PD), and type 2 diabetic (T2D) individuals in Durban, South Africa. The mean $\pm$ SEM IL-6 concentrations were 3.28 $\pm$ 0.59 pg/mL in the NPD group, 8.06 $\pm$ 1.44 pg/mL in the PD group and 14.61 $\pm$ 2.49 pg/mL in the T2D group. Statistical comparisons test revealed that IL-6 levels in the PD group were significantly higher than those in the NPD group ($p = 0.0369$), while concentrations in the T2D group were significantly elevated compared to the NPD group ($p < 0.0001$). However, the difference between the PD and T2D groups was not statistically significant ($p = 0.1648$) (see Figure 2). These results indicate a trend of increasing IL-6 concentrations across the glycaemic spectrum. For reference, normal IL-6 concentrations in healthy individuals typically range between <1 to 5 pg/mL, suggesting that both the PD and T2D groups exhibited elevated IL-6 levels consistent with systemic inflammation [36].

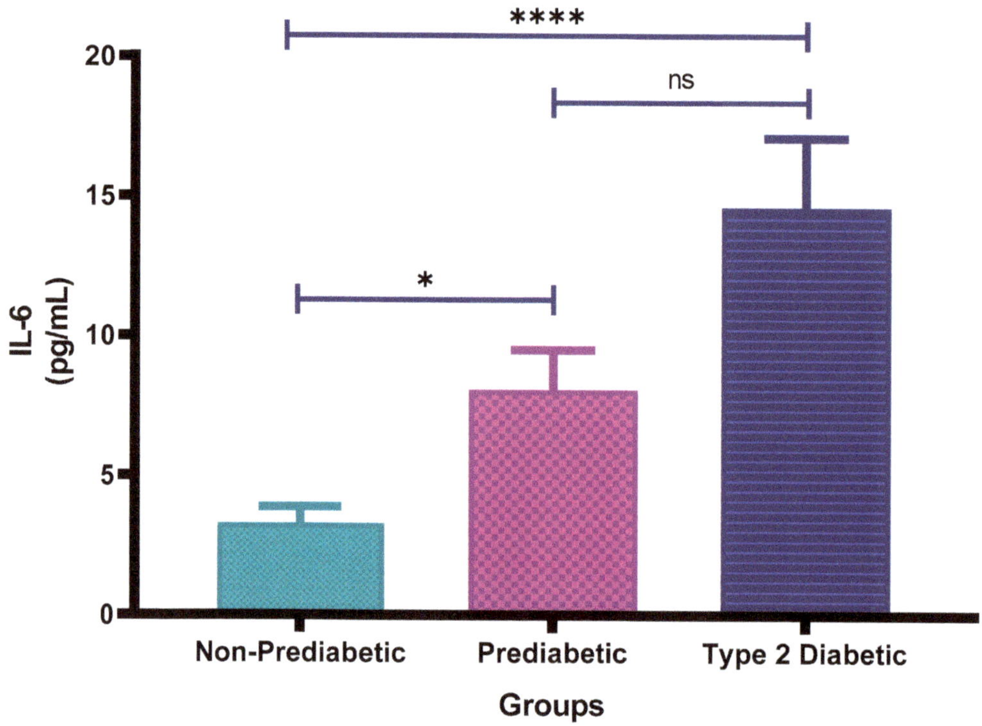

Figure 2. Interleukin-6 (IL-6) concentration in the non-prediabetic (NPD) group ($n = 10$), prediabetic (PD) group ($n = 15$) and type 2 diabetic (T2D) group ($n = 15$). Values are presented as means ± SEM. Asterisk Indicates the significant (* $p < 0.05$, **** $p < 0.0001$) differences between the groups. ns: Non-Significant.

3.4. Fibrinogen Concentration

Figure 3 shows the plasma fibrinogen concentrations in non-prediabetic (NPD), prediabetic (PD) and type 2 diabetic (T2D) individuals in Durban, South Africa. The mean ± SEM fibrinogen concentrations were 1.823 ± 0.023 g/L in the NPD group, 2.938 ± 0.018 g/L in the PD group, and 4.060 ± 0.019 g/L in the T2D group. Statistical comparisons test showed that fibrinogen levels in the PD group were significantly higher than those in the NPD group ($p = 0.0008$) and significantly lower than those in the T2D group ($p = 0.0008$). Additionally, the T2D group had significantly higher fibrinogen concentrations compared to the NPD group ($p < 0.0001$) (see Figure 3). These results indicate a clear increase in fibrinogen levels across the glycaemic spectrum. For comparison, the typical plasma fibrinogen reference range in healthy adults is approximately 2–4 g/L, indicating that all study groups exhibited markedly elevated fibrinogen levels consistent with a pro-thrombotic or inflammatory state [35].

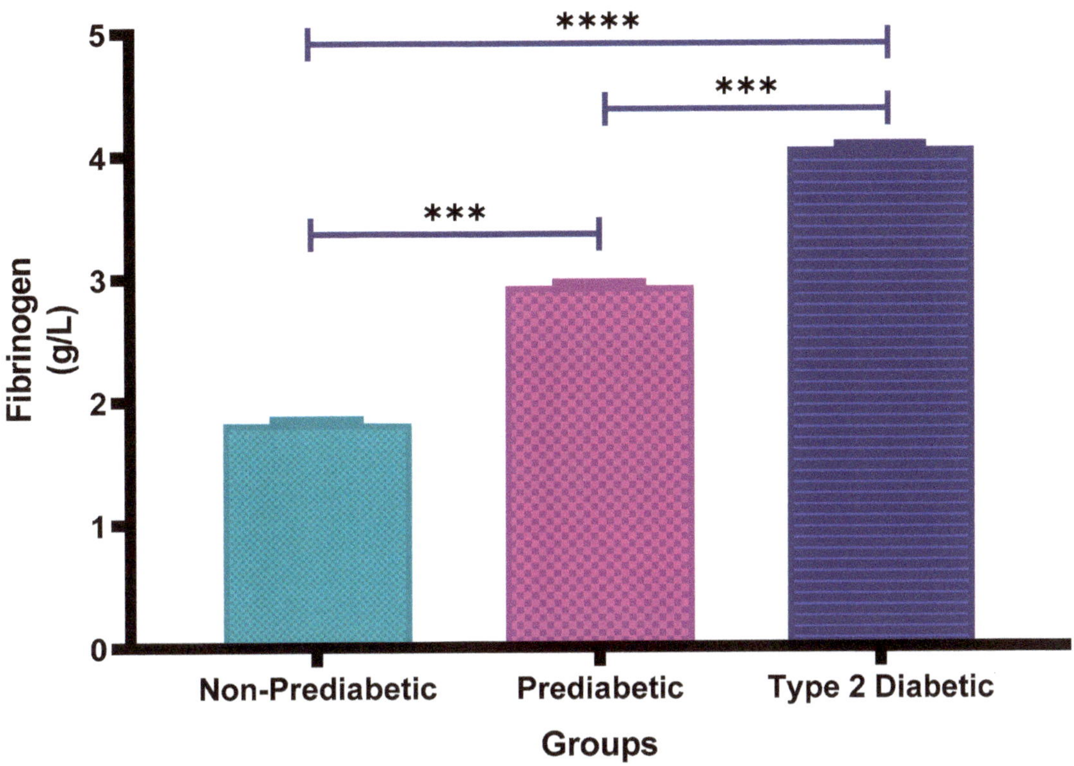

Figure 3. Plasma fibrinogen concentration in the non-prediabetic (NPD) group (n = 20), prediabetic (PD) group (n = 20) and type 2 diabetic (T2D) group (n = 20). Values are presented as means ± SEM. Asterisk Indicates the significant (*** $p < 0.001$, **** $p < 0.0001$) differences between the groups.

3.5. D-Dimer (D2D) Concentration

Figure 4 shows the plasma D-dimer concentrations in non-prediabetic (NPD), prediabetic (PD), and type 2 diabetic (T2D) individuals in Durban, South Africa. The mean ± SEM D-dimer concentrations were 68.40 ± 22.88 ng/mL in the NPD group, 265.1 ± 45.11 ng/mL in the PD group, and 429.8 ± 29.82 ng/mL in the T2D group. Statistical comparisons test indicated that D-dimer levels in the PD group were significantly higher than those in the NPD group (p = 0.0451), while levels in the T2D group were significantly elevated compared to the NPD group ($p < 0.0001$). However, the difference between the PD and T2D groups was not statistically significant (p = 0.1009) (See Figure 4). These findings suggest a progressive increase in fibrinolytic activity across the glycaemic spectrum. For reference, normal plasma D-dimer concentrations are typically below 500 ng/mL [37]. Although the D-dimer levels in all groups in this study remained below this clinical threshold, individuals with T2D exhibited the highest concentrations, suggesting a trend toward increased coagulation and fibrinolytic activity.

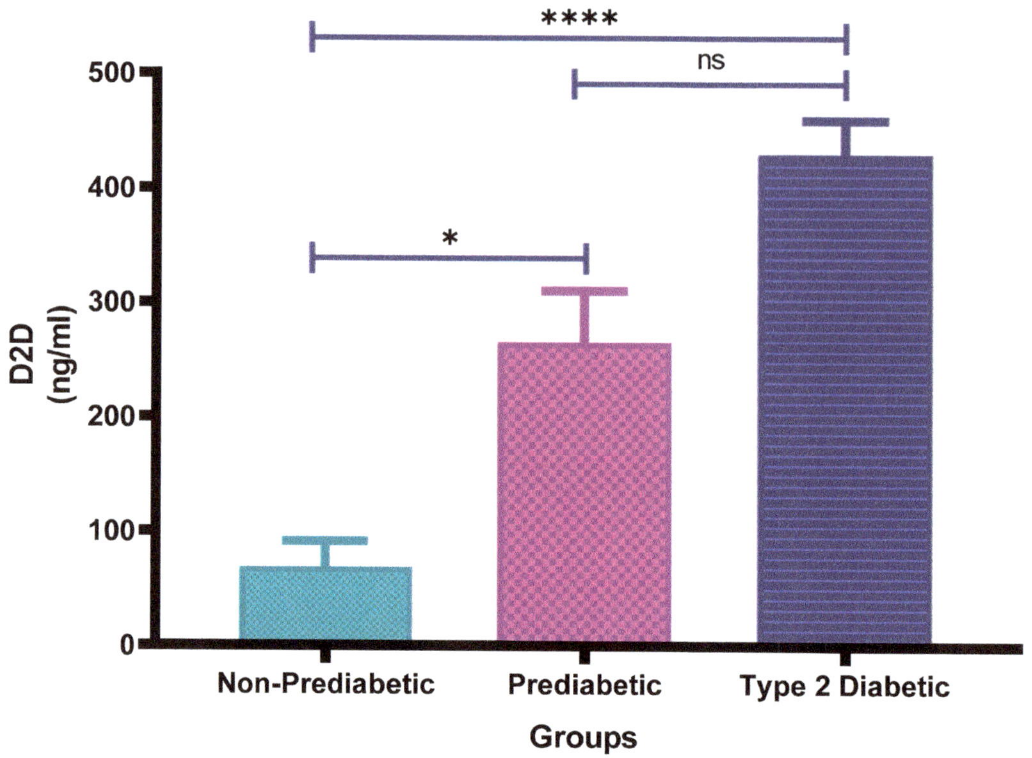

Figure 4. Plasma D-dimer (D2D) concentration in the non-prediabetic (NPD) group (n = 10), prediabetic (PD) group (n = 15) and type 2 diabetic (T2D) group (n = 15). Values are presented as means ± SEM. Asterisk Indicates the significant (* $p < 0.05$, **** $p < 0.0001$) differences between the groups. ns: Non-Significant.

3.6. Calcium Binding Protein (S100B) Concentration

Figure 5 shows the plasma calcium-binding protein S100B concentrations in non-prediabetic (NPD), prediabetic (PD) and type 2 diabetic (T2D) individuals in Durban, South Africa. The mean ± SEM S100B levels were 0.124 ± 0.0031 Ug/L in the NPD group, 0.221 ± 0.0025 Ug/L in the PD group, and 0.327 ± 0.0033 Ug/L in the T2D group. Statistical comparisons test revealed that S100B concentrations were significantly higher in the PD group compared to the NPD group (p = 0.0008) and significantly lower in the PD group compared to the T2D group (p = 0.0008). Furthermore, the T2D group showed significantly higher S100B levels than the NPD group ($p < 0.0001$) (See Figure 5). These findings suggest a progressive increase in plasma S100B concentration with worsening glycaemic status. For reference, typical plasma S100B levels in healthy adults are below 0.15 Ug/L, indicating that individuals with PD or T2D exhibited elevated S100B consistent with possible neurovascular injury or inflammation [38].

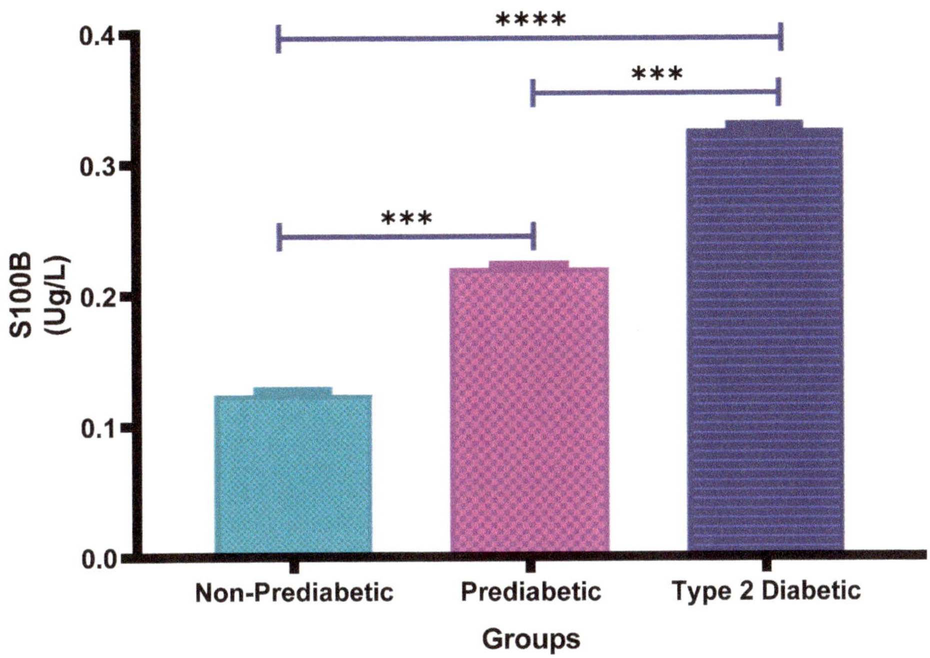

Figure 5. Plasma calcium binding protein (S100B) concentration in the non-prediabetic (NPD) group ($n = 20$), prediabetic (PD) group ($n = 20$) and type 2 diabetic (T2D) group ($n = 20$). Values are presented as means ± SEM. Asterisk Indicates the significant (*** $p < 0.001$, **** $p < 0.0001$) differences between the groups.

3.7. Correlation Analysis

Analyses were conducted using Pearson's correlation coefficients on the pooled dataset of all participants with available data ($n = 60$), combining the NPD, PD and T2D groups. Both impaired fasting glucose and HbA1c showed strong, positive correlations with S100B concentrations, with correlation coefficients of $r = 0.75$ and $r = 0.75$, respectively ($p < 0.0001$ for both; 95% CI: 0.62–0.84), as shown in Table 3. These significant positive correlations indicate that as blood glucose levels and long-term glycaemic control worsen, plasma S100B concentrations tend to increase. This suggests a link between glycaemic dysregulation and elevated S100B, which may reflect increased neurovascular injury or inflammatory processes in individuals progressing from non-prediabetic to diabetic states.

Table 3. Correlation between glucose and glycated haemoglobin (HbA1c) with calcium-binding protein (S100B) in the non-prediabetic (NPD), prediabetic (PD) and type 2 diabetic (T2D) groups. r = Pearson's correlation coefficient, n = sample size. 95% confidence intervals (CI) for each correlation are included.

Correlation Analysis	Impaired Fasting Glucose vs. S100B	HbA1c vs. S100B
r	0.7475	0.7458
95% CI	0.6093–0.8417	0.6068–0.8406
p value	0.0001	0.0001
n	60	60

3.8. Glial Fibrillary Acidic Protein (GFAP) Concentration

Figure 6 shows the plasma glial fibrillary acidic protein (GFAP) concentrations in non-prediabetic (NPD), prediabetic (PD) and type 2 diabetic (T2D) individuals in Durban, South Africa. The mean ± SEM GFAP concentrations were 80.3 ± 11.3 ng/L in the NPD group, 210.1 ± 46.9 ng/L in the PD group and 251.7 ± 52.7 ng/L in the T2D group. Statistical comparisons test revealed that GFAP levels were significantly higher in the PD group compared to the NPD group (p = 0.0082) and significantly elevated in the T2D group compared to the NPD group (p = 0.0071). However, there was no significant difference between the PD and T2D groups (p > 0.9999) (See Figure 6). The reference range for serum GFAP varies by age and assay, but generally healthy adults (20–49) have levels ranging from 25 to 242 ng/L [39]. These findings suggest a rising trend in GFAP concentration with worsening glycaemic status. Notably, mean GFAP levels in the PD and T2D groups exceed the general adult reference range, providing supporting evidence of astrocyte activation and possible neurovascular impairment associated with dysglycaemia.

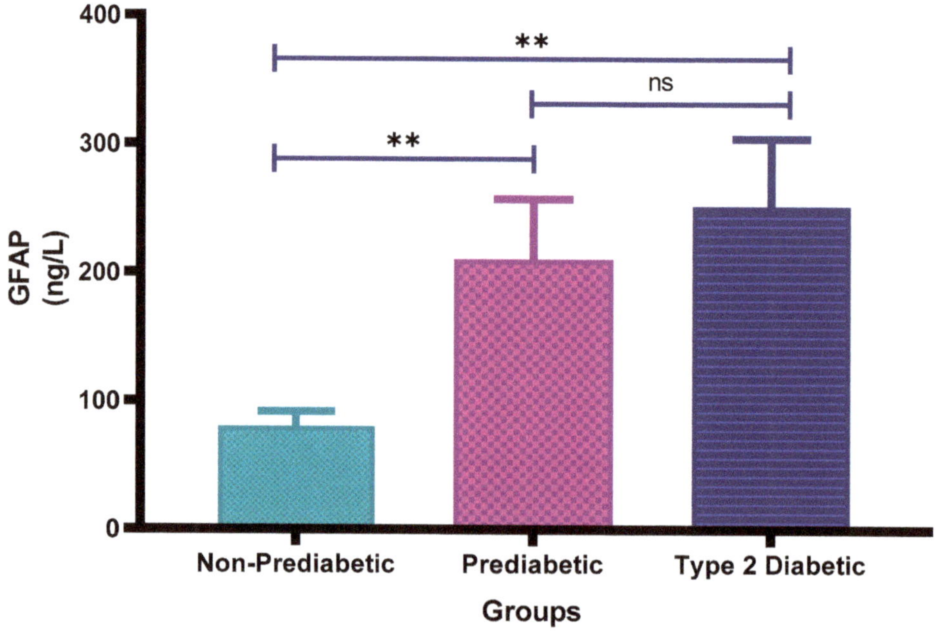

Figure 6. Plasma glial fibrillary acidic protein (GFAP) concentration in the non-prediabetic (NPD) group (n = 10), prediabetic (PD) group (n = 15) and type 2 diabetic (T2D) group (n = 15). Values are presented as means ± SEM. Asterisk Indicates the significant (** p < 0.01) differences between the groups. ns: Non-Significant.

3.9. Neuron Specific Enolase (NSE) Concentration

Figure 7 shows the plasma neuron-specific enolase (NSE) concentrations in non-prediabetic (NPD), prediabetic (PD) and type 2 diabetic (T2D) individuals in Durban, South Africa. The mean ± SEM NSE concentrations were 4.06 ± 0.73 ng/mL in the NPD group, 21.57 ± 7.39 ng/mL in the PD group and 31.16 ± 4.74 ng/mL in the T2D group. Statistical comparisons test revealed that NSE levels were significantly higher in the PD group compared to the NPD group (p = 0.0118), significantly elevated in the T2D group compared to the NPD group (p < 0.0001) and also significantly increased in the T2D group compared to the PD group (p = 0.0105) (see Figure 7). The reference range for serum NSE is typically 0–18.9 ng/mL [40]. The values observed in the PD and T2D groups exceed this range, suggesting progressive neuronal injury or neurovascular stress associated with dysglycaemia.

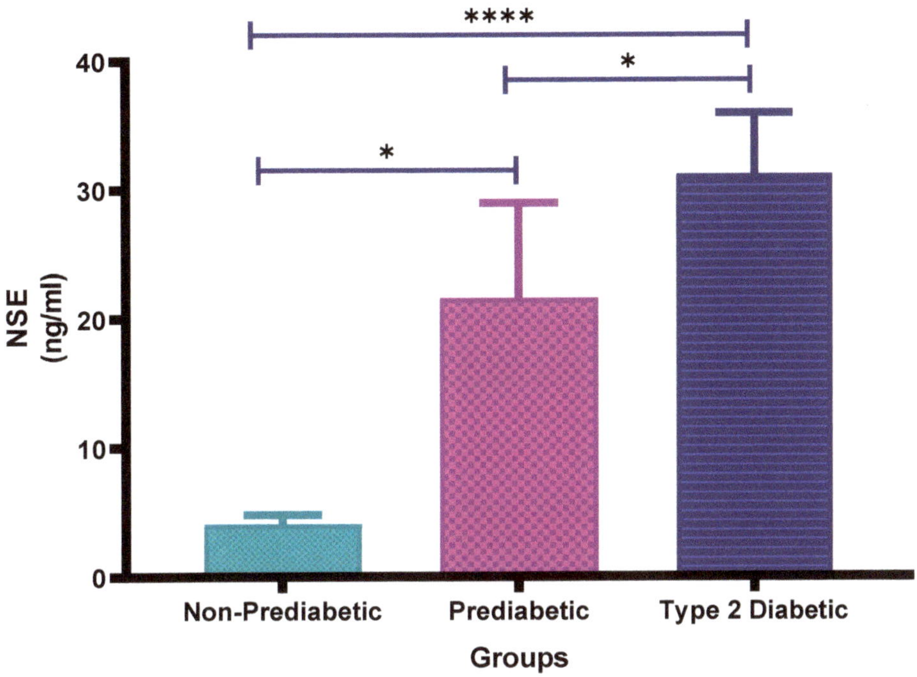

Figure 7. Neuron-specific enolase (NSE) concentration in the non-prediabetic (NPD) group ($n = 10$), prediabetic (PD) group ($n = 15$) and type 2 diabetic (T2D) group ($n = 15$). Values are presented as means ± SEM. Asterisk Indicates the significant (* $p < 0.05$, **** $p < 0.0001$) differences between the groups.

4. Discussion

The increasing prevalence of T2DM has been strongly linked to modern lifestyle factors, including poor dietary habits, physical inactivity and rapid urbanisation associated with economic development [41]. Research has demonstrated that many of the complications related to T2DM, such as cerebrovascular diseases, specifically stroke, begin during the prediabetic stage [42]. Despite this, there is limited understanding of the specific link between prediabetes and stroke risk, especially in populations where T2DM is on the rise. This study aimed to fill this gap by investigating the relationship between prediabetes and stroke risk in individuals aged 25–45 years from Durban, South Africa. Through the retrospective analysis of biomarkers related to inflammation, coagulation and brain damage, we sought to identify early indicators that could help predict stroke risk in prediabetic individuals and contribute to better detection and intervention strategies. It is important to note that this study did not directly assess clinical stroke events. Rather, biomarkers associated with inflammation, coagulation and neurovascular injury were used as surrogate indicators to explore potential stroke risk. This distinction is critical to prevent overinterpretation of the findings.

The results presented here are descriptive and based on biomarker trends rather than confirmed clinical stroke events. Therefore, the findings should be interpreted as indicative of potential stroke risk rather than deterministic outcomes. This study provides proof-of-concept evidence of early neurovascular, inflammatory and coagulation changes in individuals with prediabetes in a South African population—a group underrepresented in current research. Despite the limited sample size, the consistent trends across multiple biomarkers support the biological relevance of these findings. These results provide initial evidence that can inform and guide larger-scale future studies.

It should be acknowledged that the number of participants in each group was small, and the sex distribution was unequal. The primary objective of this study was to assess biomarker changes associated with prediabetes, rather than to evaluate sex-specific differences. Pooling data allowed for greater statistical power in addressing this research aim. However, participants were generally healthy young adults without major comorbidities, which supports the validity of pooling results across sexes in this study.

These results of this study support the growing understanding that prediabetes is not a benign intermediate state, but a metabolically active condition marked by early pathological changes. By highlighting significant biomarker alterations in a young South African cohort, this study contributes valuable regional data to a field largely shaped by research from other countries.

Insulin resistance is a key feature of prediabetes and plays a significant role in the development of associated complications, including stroke [29]. In individuals with prediabetes, insulin resistance results from defects in the insulin signalling pathways. These defects impair glucose uptake and metabolism, which can increase the risk of cardiovascular and cerebrovascular events [43]. One of the primary markers of glucose regulation in prediabetes is HbA1c, which reflects the average blood glucose levels over 3 months [44]. In this study, individuals in the PD group showed significantly elevated ($p < 0.0001$) HbA1c levels compared to the NPD group but significantly lower levels than that of individuals in the T2D group ($p < 0.0001$). This increase in HbA1c suggests ongoing hyperglycaemia and impaired glucose control, which is consistent with the definition of prediabetes as outlined by the ADA and WHO, where the condition is characterised by intermediate insulin resistance and elevated HbA1c levels [45]. The results in this study indicate that insulin resistance, as indicated by elevated HbA1c levels, may contribute to the increased risk of stroke in individuals with prediabetes.

Inflammation is a fundamental physiological response that helps the body fight infections but when chronic, it can contribute to the development of various diseases, including cerebrovascular disease [46]. In a normal response, inflammation is acute, involving the activation of both immune and non-immune cells [46]. However, in prediabetes and T2DM, this inflammatory response becomes persistent and low-grade.

Chronic hyperglycaemia associated with prediabetes and T2DM stimulates the release of inflammatory cytokines such as CRP and IL-6. High levels of inflammatory cytokines have been known to cause injury to the vascular endothelium.

CRP is an acute-phase protein synthesised by the liver following inflammation, and it usually indicates systemic inflammation [47]. The high level of CRP was linked to an increased risk for rupture of plaque and vascular thrombosis since CRP mediates endothelial dysfunction and destabilises atherosclerotic plaques toward rupture [48]. Increased IL-6 levels are strongly associated with the development of chronic inflammation [49]. The pro-inflammatory cytokine IL-6 plays a role in several biological activities, including cell proliferation and differentiation and has been implicated in the pathogenesis of atherosclerosis and plaque formation within the vasculature [50]. In individuals with prediabetes, elevated blood glucose levels exacerbate this inflammatory response, leading to endothelial injury and the initiation of atherosclerosis, a major cause of ischaemic stroke [51]. This study's results revealed significantly higher CRP levels in the PD group compared to NPD ($p = 0.0008$), and significantly higher levels in T2D compared to both PD ($p = 0.0008$) and

NPD ($p < 0.0001$) groups. These results indicate a progressive increase in CRP across the glycaemic spectrum. For context, Hs-CRP concentrations below 3 mg/L are generally considered normal, so individuals in both PD and T2D groups showed elevated CRP relative to clinical reference values.

These CRP findings align with the conclusions drawn from previous research assessing various markers of carotid atherosclerosis in individuals with T2D [52]. The study demonstrated a significant correlation between elevated CRP levels and the presence of carotid artery disease, as well as an increased risk of cerebrovascular events in individuals with T2DM. However, by focusing on a younger South African prediabetic population, this study extends current knowledge to an underrepresented group, providing new insights into early inflammatory processes linked to stroke risk. The reviewed literature further emphasises CRP's potential as a predictive marker for stroke risk. Additionally, it highlights T2DM as an independent risk factor for both stroke and its recurrence. The high levels of CRP in the present study strongly point toward its role as an inflammatory marker related to carotid atherosclerosis, one of the major contributors to ischaemic stroke risk. The higher level of CRP among the individuals with T2D further established the role of inflammation in the pathogenesis of carotid atherosclerosis as one of the key contributors to ischaemic stroke. The increased levels of CRP in prediabetes suggest that even minor early metabolic disturbances may enhance vascular risk, hence the stratification and need for early intervention in such cases.

Statistical analysis of this study's results showed that IL-6 levels were significantly higher in the PD group compared to the NPD group ($p = 0.0369$) and significantly elevated in the T2D group compared to the NPD group ($p < 0.0001$). However, there was no significant difference between the PD and T2D groups ($p = 0.1648$). The IL-6 results further support the findings of [53,54], who found higher plasma IL-6 levels in individuals with diabetes compared to non-diabetic individuals. However, unlike the study by Pickup, which indicated no significant difference in IL-6 levels among individuals with diabetes and complications of macrovascular and microvascular nature, the study by [54], along with this current study, noted significant differences. Though this study's IL-6 results align with the findings by [54], it has to be taken into consideration that another study has linked IL-6 with increased insulin resistance, further exacerbating hyperglycaemia and vascular status [55]. This creates a vicious cycle, increasing the risk of stroke in individuals with T2DM. Overall, CRP and IL-6 levels and consequently the risk of a stroke incrementally increases from the NPD to PD to T2D group, as a result of worsening metabolic and inflammatory profiles, with prediabetes representing an intermediate stage of the inflammatory response. The CRP and IL-6 results follow a similar trend as seen in the HbA1C results, which increased incrementally from the NPD to PD to T2D group.

Although individuals with HIV were excluded from this study to reduce confounding, it is important to acknowledge that HIV is highly prevalent in KwaZulu-Natal and can independently elevate inflammatory markers such as CRP and IL-6. As such, while the exclusion enhances the specificity of our findings, it may limit the generalisability of results to populations with high HIV comorbidity. HIV and other major illnesses are not the only potential confounding factors; lifestyle-related variables such as smoking, alcohol use and poor diet may also influence inflammatory and vascular biomarkers. These factors were not accounted for in this study.

It is important to note that unassessed lifestyle factors, including smoking, alcohol consumption, diet, and body mass index (BMI), could influence inflammatory and coagulation biomarkers. Smoking and excessive alcohol intake are known to elevate CRP and fibrinogen levels, potentially confounding the observed biomarker elevations in prediabetic and diabetic individuals [56,57]. Similarly, higher BMI and poor dietary habits can exacerbate low-grade inflammation and hypercoagulability, contributing to increased vascular risk [58,59]. In future research, multivariate adjustments could be applied by including lifestyle factors as covariates in regression models (e.g., adjusting CRP levels for smoking status or BMI), which would help distinguish the independent contribution of prediabetes from these confounders. Therefore, future studies should consider incorporating these variables and performing multivariate adjustments to more accurately assess the independent effects of prediabetes on stroke-related biomarkers, particularly in populations with high HIV prevalence or diverse lifestyle exposures.

Beyond its role in driving inflammation, T2DM is also known to impair the normal functioning of the coagulation cascade, further enhancing the risk of thrombotic events. The inflammatory markers not only indicate systemic inflammation but also contribute to the activation of coagulation pathways, leading to an increased thrombotic state [60].

Fibrinogen and D-dimer are key biomarkers involved in the coagulation cascade and thrombus formation [61]. Fibrinogen is a plasma protein that plays a critical role in blood clot formation, while D-dimer is a degradation product of fibrin, typically elevated in cases of thrombosis [62]. Both are associated with increased thrombogenic activity, contributing to the risk of cardiovascular and cerebrovascular events [63,64]. Elevated fibrinogen is also associated with pro-inflammatory states, which further promotes vascular dysfunction [65]. In T2DM, insulin resistance predisposes individuals to abnormal fibrinogen metabolism, promoting clot formation and impairing fibrinolysis [66]. In addition, raised D-dimer levels in individuals with diabetes indicate ongoing thrombotic activity and enhanced vascular events [67]. The results in this study indicated that plasma fibrinogen levels were significantly higher in PD than in NPD ($p = 0.0008$) and further elevated in T2D compared to both PD ($p = 0.0008$) and NPD ($p < 0.0001$). These findings demonstrate a stepwise increase in fibrinogen along the glycaemic spectrum.

While fibrinogen levels remained within clinically accepted reference ranges (2–4 g/L), the observed upward trend across the glycaemic spectrum may represent an early, compensatory response to low-grade inflammation or endothelial activation. In the absence of acute illness or overt vascular damage, fibrinogen concentrations may rise gradually but remain subclinical, supporting its role as a sensitive marker of early prothrombotic risk in dysglycaemic individuals. Fibrinogen is an acute-phase reactant, and its levels can remain in the "normal" range until systemic inflammation or endothelial injury reaches a more advanced or clinically overt stage. In early dysglycaemia, like prediabetes, subtle inflammatory or metabolic disturbances may stimulate fibrinogen production modestly, but not enough to push levels beyond clinical thresholds.

The results in the study indicated that D-dimer levels were significantly higher in PD vs. NPD ($p = 0.0451$) and in T2D vs. NPD ($p < 0.0001$), though the PD–T2D difference was not significant ($p = 0.1009$). Similarly to fibrinogen, D-dimer levels remained within the reference range but showed a stepwise increase across the glycaemic spectrum. This concurrent elevation suggests active clot formation and breakdown, which is characteristic of a hypercoagulable state. Fibrinogen reflects the production of fibrin and the potential for clot formation, while D-dimer indicates the breakdown of fibrin clots. The simultaneous rise in both markers suggests that coagulation is being activated alongside ongoing fibrinolysis, likely driven by persistent endothelial dysfunction, vascular inflammation, and chronic metabolic stress in individuals with dysglycaemia.

The fibrinogen and D-dimer levels observed in this study were significantly elevated across the glycaemic spectrum but remained within clinically accepted reference ranges. This finding contrasts with several international studies reporting elevations beyond these limits in diabetes mellitus. By demonstrating early, subclinical increases in these coagulation markers in a younger South African prediabetic population, the study provides novel evidence of a prothrombotic state that may precede overt vascular disease, thereby enhancing understanding of coagulation dynamics in early dysglycaemia.

While inflammation and coagulation are common pathways in both diabetes and stroke pathology, these are non-specific mechanisms. As chronic inflammation and coagulation abnormalities persist in prediabetes and T2DM, they impair vascular function, heightening the risk of cerebrovascular damage. These changes contribute to the development of brain injury, particularly ischaemic stroke, by compromising cerebral circulation. Therefore, assessing brain injury provides the most accurate insight into the impact of these metabolic disturbances in this study.

S100B is a protein with diverse functions in the body's normal physiology [68]. This protein is predominantly found in the central nervous system and is released when there is damage to neural or glial cells, reflecting brain injury or neurological distress [69]. S100B is mainly produced by astrocytes and is involved in regulating processes like cell proliferation, differentiation and inflammation [70].

Hyperglycaemia accelerates intracellular glucose metabolism and mitochondrial electron transport, promotes the formation of advanced glycation end products, and activates NADPH oxidase, all of which increase the generation of reactive oxygen species (ROS) [68]. Elevated ROSs induce oxidative stress within the neurovascular unit and act as potent activators of astrocytes, triggering morphological and functional changes via intracellular signalling pathways [70]. In response to this oxidative and inflammatory environment, activated astrocytes upregulate and release calcium-binding proteins such as S100B into the extracellular space [69]. Once extracellular, S100B functions both as a damage-associated molecular pattern and as a mediator of neuroinflammation, linking hyperglycaemia-driven oxidative stress to astrocytic activation and S100B secretion [68–70].

In individuals with T2DM, S100B levels are often elevated with several contributing factors [24]. These include chronic low-grade inflammation, oxidative stress and microvascular changes, which can result from prolonged hyperglycaemia and metabolic abnormalities. Elevated S100B levels can indicate neurological complications associated with the disease, such as cognitive impairment, diabetic neuropathy or even an increased risk of stroke [24]. Statistical comparisons of results in this study revealed significantly higher S100B concentrations in the PD group compared to the NPD group ($p = 0.0008$), and significantly higher levels in the T2D group compared to both the NPD group ($p < 0.0001$) and the PD group ($p = 0.0008$). These findings demonstrate a stepwise elevation in plasma S100B concentrations with worsening glycaemic status. For reference, typical plasma S100B levels in healthy adults are below 0.15 μg/L. This significant difference can be attributed to factors such as persistent hyperglycaemia and associated conditions within the T2D group, indicating that individuals with type 2 diabetes suffered the greatest extent of brain injury, with the NPD levels signifying a healthier and less injured state. The elevated level of S100B in the PD group can be attributed to the fact that individuals with prediabetes often experience intermediate insulin resistance, chronic inflammation and metabolic dysfunction. These factors can lead to small vessel damage in the brain, which may eventually result in strokes. Elevated S100B levels may not directly predict an imminent stroke event; however, it may serve as a marker of neurological damage that has occurred, reflecting the severity of the event [71]. S100B is more indicative of a past or ongoing stroke rather than a predictive marker for future stroke risk. The incremental rise in S100B levels from the NPD to PD to T2D group reflects the increasing severity of brain injury, with prediabetes serving as an intermediary stage. This trend is consistent with the incremental increase pattern observed in the HbA1c, CRP, IL-6, fibrinogen and D-dimer levels. Moreover, the correlation analysis revealed a strong positive correlation between both impaired fasting blood glucose and HbA1c with the calcium-binding protein S100B in the NPD, PD and

T2D groups. The small p-values (0.0001) indicate that these correlations are statistically significant, implying that as glucose and HbA1c levels increase, S100B levels tend to rise.

GFAP and NSE are important biomarkers used to assess brain injury, especially in neurological disorders like stroke [72–74]. GFAP is primarily found in the astrocytes and its increased level reflects the activation or injury of astrocytes that may result from inflammation or damage to the brain [75]. As a neuroinflammatory marker, GFAP has also been used to assess the extent of brain damage in cases of stroke or neurodegenerative conditions [76,77]. NSE, in turn, is an intracellular enzyme highly expressed in neurons, where elevated levels are indicative of neuronal damage or death [78]. Both biomarkers are valuable in assessing neuronal injury and have been involved in the process of determining the effects of chronic disorders, such as T2DM, on the brain. In individuals with prediabetes or T2DM, elevated levels of GFAP and NSE can indicate early neuronal damage, providing insight into the risk of cerebrovascular events, including stroke.

The results in this study indicated that the mean $\pm$ SEM NSE concentrations were 4.06 $\pm$ 0.73 ng/mL in the NPD group, 21.57 $\pm$ 7.39 ng/mL in the PD group, and 31.16 $\pm$ 4.74 ng/mL in the T2D group. Statistical comparisons revealed that NSE concentrations were significantly higher in the PD group compared to the NPD group ($p = 0.0118$), significantly elevated in the T2D group compared to the NPD group ($p < 0.0001$) and significantly increased in the T2D group compared to the PD group ($p = 0.0105$). These results indicate a stepwise rise in NSE concentrations with worsening glycaemic status. Values observed in the PD and T2D groups exceeded the reference range, suggesting the presence of progressive neuronal injury or neurovascular stress linked to dysglycaemia.

The NSE findings of this study are consistent with the work by [23], who reported concentrations of 7.5 $\pm$ 1.5 ng/mL in controls, 15.2 $\pm$ 2.4 ng/mL in normoglycaemic stroke patients, and 19.7 $\pm$ 4.7 ng/mL in hyperglycaemic stroke patients. Quantitatively, NSE in our PD group (21.57 $\pm$ 7.39 ng/mL) was approximately 1.4-fold higher than in hyperglycaemic stroke patients, while concentrations in the T2D group (31.16 $\pm$ 4.74 ng/mL) were nearly 1.6-fold greater. Even at the prediabetic stage, NSE levels were almost three times higher than the non-stroke control group reported by [23] (21.57 vs. 7.5 ng/mL). These findings demonstrate a consistent pattern of increased neuronal injury associated with glycaemic dysregulation and suggest that neuronal stress of a magnitude similar to, or greater than, that observed in stroke may already be present in prediabetes. This supports the hypothesis that hyperglycaemia is not only a consequence of acute cerebrovascular events but may act as an early driver of neurovascular injury, thereby predisposing individuals to future stroke.

These results further support NSE as a reliable biomarker of neuronal injury, indicating that hyperglycaemia in prediabetic and diabetic individuals can induce metabolic stress and early neurovascular damage, even in the absence of clinical stroke.

The results in this study revealed that GFAP levels were significantly higher in the PD group compared to the NPD group ($p = 0.0082$) and significantly elevated in the T2D group compared to the NPD group ($p = 0.0071$). However, there was no significant difference between the PD and T2D groups ($p > 0.9999$). The reference range for serum GFAP varies by age and assay, but generally, healthy adults aged 20 to 49 years have levels ranging from 25 to 242 ng/L. The GFAP findings of this study are consistent with the work by [22], who examined serum levels of GFAP in individuals with T2DM and cognitive impairments. The study found that individuals with T2DM, particularly those with cognitive impairments, had significantly higher GFAP levels compared to healthy controls. While [22] focused on GFAP and cognitive impairments, GFAP is also considered an important marker in stroke assessment. This supports the notion that elevated GFAP reflects astrocyte activation and neurovascular injury that may underlie both cognitive decline and increased stroke risk in dysglycaemic states. In this study, which investigated stroke risk specifically in individuals with prediabetes and no other underlying disorders or diseases, similar elevated GFAP levels were observed. This supports the relevance of GFAP as a biomarker in assessing brain injury and differentiating between stroke types, highlighting its significance in individuals with diabetes at risk for cerebrovascular events.

The neurovascular injury markers S100B, GFAP and NSE, which were the central focus of this study, showed significant elevations across the glycaemic spectrum, providing strong evidence of early neurovascular compromise. These findings suggest that subclinical blood–brain barrier disruption and neuronal stress may already be present in the prediabetic state. Taken together with the elevations in inflammation and coagulation, these results suggest that even in the absence of clinical stroke, individuals with prediabetes may exhibit early signs of neuroinflammation and neuronal injury. Supporting evidence from previous research shows that white matter alterations are significantly correlated with cognitive impairment, particularly in processing speed, in individuals with Metabolic Syndrome [79]. These findings indicate that a cluster of silent vascular risk factors may contribute to progressive cerebrovascular pathology and cognitive decline as patients age. This underscores the relevance of our study, as identifying early biomarker changes in the prediabetic stage could help detect and mitigate subclinical neurovascular alterations before irreversible structural or cognitive damage occurs. This provides a biological rationale for incorporating neurological screening in metabolic disease monitoring, particularly in high-risk populations. If left untreated, this covert neurovascular injury could potentiate cumulative damage, ultimately elevating the risk of symptomatic cerebrovascular events such as stroke as metabolic dysfunction worsens. Furthermore, given South Africa's rising burden of non-communicable diseases and the relatively young age of onset observed in this cohort (25 to 45 years), these findings highlight the urgent need for early detection strategies that integrate neurological risk profiling into routine metabolic health assessments.

The neurovascular injury markers, taken together with inflammatory and coagulation markers, suggest that biological processes linked to stroke pathogenesis may begin much earlier than traditionally expected. The findings may inform future research on predictive biomarker panels for stroke risk stratification in younger adults, particularly in regions like South Africa, where the burden of diabetes-related complications is rising.

5. Conclusions

Intermediate insulin resistance and intermediate hyperglycaemia associated with prediabetes lead to inflammation and endothelial dysfunction, which are known risk factors for stroke. There is a positive correlation between insulin resistance and inflammatory markers, such as CRP and IL-6, suggesting that prediabetes increases the risk of developing cerebrovascular events. While there are several other markers of brain injury, this study demonstrated that S100B, GFAP and NSE have potential for use as biomarkers in the development of prediabetes-associated stroke events. The elevated neurodegenerative markers (S100B, GFAP, and NSE) in both the prediabetes (PD) and T2D groups indicate ongoing neurovascular damage and an increased stroke risk that begins to emerge in the prediabetic stage and intensifies with progression to T2DM. These results suggest that individuals with prediabetes and T2DM experience early stages of neurodegeneration even in the absence of a stroke. Importantly, prediabetes is a reversible condition, and early identification of at-risk individuals provides a critical window for intervention through lifestyle modification and glycaemic control. The panel of biomarkers used, while requiring further independent validation, demonstrates potential as a cost-effective, non-invasive tool for early stroke risk monitoring by detecting intermediate metabolic and vascular alterations that precede overt cerebrovascular events. These findings provide novel, region-specific insights from a South African cohort, highlighting the burden of prediabetes in local populations and emphasising the potential for early detection, preventive strategies, and biomarker-based monitoring in communities underrepresented in current research. The current study, therefore, warrants further investigation on these markers for the early detection of stroke risk in prediabetes. However, it must be emphasised that this study did not include participants with confirmed stroke diagnoses, and the findings are based solely on biomarker trends suggestive of elevated risk; therefore, findings should be interpreted as indicative of potential stroke risk rather than deterministic outcomes. The study's sample pool was limited to two hospitals in Durban, so findings may not be generalisable across the entire South African population. Due to the retrospective nature of the study, it also did not account for lifestyle factors like smoking, alcohol or diet. Although the sample size was limited, the findings remain clinically significant and highlight the potential utility of the biomarker panel for early stroke risk monitoring and preventive strategies. The results provide valuable initial evidence to inform and guide larger-scale, prospective investigations in the future. Future research should include these variables, expand the sample size and include diverse populations. Longitudinal studies are needed to clarify the cause-and-effect relationship between prediabetes and stroke risk. A key recommendation is to incorporate advanced diagnostic techniques, such as MRI and Transcranial Doppler Ultrasound, to improve the accuracy of stroke detection and validate biomarkers like S100B, GFAP and NSE as non-invasive stroke indicators. Additionally, studies should investigate individuals who have already experienced a stroke, monitoring their glucose and insulin levels to assess whether prediabetes or insulin resistance contributed to the stroke. This could provide valuable insights into the role of metabolic disturbances in stroke development and recurrence, ultimately informing targeted prevention and treatment strategies.

Furthermore, South Africa as a whole needs to raise awareness about diabetes, particularly prediabetes and its associated risks. While it may not receive the same attention as infectious diseases, such as HIV/AIDS or tuberculosis, diabetes complications, such as stroke, cardiovascular disease and kidney failure, can be just as fatal. Public health campaigns should educate the population about early detection, lifestyle changes and diabetes management to reduce its impact, ultimately improving health outcomes across the country.

Author Contributions: Y.N. and A.K. conceptualised and designed the study and drafted the manuscript. Both authors reviewed and edited the final draft. All authors have read and agreed to the published version of the manuscript.

Funding: This study was funded by the College of Health Science (CHS); however, the funders had no role in the study design and manuscript preparation.

Institutional Review Board Statement: This study was conducted with ethical approval that was obtained from the University of KwaZulu-Natal (UKZN) Biomedical Research Ethics Committee (BREC) (approval number: BE266/19; approval date: 4 May 2024) as well as the Provincial Health Research Committee from the KZN Department of Health.

Informed Consent Statement: Written informed consent was obtained from patients prior to sample collection.

Data Availability Statement: Upon request, the raw data supporting the conclusions of this article will be made available by the authors.

Acknowledgments: The authors would like to acknowledge the Department of Human Physiology at UKZN, Westville, for their support and academic contributions to this research.

Conflicts of Interest: The authors declare no conflicts of interest. The funders had no role in the design of the study, in the collection, analyses, or interpretation of data, in the writing of the manuscript, or in the decision to publish the results.

Abbreviations

The following abbreviations are used in this manuscript:

ADA	American Diabetes Association
BREC	Biomedical Research Ethics Committee
CI	Confidence Level
CRP	C-reactive protein
CT	Computed tomography
ELISA	Enzyme-Linked Immunosorbent Assay
GFAP	Glial fibrillary acidic protein
HbA1c	Glycated Haemoglobin
IFG	Impaired fasting glucose
IGT	Impaired glucose tolerance
IL-6	Interleukin 6
IQR	Interquartile Range
MRI	Magnetic resonance imaging
NPD	Non Prediabetic
NSE	Neuron-specific enolase
PD	Prediabetic
ROS	Reactive oxygen species
S100B	Calcium-binding protein B
SEM	Standard error of mean
T2D	Type 2 diabetic
T2DM	Type 2 diabetes mellitus
UKZN	University of KwaZulu-Natal
WHO	World Health Organization

References

1. Goyal, R.; Singhal, M.; Jialal, I. Type 2 Diabetes. In *StatPearls*; StatPearls Publishing: Treasure Island, FL, USA, 2023.
2. Pheiffer, C.; Pillay-van Wyk, V.; Joubert, J.D.; Levitt, N.; Nglazi, M.D.; Bradshaw, D. The prevalence of type 2 diabetes in South Africa: A systematic review protocol. *BMJ Open* **2018**, *8*, e021029. [CrossRef]
3. Pheiffer, C.; Pillay-van Wyk, V.; Turawa, E.; Levitt, N.; Kengne, A.P.; Bradshaw, D. Prevalence of Type 2 Diabetes in South Africa: A Systematic Review and Meta-Analysis. *Int. J. Environ. Res. Public Health* **2021**, *18*, 5868. [CrossRef]
4. Kone, N.; Cassim, N.; Maposa, I.; George, J.A. Diabetic control and compliance using glycated haemoglobin (HbA1C) testing guidelines in public healthcare facilities of Gauteng province, South Africa. *PLoS ONE* **2023**, *18*, e0278789. [CrossRef]
5. Kenny, H.C.; Abel, E.D. Heart Failure in Type 2 Diabetes Mellitus. *Circ. Res.* **2019**, *124*, 121–141. [CrossRef] [PubMed]
6. Hastrup, S.; Hedegaard, J.N.; Andersen, G.; Rungby, J.; Johnsen, S.P. Type 2 diabetes mellitus in patients with ischemic stroke—A nationwide study. *Diabet. Med.* **2024**, *41*, e15337. [CrossRef]
7. Mosenzon, O.; Cheng, A.Y.; Rabinstein, A.A.; Sacco, S. Diabetes and Stroke: What are the connections? *J. Stroke* **2023**, *25*, 26–38. [CrossRef]
8. Murphy, S.J.; Werring, D.J. Stroke: Causes and clinical features. *Medicine* **2020**, *48*, 561–566. [CrossRef]
9. Bradshaw, D.; Norman, R.; Pieterse, D.; Levitt, N.S. Estimating the burden of disease attributable to diabetes in South Africa in 2000. *S. Afr. Med. J.* **2007**, *97*, 700–706. [CrossRef]
10. Chen, R.; Ovbiagele, B.; Feng, W. Diabetes and Stroke: Epidemiology, Pathophysiology, Pharmaceuticals and Outcomes. *Am. J. Med. Sci.* **2016**, *351*, 380–386. [CrossRef]
11. Arboix, A.; Milian, M.; Oliveres, M.; García-Eroles, L.; Massons, J. Impact of female gender on prognosis in type 2 diabetic patients with ischemic stroke. *Eur. Neurol.* **2006**, *56*, 6–12. [CrossRef]
12. Kuriakose, D.; Xiao, Z. Pathophysiology and Treatment of Stroke: Present Status and Future Perspectives. *Int. J. Mol. Sci.* **2020**, *21*, 7609. [CrossRef]

13. Ménégaut, L.; Laubriet, A.; Crespy, V.; Leleu, D.; Pilot, T.; Van Dongen, K.; de Barros, J.-P.P.; Gautier, T.; Petit, J.-M.; Thomas, C.; et al. Inflammation and oxidative stress markers in type 2 diabetes patients with Advanced Carotid atherosclerosis. *Cardiovasc. Diabetol.* **2023**, *22*, 248. [CrossRef]
14. Galicia-Garcia, U.; Benito-Vicente, A.; Jebari, S.; Larrea-Sebal, A.; Siddiqi, H.; Uribe, K.B.; Ostolaza, H.; Martín, C. Pathophysiology of Type 2 Diabetes Mellitus. *Int. J. Mol. Sci.* **2020**, *21*, 6275. [CrossRef] [PubMed]
15. Jin, R.; Yang, G.; Li, G. Inflammatory mechanisms in ischemic stroke: Role of inflammatory cells. *J. Leukoc. Biol.* **2010**, *87*, 779–789. [CrossRef] [PubMed]
16. Parikh, N.S.; Merkler, A.E.; Iadecola, C. Inflammation, Autoimmunity, Infection, and Stroke: Epidemiology and Lessons from Therapeutic Intervention. *Stroke* **2020**, *51*, 711–718. [CrossRef]
17. den Hertog, H.M.; van Rossum, J.A.; van der Worp, H.B.; van Gemert, H.M.; de Jonge, R.; Koudstaal, P.J.; Dippel, D.W. C-reactive protein in the very early phase of acute ischemic stroke: Association with poor outcome and death. *J. Neurol.* **2009**, *256*, 2003–2008. [CrossRef] [PubMed]
18. Aref, H.M.A.; Fahmy, N.A.; Khalil, S.H.; Ahmed, M.F.; ElSadek, A.; Abdulghani, M.O. Role of interleukin-6 in ischemic stroke outcome. *Egypt. J. Neurol. Psychiatry Neurosurg.* **2020**, *56*, 12. [CrossRef]
19. Nam, K.-W.; Kwon, H.-M.; Lee, Y.-S. Clinical significance of D-dimer levels during acute period in ischemic stroke. *Thromb. J.* **2023**, *21*, 55. [CrossRef]
20. Peycheva, M.; Deneva, T.; Zahariev, Z. The role of fibrinogen in acute ischaemic stroke. *Neurol. Neurochir. Pol.* **2021**, *55*, 74–80. [CrossRef]
21. Prasad, M.K.; Marandi, S.; Mishra, B.; Guria, R.T.; Kumar, A.; Birua, H.; Ray, H.N.; Dungdung, A.; Kumar, D.; Maitra, S. Association of Fibrinogen with Ischemic Stroke: A Systematic Review and Meta-Analysis. *Cureus* **2023**, *15*, e34335. [CrossRef]
22. Ayala-Guerrero, L.; García-delaTorre, P.; Sánchez-García, S.; Guzmán-Ramos, K. Serum Levels of Glial Fibrillary Acidic Protein Association with Cognitive Impairment and Type 2 Diabetes. *Arch. Med. Res.* **2022**, *53*, 501–507. [CrossRef] [PubMed]
23. Pandey, A.; Saxena, K.; Verma, M.; Bharosay, A. Correlative study between neuron-specific enolase and blood sugar level in ischemic stroke patients. *J. Neurosci. Rural. Pract.* **2011**, *2*, 50–54. [CrossRef]
24. Yu, H.; Li, H.; Liu, X.; Du, X.; Deng, B. Levels of serum S100B are associated with cognitive dysfunction in patients with type 2 diabetes. *Aging* **2020**, *12*, 4193–4203. [CrossRef]
25. Sosibo, A.M.; Mzimela, N.C.; Ngubane, P.S.; Khathi, A. Prevalence and correlates of pre-diabetes in adults of mixed ethnicities in the South African population: A systematic review and meta-analysis. *PLoS ONE* **2022**, *17*, e0278347. [CrossRef]
26. Tabák, A.G.; Herder, C.; Rathmann, W.; Brunner, E.J.; Kivimäki, M. Prediabetes: A high-risk state for diabetes development. *Lancet* **2012**, *379*, 2279–2290. [CrossRef]
27. Gidey, G.; Hiruy, M.; Teklu, D.; Ramanathan, K.; Amare, H. Prevalence of Prediabetes and Related Modifiable Cardiovascular Risk Factors Among Employees of Ayder Comprehensive Specialized Hospital, Tigray, Northern Ethiopia. *Diabetes Metab. Syndr. Obes.* **2023**, *16*, 643–652. [CrossRef] [PubMed]
28. Sosibo, A.M.; Mzimela, N.C.; Ngubane, P.S.; Khathi, A. Prevalence of pre-diabetes in adults aged 25–45 years in a Durban-based clinical setting, South Africa: A retrospective study. *Prim. Care Diabetes* **2023**, *17*, 650–654. [CrossRef] [PubMed]
29. Mijajlović, M.D.; Aleksić, V.M.; Šternić, N.M.; Mirković, M.M.; Bornstein, N.M. Role of prediabetes in stroke. *Neuropsychiatr. Dis. Treat.* **2017**, *13*, 259–267. [CrossRef]
30. Luc, K.; Schramm-Luc, A.; Guzik, T.J.; Mikolajczyk, T.P. Oxidative stress and inflammatory markers in prediabetes and diabetes. *J. Physiol. Pharmacol.* **2019**, *70*, 809–824. [CrossRef]
31. Feng, S.; Yang, M.; Liu, S.; He, Y.; Deng, S.; Gong, Y. Oxidative stress as a bridge between age and stroke: A narrative review. *J. Intensive Med.* **2023**, *3*, 313–319. [CrossRef]
32. Kezerle, L.; Tsadok, M.A.; Akriv, A.; Senderey, A.B.; Bachrach, A.; Leventer-Roberts, M.; Haim, M. Pre-Diabetes Increases Stroke Risk in Patients with Nonvalvular Atrial Fibrillation. *J. Am. Coll. Cardiol.* **2021**, *77*, 875–884. [CrossRef]
33. Jansen van Vuuren, J.; Pillay, S.; Naidoo, A. Circulating Biomarkers in Long-Term Stroke Prognosis: A Scoping Review Focusing on the South African Setting. *Cureus* **2022**, *14*, e23971. [CrossRef]
34. Kang, H. Sample size determination and power analysis using the G*Power software. *J. Educ. Eval. Health Prof.* **2021**, *18*, 17. [CrossRef] [PubMed]
35. Pagana, K.; Pagana, T.; Pagana, T. *Mosby's Diagnostic & Laboratory Test Reference*, 14th ed.; Elsevier: St. Louis, MO, USA, 2019.
36. Yin, J.-X.; Agbana, Y.L.; Sun, Z.-S.; Fei, S.-W.; Zhao, H.-Q.; Zhou, X.-N.; Chen, J.-H.; Kassegne, K. Increased interleukin-6 is associated with long COVID-19: A systematic review and meta-analysis. *Infect. Dis. Poverty* **2023**, *12*, 43. [CrossRef] [PubMed]
37. Killeen, R.B.; Kok, S.J. D-Dimer Test. In *StatPearls*; StatPearls Publishing: Treasure Island, FL, USA, 2025.
38. Ben Abdesselam, O.; Vally, J.; Adem, C.; Foglietti, M.-J.; Beaudeux, J.-L. Reference Values for Serum S-100B Protein Depend on the Race of Individuals. *Clin. Chem.* **2003**, *49*, 836–837. [CrossRef]
39. Tybirk, L.; Hviid, C.V.B.; Knudsen, C.S.; Parkner, T. Serum GFAP-reference interval and preanalytical properties in Danish adults. *Clin. Chem. Lab. Med.* **2022**, *60*, 1830–1838. [CrossRef] [PubMed]

40. Xie, E.; Zhang, W.; Xu, H.; Ling, Y.; Zhang, Q.; Pan, S. Correction of serum NSE reference intervals includes the unidentified hemolysis sample: 1-year data analysis from healthcare individuals. *J. Clin. Lab. Anal.* **2019**, *33*, e22997. [CrossRef]
41. Gassasse, Z.; Smith, D.; Finer, S.; Gallo, V. Association between urbanisation and type 2 diabetes: An ecological study. *BMJ Glob. Health* **2017**, *2*, e000473. [CrossRef]
42. van Sloten, T.T.; Sedaghat, S.; Carnethon, M.R.; Launer, L.J.; Stehouwer, C.D.A. Cerebral microvascular complications of type 2 diabetes: Stroke, cognitive dysfunction, and depression. *Lancet Diabetes Endocrinol.* **2020**, *8*, 325–336. [CrossRef]
43. Freeman, A.M.; Acevedo, L.A.; Pennings, N. Insulin Resistance. In *StatPearls*; StatPearls Publishing: Treasure Island, FL, USA, 2024.
44. Sherwani, S.I.; Khan, H.A.; Ekhzaimy, A.; Masood, A.; Sakharkar, M.K. Significance of HbA1c Test in Diagnosis and Prognosis of Diabetic Patients. *Biomark Insights* **2016**, *11*, 95–104. [CrossRef]
45. Echouffo-Tcheugui, J.B.; Selvin, E. Prediabetes and What It Means: The Epidemiological Evidence. *Annu. Rev. Public Health* **2021**, *42*, 59–77. [CrossRef] [PubMed]
46. Chen, L.; Deng, H.; Cui, H.; Fang, J.; Zuo, Z.; Deng, J.; Li, Y.; Wang, X.; Zhao, L. Inflammatory responses and inflammation-associated diseases in organs. *Oncotarget* **2018**, *9*, 7204–7218. [CrossRef]
47. Stanimirovic, J.; Radovanovic, J.; Banjac, K.; Obradovic, M.; Essack, M.; Zafirovic, S.; Gluvic, Z.; Gojobori, T.; Isenovic, E.R. Role of C-Reactive Protein in Diabetic Inflammation. *Mediat. Inflamm.* **2022**, *2022*, 3706508. [CrossRef]
48. Badimon, L.; Peña, E.; Arderiu, G.; Padró, T.; Slevin, M.; Vilahur, G.; Chiva-Blanch, G. C-Reactive Protein in Atherothrombosis and Angiogenesis. *Front. Immunol.* **2018**, *9*, 430. [CrossRef]
49. Fonseca, J.E.; Santos, M.J.; Canhão, H.; Choy, E. Interleukin-6 as a key player in systemic inflammation and joint destruction. *Autoimmun. Rev.* **2009**, *8*, 538–542. [CrossRef]
50. Zhang, H.; Dhalla, N.S. The Role of Pro-Inflammatory Cytokines in the Pathogenesis of Cardiovascular Disease. *Int. J. Mol. Sci.* **2024**, *25*, 1082. [CrossRef]
51. Liang, Y.; Wang, M.; Wang, C.; Liu, Y.; Naruse, K.; Takahashi, K. The Mechanisms of the Development of Atherosclerosis in Prediabetes. *Int. J. Mol. Sci.* **2021**, *22*, 4108. [CrossRef]
52. Kadoglou, N.P.E.; Avgerinos, E.D.; Liapis, C.D. An Update on Markers of Carotid Atherosclerosis in Patients with Type 2 Diabetes. *Biomark. Med.* **2010**, *4*, 601–609. [CrossRef] [PubMed]
53. Pickup, J.C.; Chusney, G.D.; Thomas, S.M.; Burt, D. Plasma interleukin-6, tumour necrosis factor α and blood cytokine production in type 2 diabetes. *Life Sci.* **2000**, *67*, 291–300. [CrossRef] [PubMed]
54. Wu, W.; Wang, M.; Sun, Z.; Wang, X.; Miao, J.; Zheng, Z. The predictive value of TNF-α and IL-6 and the incidence of macrovascular complications in patients with type 2 diabetes. *Acta Diabetol.* **2012**, *49*, 3–7. [CrossRef]
55. Rehman, K.; Akash, M.S.H.; Liaqat, A.; Kamal, S.; Qadir, M.I.; Rasul, A. Role of Interleukin-6 in Development of Insulin Resistance and Type 2 Diabetes Mellitus. *Crit. Rev. Eukaryot. Gene Expr.* **2017**, *27*, 229–236. [CrossRef]
56. Dijk, W.; Akkermans, R.; Heijdra, Y.; Weel, C.; Schermer, T.; Scheepers, P.; Lenders, J. The Acute Effect of Cigarette Smoking on the High-Sensitivity CRP and Fibrinogen Biomarkers in Chronic Obstructive Pulmonary Disease Patients. *Biomark. Med.* **2013**, *7*, 211–219. [CrossRef]
57. Feng, D.; Liu, T.; Wang, H.; Karp, E.; Ling, W.; Chen, W.Q. Mediation role of C-reactive protein on the association between smoking quantity and type 2 diabetes in current Chinese smokers. *J. Diabetes Res.* **2014**, *2014*, 171538. [CrossRef]
58. Dina, C.; Tit, D.M.; Radu, A.; Bungau, G.; Radu, A.F. Obesity, Dietary Patterns, and Cardiovascular Disease: A Narrative Review of Metabolic and Molecular Pathways. *Curr. Issues Mol. Biol.* **2025**, *47*, 440. [CrossRef] [PubMed]
59. Henning, R.J. Obesity and obesity-induced inflammatory disease contribute to atherosclerosis: A review of the pathophysiology and treatment of obesity. *Am. J. Cardiovasc. Dis.* **2021**, *11*, 504–529. [PubMed]
60. Stark, K.; Massberg, S. Interplay between inflammation and thrombosis in cardiovascular pathology. *Nat. Rev. Cardiol.* **2021**, *18*, 666–682. [CrossRef] [PubMed]
61. Nakae, R.; Murai, Y.; Yokobori, S. Biomarkers in Neurological Injury: Fibrinogen, Fibrinogen/Fibrin Degradation Products (FDPs), and D-dimer. In *Biomarkers in Trauma, Injury and Critical Care*; Rajendram, R., Preedy, V.R., Patel, V.B., Eds.; Springer International Publishing: Cham, Switzerland, 2023; pp. 43–57.
62. Zhou, W.; Qu, C.; Liu, X.; Huang, J. Diagnostic value of D-dimer to fibrinogen ratio for pulmonary embolism in postpartum women. *BMC Pregnancy Childbirth* **2024**, *24*, 482. [CrossRef] [PubMed]
63. Surma, S.; Banach, M. Fibrinogen and Atherosclerotic Cardiovascular Diseases-Review of the Literature and Clinical Studies. *Int. J. Mol. Sci.* **2021**, *23*, 193. [CrossRef]
64. Hou, H.; Xiang, X.; Pan, Y.; Li, H.; Meng, X.; Wang, Y. Association of Level and Increase in D-Dimer with All-Cause Death and Poor Functional Outcome After Ischemic Stroke or Transient Ischemic Attack. *J. Am. Heart Assoc.* **2021**, *10*, e018600. [CrossRef]
65. Vilar, R.; Fish, R.J.; Casini, A.; Neerman-Arbez, M. Fibrin(ogen) in human disease: Both friend and foe. *Haematologica* **2020**, *105*, 284–296. [CrossRef]

66. Bryk-Wiązania, A.H.; Undas, A. Hypofibrinolysis in type 2 diabetes and its clinical implications: From mechanisms to pharmacological modulation. *Cardiovasc. Diabetol.* **2021**, *20*, 191. [CrossRef] [PubMed]
67. Cheng, L.; Fu, Q.; Zhou, L.; Fan, Y.; Liu, F.; Fan, Y.; Zhang, X.; Lin, W.; Wu, X. D-dimer as a predictor of cardiovascular outcomes in patients with diabetes mellitus. *BMC Cardiovasc. Disord.* **2022**, *22*, 82. [CrossRef] [PubMed]
68. Arrais, A.; Melo, L.; Norrara, B.; Almeida, M.; Freire, K.; Melo, A.; Oliveira, L.; Vieira Lima, F.; Engelberth, R.; Cavalcante, J.; et al. S100B protein: General characteristics and pathophysiological implications in the Central Nervous System. *Int. J. Neurosci.* **2022**, *132*, 313–321. [CrossRef] [PubMed]
69. Michetti, F.; D'Ambrosi, N.; Toesca, A.; Puglisi, M.A.; Serrano, A.; Marchese, E.; Corvino, V.; Geloso, M.C. The S100B story: From biomarker to active factor in neural injury. *J. Neurochem.* **2019**, *148*, 168–187. [CrossRef]
70. Brozzi, F.; Arcuri, C.; Giambanco, I.; Donato, R. S100B Protein Regulates Astrocyte Shape and Migration via Interaction with Src Kinase: Implications for Astrocyte Development, Activation, and Tumor Growth. *J. Biol. Chem.* **2009**, *284*, 8797–8811. [CrossRef]
71. Elting, J.W.; de Jager, A.E.; Teelken, A.W.; Schaaf, M.J.; Maurits, N.M.; van der Naalt, J.; Sibinga, C.T.; Sulter, G.A.; De Keyser, J. Comparison of serum S-100 protein levels following stroke and traumatic brain injury. *J. Neurol. Sci.* **2000**, *181*, 104–110. [CrossRef]
72. Kawata, K.; Liu, C.Y.; Merkel, S.F.; Ramirez, S.H.; Tierney, R.T.; Langford, D. Blood biomarkers for brain injury: What are we measuring? *Neurosci. Biobehav. Rev.* **2016**, *68*, 460–473. [CrossRef]
73. González-García, S.; González-Quevedo, A.; Fernández-Concepción, O.; Peña-Sánchez, M.; Menéndez-Saínz, C.; Hernández-Díaz, Z.; Arteche-Prior, M.; Pando-Cabrera, A.; Fernández-Novales, C. Short-term prognostic value of serum neuron specific enolase and S100B in acute stroke patients. *Clin. Biochem.* **2012**, *45*, 1302–1307. [CrossRef]
74. Singh, H.V.; Pandey, A.; Shrivastava, A.K.; Raizada, A.; Singh, S.K.; Singh, N. Prognostic value of neuron specific enolase and IL-10 in ischemic stroke and its correlation with degree of neurological deficit. *Clin. Chim. Acta* **2013**, *419*, 136–138. [CrossRef]
75. Brahmachari, S.; Fung, Y.K.; Pahan, K. Induction of glial fibrillary acidic protein expression in astrocytes by nitric oxide. *J. Neurosci.* **2006**, *26*, 4930–4939. [CrossRef] [PubMed]
76. Jæger, H.S.; Tranberg, D.; Larsen, K.; Valentin, J.B.; Blauenfeldt, R.A.; Luger, S.; Bache, K.G.; Gude, M.F. Diagnostic performance of Glial Fibrillary Acidic Protein and Prehospital Stroke Scale for identification of stroke and stroke subtypes in an unselected patient cohort with symptom onset < 4.5 h. *Scand. J. Trauma Resusc. Emerg. Med.* **2023**, *31*, 1. [CrossRef] [PubMed]
77. Peretti, D.E.; Boccalini, C.; Ribaldi, F.; Scheffler, M.; Marizzoni, M.; Ashton, N.J.; Zetterberg, H.; Blennow, K.; Frisoni, G.B.; Garibotto, V. Association of glial fibrillary acid protein, Alzheimer's disease pathology and cognitive decline. *Brain* **2024**, *147*, 4094–4104. [CrossRef]
78. Bezek, S.; Biberthaler, P.; Martinez-Espina, I.; Bogner-Flatz, V. Chapter 11—Pathophysiology and clinical implementation of traumatic brain injury biomarkers: Neuron-specific enolase. In *Biomarkers for Traumatic Brain Injury*; Wu, A.H.B., Peacock, W.F., Eds.; Academic Press: Cambridge, MA, USA, 2020; pp. 169–182.
79. Segura, B.; Jurado, M.A.; Freixenet, N.; Bargalló, N.; Junqué, C.; Arboix, A. White matter fractional anisotropy is related to processing speed in metabolic syndrome patients: A case-control study. *BMC Neurol.* **2010**, *10*, 64. [CrossRef] [PubMed]

MDPI AG
Grosspeteranlage 5
4052 Basel
Switzerland
Tel.: +41 61 683 77 34

Current Issues in Molecular Biology Editorial Office
E-mail: cimb@mdpi.com
www.mdpi.com/journal/cimb

www.ingramcontent.com/pod-product-compliance
Lightning Source LLC
Chambersburg PA
CBHW061511210926
51850CB00008B/1878
* 9 7 8 3 7 2 5 8 6 4 2 6 3 *